PENGUIN BOOKS

The Book of Why

'Pearl's accomplishments over the last 30 years have provided the theoretical basis for progress in artificial intelligence ... and they have redefined the term "thinking machine"'
Vint Cerf, Google, Inc.

'Have you ever wondered about the puzzles of correlation and causation? This wonderful book has illuminating answers and it is fun to read' Daniel Kahneman, author of *Thinking, Fast and Slow*

'"Correlation is not causation". That scientific refrain has had social consequences – not least, the long debate over links between tobacco and lung cancer. Judea Pearl proposes a radical mathematical solution. A prominent researcher in AI, Pearl teams up with writer Dana Mackenzie to explore the science of causation, now bearing fruit in biology, medicine, social science and AI' Barbara Kiser, *Nature*

'Judea Pearl has been the heart and soul of a revolution in artificial intelligence and in computer science more broadly. He is an Alan Turing of our time' Eric Horvitz, Managing Director, Microsoft Research

'Modern applications of AI, such as robotics, self-driving cars, speech recognition, and machine translation deal with uncertainty. Pearl has been instrumental in supplying the rationale and much valuable technology that allow these applications to flourish'
Alfred Spector, Vice President of Research, Google, Inc.

Judea Pearl is a world-renowned Israeli-American computer scientist and philosopher, known for his world-leading work in AI and the development of Bayesian networks, as well as his theory of causal and counterfactual inference. In 2011, he won the most prestigious award in computer science, the Alan Turing Award. He has also received the Rumelhart Prize (Cognitive Science Society), the Benjamin Franklin Medal (Franklin Institute) and the Lakatos Award (London School of Economics) and he is the founder and president of the Daniel Pearl Foundation.

Dana Mackenzie, a PhD mathematician turned science writer, has written for such magazines as *Science*, *New Scientist* and *Discover*.

JUDEA PEARL
AND DANA MACKENZIE

The Book of Why
The New Science of Cause and Effect

PENGUIN BOOKS

PENGUIN BOOKS

UK | USA | Canada | Ireland | Australia
India | New Zealand | South Africa

Penguin Books is part of the Penguin Random House group of companies
whose addresses can be found at global.penguinrandomhouse.com.

First published in the United States of America by Basic Books 2018
First published in Great Britain by Allen Lane 2018
Published in Penguin Books 2019
001

Copyright © Judea Pearl and Dana Mackenzie, 2018

The moral rights of the authors have been asserted

Printed and bound in Great Britain by Clays Ltd, Elcograf S.p.A.

A CIP catalogue record for this book is available from the British Library

ISBN: 978–0–141–98241–0

To Ruth

CONTENTS

PREFACE

ALMOST two decades ago, when I wrote the preface to my book *Causality* (2000), I made a rather daring remark that friends advised me to tone down. "Causality has undergone a major transformation," I wrote, "from a concept shrouded in mystery into a mathematical object with well-defined semantics and well-founded logic. Paradoxes and controversies have been resolved, slippery concepts have been explicated, and practical problems relying on causal information that long were regarded as either metaphysical or unmanageable can now be solved using elementary mathematics. Put simply, causality has been mathematized."

Reading this passage today, I feel I was somewhat shortsighted. What I described as a "transformation" turned out to be a "revolution" that has changed the thinking in many of the sciences. Many now call it "the Causal Revolution," and the excitement that it has generated in research circles is spilling over to education and applications. I believe the time is ripe to share it with a broader audience.

This book strives to fulfill a three-pronged mission: first, to lay before you in nonmathematical language the intellectual content of the Causal Revolution and how it is affecting our lives as well as our future; second, to share with you some of the heroic journeys, both successful and failed, that scientists have embarked on when confronted by critical cause-effect questions.

Finally, returning the Causal Revolution to its womb in artificial intelligence, I aim to describe to you how robots can be constructed that learn to communicate in our mother tongue—the language of cause and effect. This new generation of robots should explain to us why things happened, why they responded the way they did, and why nature operates one way and not another. More ambitiously, they should also teach us about ourselves: why our mind clicks the way it does and what it means to think rationally about cause and effect, credit and regret, intent and responsibility.

When I write equations, I have a very clear idea of who my readers are. Not so when I write for the general public—an entirely new adventure for me. Strange, but this new experience has been one of the most rewarding educational trips of my life. The need to shape ideas in your language, to guess your background, your questions, and your reactions, did more to sharpen my understanding of causality than all the equations I have written prior to writing this book.

For this I will forever be grateful to you. I hope you are as excited as I am to see the results.

Judea Pearl
Los Angeles, October 2017

INTRODUCTION: MIND OVER DATA

Every science that has thriven has thriven upon its own symbols.

—Augustus de Morgan (1864)

THIS book tells the story of a science that has changed the way we distinguish facts from fiction and yet has remained under the radar of the general public. The consequences of the new science are already impacting crucial facets of our lives and have the potential to affect more, from the development of new drugs to the control of economic policies, from education and robotics to gun control and global warming. Remarkably, despite the diversity and apparent incommensurability of these problem areas, the new science embraces them all under a unified framework that was practically nonexistent two decades ago.

The new science does not have a fancy name: I call it simply "causal inference," as do many of my colleagues. Nor is it particularly high-tech. The ideal technology that causal inference strives to emulate resides within our own minds. Some tens of thousands of years ago, humans began to realize that certain things cause other things and that tinkering with the former can change the latter. No other species grasps this, certainly not to the extent that we do. From this discovery came organized societies, then towns and cities, and eventually the science- and technology-based civilization we enjoy today. All because we asked a simple question: Why?

Causal inference is all about taking this question seriously. It posits that the human brain is the most advanced tool ever devised for managing causes and effects. Our brains store an incredible amount of causal knowledge which, supplemented by data, we could harness to answer some of the most pressing questions of our time. More ambitiously, once we really understand the logic behind causal thinking, we could emulate it on modern computers and create an "artificial scientist." This smart robot would discover yet unknown phenomena, find explanations to pending scientific dilemmas, design new experiments, and continually extract more causal knowledge from the environment.

But before we can venture to speculate on such futuristic developments, it is important to understand the achievements that causal inference has tallied thus far. We will explore the way that it has transformed the thinking of scientists in almost every data-informed discipline and how it is about to change our lives.

The new science addresses seemingly straightforward questions like these:

- How effective is a given treatment in preventing a disease?
- Did the new tax law cause our sales to go up, or was it our advertising campaign?
- What is the health-care cost attributable to obesity?
- Can hiring records prove an employer is guilty of a policy of sex discrimination?
- I'm about to quit my job. Should I?

These questions have in common a concern with cause-and-effect relationships, recognizable through words such as "preventing," "cause," "attributable to," "policy," and "should I." Such words are common in everyday language, and our society constantly demands answers to such questions. Yet, until very recently, science gave us no means even to articulate, let alone answer, them.

By far the most important contribution of causal inference to mankind has been to turn this scientific neglect into a thing of the past. The new science has spawned a simple mathematical language to articulate causal relationships that we know as well as those we wish to find out about. The ability to express this information in mathematical form has unleashed a wealth of powerful and principled methods for combining our knowledge with data and answering causal questions like the five above.

I have been lucky to be part of this scientific development for the past quarter century. I have watched its progress take shape in students' cubicles and research laboratories, and I have heard its breakthroughs resonate in somber scientific conferences, far from the limelight of public attention. Now, as we enter the era of strong artificial intelligence (AI) and many tout the endless possibilities of Big Data and deep learning, I find it timely and exciting to present to the reader some of the most adventurous paths that the new science is taking, how it impacts data science, and the many ways in which it will change our lives in the twenty-first century.

When you hear me describe these achievements as a "new science," you may be skeptical. You may even ask, Why wasn't this done a long time ago? Say when Virgil first proclaimed, "Lucky is he who has been able to understand the causes of things" (29 BC). Or when the founders of modern statistics, Francis Galton and Karl Pearson, first discovered that population data can shed light on scientific questions. There is a long tale behind their unfortunate failure to embrace causation at this juncture, which the historical sections of this book will relate. But the most serious impediment, in my opinion, has been the fundamental gap between the vocabulary in which we cast causal questions and the traditional vocabulary in which we communicate scientific theories.

To appreciate the depth of this gap, imagine the difficulties that a scientist would face in trying to express some obvious causal relationships—say, that the barometer reading B tracks the atmospheric pressure P. We can easily write down this relationship in an equation such as $B = kP$, where k is some constant of proportionality. The rules of algebra now permit us to rewrite this same equation in a wild variety of forms, for example, $P = B/k$, $k = B/P$, or $B - kP = 0$. They all mean the same thing—that if we know any two of the three quantities, the third is determined. None of the letters k, B, or P is in any mathematical way privileged over any of the others. How then can we express our strong conviction that it is the pressure that causes the barometer to change and not the other way around? And if we cannot express even this, how can we hope to express the many other causal convictions that do not have mathematical formulas, such as that the rooster's crow does not cause the sun to rise?

My college professors could not do it and never complained. I would be willing to bet that none of yours ever did either. We now understand why: never were they shown a mathematical language of causes; nor were they shown its benefits. It is in fact an indictment of science that it has neglected to develop such a language for so many generations. Everyone knows that flipping a switch will cause a light to turn on or off and that a hot, sultry summer afternoon will cause sales to go up at the local ice-cream parlor. Why then have scientists not captured such obvious facts in formulas, as they did with the basic laws of optics, mechanics, or geometry? Why have they allowed these facts to languish in bare intuition, deprived of mathematical tools that have enabled other branches of science to flourish and mature?

Part of the answer is that scientific tools are developed to meet scientific needs. Precisely because we are so good at handling questions about switches, ice cream, and barometers, our need for special mathematical machinery to handle them was not obvious. But as scientific curiosity increased and we began posing causal questions in complex legal, business, medical, and policy-making situations, we found ourselves lacking the tools and principles that mature science should provide.

Belated awakenings of this sort are not uncommon in science. For example, until about four hundred years ago, people were quite happy with their natural ability to manage the uncertainties in daily life, from crossing a street to risking a fistfight. Only after gamblers invented intricate games of chance, sometimes carefully designed to trick us into making bad choices, did mathematicians like Blaise Pascal (1654), Pierre de Fermat (1654), and Christiaan Huygens (1657) find it necessary to develop what we today call probability theory. Likewise, only when insurance organizations demanded accurate estimates of life annuity did mathematicians like Edmond Halley (1693) and Abraham de Moivre (1725) begin looking at mortality tables to calculate life expectancies. Similarly, astronomers' demands for accurate predictions of celestial motion led Jacob Bernoulli, Pierre-Simon Laplace, and Carl Friedrich Gauss to develop a theory of errors to help us extract signals from noise. These methods were all predecessors of today's statistics.

Ironically, the need for a theory of causation began to surface at the same time that statistics came into being. In fact, modern statistics hatched from the causal questions that Galton and Pearson asked about heredity and their ingenious attempts to answer them using cross-generational data. Unfortunately, they failed in this endeavor, and rather than pause to ask why, they declared those questions off limits and turned to developing a thriving, causality-free enterprise called statistics.

This was a critical moment in the history of science. The opportunity to equip causal questions with a language of their own came very close to being realized but was squandered. In the following years, these questions were declared unscientific and went underground. Despite heroic efforts by the geneticist Sewall Wright (1889–1988), causal vocabulary was virtually prohibited for more than half a century. And when you prohibit speech, you prohibit thought and stifle principles, methods, and tools.

Readers do not have to be scientists to witness this prohibition. In Statistics 101, every student learns to chant, "Correlation is not causation." With good reason! The rooster's crow is highly correlated with the sunrise; yet it does not cause the sunrise.

Unfortunately, statistics has fetishized this commonsense observation. It tells us that correlation is not causation, but it does not tell us what causation is. In vain will you search the index of a statistics textbook for an entry on "cause." Students are not allowed to say that X is the cause of Y—only that X and Y are "related" or "associated."

Because of this prohibition, mathematical tools to manage causal questions were deemed unnecessary, and statistics focused exclusively on how to summarize data, not on how to interpret it. A shining exception was path analysis, invented by geneticist Sewall Wright in the 1920s and a direct ancestor of the methods we will entertain in this book. However, path analysis was badly underappreciated in statistics and its satellite communities and languished for decades in its embryonic status. What should have been the first step toward causal inference remained the only step until the 1980s. The rest of statistics, including the many disciplines that looked to it for guidance, remained in the Prohibition era, falsely believing that the answers to all scientific questions reside in the data, to be unveiled through clever data-mining tricks.

Much of this data-centric history still haunts us today. We live in an era that presumes Big Data to be the solution to all our problems. Courses in "data science" are proliferating in our universities, and jobs for "data scientists" are lucrative in the companies that participate in the "data economy." But I hope with this book to convince you that data are profoundly dumb. Data can tell you that the people who took a medicine recovered faster than those who did not take it, but they can't tell you why. Maybe those who took the medicine did so because they could afford it and would have recovered just as fast without it.

Over and over again, in science and in business, we see situations where mere data aren't enough. Most big-data enthusiasts, while somewhat aware of these limitations, continue the chase after data-centric intelligence, as if we were still in the Prohibition era.

As I mentioned earlier, things have changed dramatically in the past three decades. Nowadays, thanks to carefully crafted causal models, contemporary scientists can address problems that would have once been considered unsolvable or even beyond the pale of scientific inquiry. For example, only a hundred years ago, the question of whether cigarette smoking causes a health hazard would have been considered unscientific. The mere mention of the words "cause" or "effect" would create a storm of objections in any reputable statistical journal.

Even two decades ago, asking a statistician a question like "Was it the aspirin that stopped my headache?" would have been like asking if he believed in voodoo. To quote an esteemed colleague of mine, it would be "more of a cocktail conversation topic than a scientific inquiry." But today, epidemiologists, social scientists, computer scientists, and at least some enlightened economists and statisticians pose such questions routinely and answer them with mathematical precision. To me, this change is nothing short of a revolution. I dare to call it the Causal Revolution, a scientific shakeup that embraces rather than denies our innate cognitive gift of understanding cause and effect.

The Causal Revolution did not happen in a vacuum; it has a mathematical secret behind it which can be best described as a calculus of causation, which answers some of the hardest problems ever asked about cause-effect relationships. I am thrilled to unveil this calculus not only because the turbulent history of its development is intriguing but even more because I expect that its full potential will be developed one day beyond what I can imagine . . . perhaps even by a reader of this book.

The calculus of causation consists of two languages: causal diagrams, to express what we know, and a symbolic language, resembling algebra, to express what we want to know. The causal diagrams are simply dot-and-arrow pictures that summarize our existing scientific knowledge. The dots represent quantities of interest, called "variables," and the arrows represent known or suspected causal relationships between those variables—namely, which variable "listens" to which others. These diagrams are extremely easy to draw, comprehend, and use, and the reader will find dozens of them in the pages of this book. If you can navigate using a map of one-way streets, then you can understand causal diagrams, and you can solve the type of questions posed at the beginning of this introduction.

Though causal diagrams are my tool of choice in this book, as in the last thirty-five years of my research, they are not the only kind of causal model possible. Some scientists (e.g., econometricians) like to work with mathematical equations; others (e.g., hard-core statisticians) prefer a list of assumptions that ostensibly summarizes the structure of the diagram. Regardless of language, the model should depict, however qualitatively, the process that generates the data—in other words, the cause-effect forces that operate in the environment and shape the data generated.

Side by side with this diagrammatic "language of knowledge," we also have a symbolic "language of queries" to express the questions we want answers to. For example, if we are interested in the effect of a drug (D) on lifespan (L), then our query might be written symbolically as: $P(L \mid do(D))$. In other words, what is the probability (P) that a typical patient would survive L years if made to take the drug? This question describes what epidemiologists would call an *intervention* or a *treatment* and corresponds to what we measure in a clinical trial. In many cases we may also wish to compare $P(L \mid do(D))$ with $P(L \mid do(\text{not-}D))$; the latter describes patients denied treatment, also called the "control" patients. The *do*-operator signifies that we are dealing with an intervention rather than a passive observation; classical statistics has nothing remotely similar to this operator.

We must invoke an intervention operator $do(D)$ to ensure that the observed change in Lifespan L is due to the drug itself and is not confounded with other factors that tend to shorten or lengthen life. If, instead of intervening, we let the patient himself decide whether to take the drug, those other factors might influence his decision, and lifespan differences between taking and not taking the drug would no longer be solely due to the drug. For example, suppose only those who were terminally ill took the drug. Such persons would surely differ from those who did not take the drug, and a comparison of the two groups would reflect differences in the severity of their disease rather than the effect of the drug. By contrast, forcing patients to take or refrain from taking the drug, regardless of preconditions, would wash away preexisting differences and provide a valid comparison.

Mathematically, we write the observed frequency of Lifespan L among patients who voluntarily take the drug as $P(L \mid D)$, which is the standard conditional probability used in statistical textbooks. This expression stands for the probability (P) of Lifespan L conditional on seeing the patient take Drug D. Note that $P(L \mid D)$ may be totally different from $P(L \mid do(D))$. This difference between seeing and doing is fundamental and explains why we do not regard the falling barometer to be a cause of the coming storm. Seeing the barometer fall increases the probability of the storm, while forcing it to fall does not affect this probability.

This confusion between seeing and doing has resulted in a fountain of paradoxes, some of which we will entertain in this book. A world devoid of $P(L \mid do(D))$ and governed solely by $P(L \mid D)$ would be a strange one indeed. For example, patients would avoid going to the doctor to reduce the probability of being seriously ill; cities would dismiss their firefighters to reduce the incidence of fires; doctors would recommend a drug to male and female patients but not to patients with undisclosed gender; and so on. It is hard to believe that less than three decades ago science did operate in such a world: the *do*-operator did not exist.

One of the crowning achievements of the Causal Revolution has been to explain how to predict the effects of an intervention without actually enacting it. It would never have been possible if we had not, first of all, defined the *do*-operator so that we can ask the right question and, second, devised a way to emulate it by non-invasive means.

When the scientific question of interest involves retrospective thinking, we call on another type of expression unique to causal reasoning called a counterfactual. For example, suppose that Joe took Drug D and died a month later; our question of interest is whether the drug might have caused his death. To answer this question, we need to imagine a scenario in which Joe was about to take the drug but changed his mind. Would he have lived?

Again, classical statistics only summarizes data, so it does not provide even a language for asking that question. Causal inference provides a notation and, more importantly, offers a solution. As with predicting the effect of interventions (mentioned above), in many cases we can emulate human retrospective thinking with an algorithm that takes what we know about the observed world and produces an answer about the counterfactual world. This "algorithmization of counterfactuals" is another gem uncovered by the Causal Revolution.

Counterfactual reasoning, which deals with what-ifs, might strike some readers as unscientific. Indeed, empirical observation can never confirm or refute the answers to such questions. Yet our minds make very reliable and reproducible judgments all the time about what might be or might have been. We all understand, for instance, that had the rooster been silent this morning, the sun would have risen just as well. This consensus stems from the fact that counterfactuals are not products of whimsy but reflect the very structure of our world model. Two people who share the same causal model will also share all counterfactual judgments.

Counterfactuals are the building blocks of moral behavior as well as scientific thought. The ability to reflect on one's past actions and envision alternative scenarios is the basis of free will and social responsibility. The algorithmization of counterfactuals invites thinking machines to benefit from this ability and participate in this (until now) uniquely human way of thinking about the world.

My mention of thinking machines in the last paragraph is intentional. I came to this subject as a computer scientist working in the area of artificial intelligence, which entails two points of departure from most of my colleagues in the causal inference arena. First, in the world of AI, you do not really understand a topic until you can teach it to a mechanical robot. That is why you will find me emphasizing and reemphasizing notation, language, vocabulary, and grammar. For example, I obsess over whether we can express a certain claim in a given language and whether one claim follows from others. It is amazing how much one can learn from just following the grammar of scientific utterances. My emphasis on language also comes from a deep conviction that language shapes our thoughts. You cannot answer a question that you cannot ask, and you cannot ask a question that you have no words for. As a student of philosophy and computer science, my attraction to causal inference has largely been triggered by the excitement of seeing an orphaned scientific language making it from birth to maturity.

My background in machine learning has given me yet another incentive for studying causation. In the late 1980s, I realized that machines' lack of understanding of causal relations was perhaps the biggest roadblock to giving them human-level intelligence. In the last chapter of this book, I will return to my roots, and together we will explore the implications of the Causal Revolution for artificial intelligence. I believe that strong AI is an achievable goal and one not to be feared precisely because causality is part of the solution. A causal reasoning module will give machines the ability to reflect on their mistakes, to pinpoint weaknesses in their software, to function as moral entities, and to converse naturally with humans about their own choices and intentions.

A BLUEPRINT OF REALITY

In our era, readers have no doubt heard terms like "knowledge," "information," "intelligence," and "data," and some may feel confused about the differences between them or how they interact. Now I am proposing to throw another term, "causal model," into the mix, and the reader may justifiably wonder if this will only add to the confusion.

It will not! In fact, it will anchor the elusive notions of science, knowledge, and data in a concrete and meaningful setting, and will enable us to see how the three work together to produce answers to difficult scientific questions. Figure I.1 shows a blueprint for a "causal inference engine" that might handle causal reasoning for a future artificial intelligence. It's important to realize that this is not only a blueprint for the future but also a guide to how causal models work in scientific applications today and how they interact with data.

The inference engine is a machine that accepts three different kinds of inputs—Assumptions, Queries, and Data—and produces three kinds of outputs. The first of the outputs is a Yes/No decision as to whether the given query can in theory be answered under the existing causal model, assuming perfect and unlimited data. If the answer is Yes, the inference engine next produces an Estimand. This is a mathematical formula that can be thought of as a recipe for generating the answer from any hypothetical data, whenever they are available. Finally, after the inference engine has received

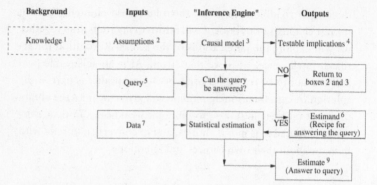

FIGURE I. How an "inference engine" combines data with causal knowledge to produce answers to queries of interest. The dashed box is not part
of the engine but is required for building it. Arrows could also be drawn
from boxes 4 and 9 to box 1, but I have opted to keep the diagram simple.

the Data input, it will use the recipe to produce an actual Estimate
for the answer, along with statistical estimates of the amount of
uncertainty in that estimate. This uncertainty reflects the limited
size of the data set as well as possible measurement errors or missing data.

To dig more deeply into the chart, I have labeled the boxes 1
through 9, which I will annotate in the context of the query "What
is the effect of Drug D on Lifespan L?"

1. "Knowledge" stands for traces of experience the reasoning
   agent has had in the past, including past observations, past
   actions, education, and cultural mores, that are deemed
   relevant to the query of interest. The dotted box around
   "Knowledge" indicates that it remains implicit in the mind
   of the agent and is not explicated formally in the model.
2. Scientific research always requires simplifying assumptions,
   that is, statements which the researcher deems worthy of
   making explicit on the basis of the available Knowledge.
   While most of the researcher's knowledge remains implicit
   in his or her brain, only Assumptions see the light of day and

are encapsulated in the model. They can in fact be read from the model, which has led some logicians to conclude that a model is nothing more than a list of assumptions. Computer scientists take exception to this claim, noting that how assumptions are represented can make a profound difference in one's ability to specify them correctly, draw conclusions from them, and even extend or modify them in light of compelling evidence.

3. Various options exist for causal models: causal diagrams, structural equations, logical statements, and so forth. I am strongly sold on causal diagrams for nearly all applications, primarily due to their transparency but also due to the explicit answers they provide to many of the questions we wish to ask. For the purpose of constructing the diagram, the definition of "causation" is simple, if a little metaphorical: a variable X is a cause of Y if Y "listens" to X and determines its value in response to what it hears. For example, if we suspect that a patient's Lifespan L "listens" to whether Drug D was taken, then we call D a cause of L and draw an arrow from D to L in a causal diagram. Naturally, the answer to our query about D and L is likely to depend on other variables as well, which must also be represented in the diagram along with their causes and effects. (Here, we will denote them collectively by Z.)

4. The listening pattern prescribed by the paths of the causal model usually results in observable patterns or dependencies in the data. These patterns are called "testable implications" because they can be used for testing the model. These are statements like "There is no path connecting D and L," which translates to a statistical statement, "D and L are independent," that is, finding D does not change the likelihood of L. If the data contradict this implication, then we need to revise our model. Such revisions require another engine, which obtains its inputs from boxes 4 and 7 and computes the "degree of fitness," that is, the degree to which the Data are compatible with the model's assumptions. For simplicity, I did not show this second engine in Figure I.1.

5. Queries submitted to the inference engine are the scientific questions that we want to answer. They must be formulated in causal vocabulary. For example, what is $P(L \mid do(D))$? One of the main accomplishments of the Causal Revolution has been to make this language scientifically transparent as well as mathematically rigorous.

6. "Estimand" comes from Latin, meaning "that which is to be estimated." This is a statistical quantity to be estimated from the data that, once estimated, can legitimately represent the answer to our query. While written as a probability formula—for example, $P(L \mid D, Z) \times P(Z)$—it is in fact a recipe for answering the causal query from the type of data we have, once it has been certified by the engine.

   It's very important to realize that, contrary to traditional estimation in statistics, some queries may not be answerable under the current causal model, even after the collection of any amount of data. For example, if our model shows that both D and L depend on a third variable Z (say, the stage of a disease), and if we do not have any way to measure Z, then the query $P(L \mid do(D))$ cannot be answered. In that case it is a waste of time to collect data. Instead we need to go back and refine the model, either by adding new scientific knowledge that might allow us to estimate Z or by making simplifying assumptions (at the risk of being wrong)—for example, that the effect of Z on D is negligible.

7. Data are the ingredients that go into the estimand recipe. It is critical to realize that data are profoundly dumb about causal relationships. They tell us about quantities like $P(L \mid D)$ or $P(L \mid D, Z)$. It is the job of the estimand to tell us how to bake these statistical quantities into one expression that, based on the model assumptions, is logically equivalent to the causal query—say, $P(L \mid do(D))$.

   Notice that the whole notion of estimands and in fact the whole top part of Figure I does not exist in traditional methods of statistical analysis. There, the estimand and the query coincide. For example, if we are interested in the proportion of people among those with Lifespan L who took the Drug D, we simply write this query as $P(D \mid L)$. The same quantity would be our estimand. This already specifies what proportions in the data need to be estimated and requires no causal knowledge. For this reason, some statisticians to this day find it extremely hard to understand why some knowledge lies outside the province of statistics and why data alone cannot make up for lack of scientific knowledge.

8. The estimate is what comes out of the oven. However, it is only approximate because of one other real-world fact about data: they are always only a finite sample from a theoretically infinite population. In our running example, the sample consists of the patients we choose to study. Even if we choose them at random, there is always some chance that the proportions measured in the sample are not representative of the proportions in the population at large. Fortunately, the discipline of statistics, nowadays empowered by advanced techniques of machine learning, gives us many, many ways to manage this uncertainty—parametric and semi-parametric models, maximum likelihood methods, and propensity scores are often used to smooth the sparse data.

9. In the end, if our model is correct and our data are sufficient, we get an answer to our causal query, such as "Drug D increases the Lifespan L of diabetic Patients Z by 30 percent, plus or minus 20 percent." Hooray! The answer will also add to our scientific knowledge (box 1) and, if things did not go the way we expected, might suggest some improvements to our causal model (box 3).

This flowchart may look complicated at first, and you might wonder whether it is really necessary. Indeed, in our ordinary lives, we are somehow able to make causal judgments without consciously going through such a complicated process and certainly without resorting to the mathematics of probabilities and proportions. Our causal intuition alone is usually sufficient for handling the kind of uncertainty we find in household routines or even in our professional lives. But if we want to teach a dumb robot to think causally, or if we are pushing the frontiers of scientific knowledge, where we do not have intuition to guide us, then a carefully structured procedure like this is mandatory.

I especially want to highlight the role of data in the above process. First, notice that we collect data only after we posit the causal model, after we state the scientific query we wish to answer, and after we derive the estimand. This contrasts with the traditional statistical approach, mentioned above, which does not even have a causal model.

But our present-day scientific world presents a new challenge to sound reasoning about causes and effects. While awareness of the need for a causal model has grown by leaps and bounds among the sciences, many researchers in artificial intelligence would like to skip the hard step of constructing or acquiring a causal model and rely solely on data for all cognitive tasks. The hope—and at present, it is usually a silent one—is that the data themselves will guide us to the right answers whenever causal questions come up.

I am an outspoken skeptic of this trend because I know how profoundly dumb data are about causes and effects. For example, information about the effects of actions or interventions is simply not available in raw data, unless it is collected by controlled experimental manipulation. By contrast, if we are in possession of a causal model, we can often predict the result of an intervention from hands-off, intervention-free data.

The case for causal models becomes even more compelling when we seek to answer counterfactual queries such as "What would have happened had we acted differently?" We will discuss counterfactuals in great detail because they are the most challenging queries for any artificial intelligence. They are also at the core of the cognitive advances that made us human and the imaginative abilities that have made science possible. We will also explain why any query about the mechanism by which causes transmit their effects—the most prototypical "Why?" question—is actually a counterfactual question in disguise. Thus, if we ever want robots to answer "Why?" questions or even understand what they mean, we must equip them with a causal model and teach them how to answer counterfactual queries, as in Figure I.1.

Another advantage causal models have that data mining and deep learning lack is adaptability. Note that in Figure I.1, the estimand is computed on the basis of the causal model alone, prior to an examination of the specifics of the data. This makes the causal inference engine supremely adaptable, because the estimand computed is good for any data that are compatible with the qualitative model, regardless of the numerical relationships among the variables.

To see why this adaptability is important, compare this engine with a learning agent—in this instance a human, but in other cases perhaps a deep-learning algorithm or maybe a human using a deep-learning algorithm—trying to learn solely from the data. By observing the outcome L of many patients given Drug D, she is able to predict the probability that a patient with characteristics Z will survive L years. Now she is transferred to a different hospital, in a different part of town, where the population characteristics (diet, hygiene, work habits) are different. Even if these new characteristics merely modify the numerical relationships among the variables recorded, she will still have to retrain herself and learn a new prediction function all over again. That's all that a deep-learning program can do: fit a function to data. On the other hand, if she possessed a model of how the drug operated and its causal structure remained intact in the new location, then the estimand she obtained in training would remain valid. It could be applied to the new data to generate a new population-specific prediction function.

Many scientific questions look different "through a causal lens," and I have delighted in playing with this lens, which over the last twenty-five years has been increasingly empowered by new insights and new tools. I hope and believe that readers of this book will share in my delight. Therefore, I'd like to close this introduction with a preview of some of the coming attractions in this book.

Chapter 1 assembles the three steps of observation, intervention, and counterfactuals into the Ladder of Causation, the central metaphor of this book. It will also expose you to the basics of reasoning with causal diagrams, our main modeling tool, and set you well on your way to becoming a proficient causal reasoner—in fact, you will be far ahead of generations of data scientists who attempted to interpret data through a model-blind lens, oblivious to the distinctions that the Ladder of Causation illuminates.

Chapter 2 tells the bizarre story of how the discipline of statistics inflicted causal blindness on itself, with far-reaching effects for all sciences that depend on data. It also tells the story of one of the great heroes of this book, the geneticist Sewall Wright, who in the 1920s drew the first causal diagrams and for many years was one of the few scientists who dared to take causality seriously.

Chapter 3 relates the equally curious story of how I became a convert to causality through my work in AI and particularly on Bayesian networks. These were the first tool that allowed computers to think in "shades of gray"—and for a time I believed they held the key to unlocking AI. Toward the end of the 1980s I became convinced that I was wrong, and this chapter tells of my journey from prophet to apostate. Nevertheless, Bayesian networks remain a very important tool for AI and still encapsulate much of the mathematical foundation of causal diagrams. In addition to a gentle, causality-minded introduction to Bayes's rule and Bayesian methods of reasoning, Chapter 3 will entertain the reader with examples of real-life applications of Bayesian networks.

Chapter 4 tells about the major contribution of statistics to causal inference: the randomized controlled trial (RCT). From a causal perspective, the RCT is a man-made tool for uncovering the query $P(L \mid do(D))$, which is a property of nature. Its main purpose is to disassociate variables of interest (say, D and L) from other variables (Z) that would otherwise affect them both. Disarming the distortions, or "confounding," produced by such lurking variables has been a century-old problem. This chapter walks the reader through a surprisingly simple solution to the general confounding problem, which you will grasp in ten minutes of playfully tracing paths in a diagram.

Chapter 5 gives an account of a seminal moment in the history of causation and indeed the history of science, when statisticians struggled with the question of whether smoking causes lung cancer. Unable to use their favorite tool, the randomized controlled trial, they struggled to agree on an answer or even on how to make sense of the question. The smoking debate brings the importance of causality into its sharpest focus. Millions of lives were lost or shortened because scientists did not have an adequate language or methodology for answering causal questions.

Chapter 6 will, I hope, be a welcome diversion for the reader after the serious matters of Chapter 5. This is a chapter of paradoxes: the Monty Hall paradox, Simpson's paradox, Berkson's paradox, and others. Classical paradoxes like these can be enjoyed as brainteasers, but they have a serious side too, especially when viewed from a causal perspective. In fact, almost all of them represent clashes with causal intuition and therefore reveal the anatomy of that intuition. They were canaries in the coal mine that should have alerted scientists to the fact that human intuition is grounded in causal, not statistical, logic. I believe that the reader will enjoy this new twist on his or her favorite old paradoxes.

Chapters 7 to 9 finally take readers on a thrilling ascent of the Ladder of Causation. We start in Chapter 7 with questions about intervention and explain how my students and I went through a twenty-year struggle to automate the answers to *do*-type questions. We succeeded, and this chapter explains the guts of the "causal inference engine," which produces the yes/no answer and the estimand in Figure I.1. Studying this engine will empower the reader to spot certain patterns in the causal diagram that deliver immediate answers to the causal query. These patterns are called back-door adjustment, front-door adjustment, and instrumental variables, the workhorses of causal inference in practice.

Chapter 8 takes you to the top of the ladder by discussing counterfactuals. These have been seen as a fundamental part of causality at least since 1748, when Scottish philosopher David Hume proposed the following somewhat contorted definition of causation: "We may define a cause to be an object followed by another, and where all the objects, similar to the first, are followed by objects similar to the second. Or, in other words, where, if the first object had not been, the second never had existed." David Lewis, a philosopher at Princeton University who died in 2001, pointed out that Hume really gave two definitions, not one, the first of regularity (i.e., the cause is regularly followed by the effect) and the second of the counterfactual ("if the first object had not been . . ."). While philosophers and scientists had mostly paid attention to the regularity definition, Lewis argued that the counterfactual definition aligns more closely with human intuition: "We think of a cause as something that makes a difference, and the difference it makes must be a difference from what would have happened without it."

Readers will be excited to find out that we can now move past the academic debates and compute an actual value (or probability) for any counterfactual query, no matter how convoluted. Of special interest are questions concerning necessary and sufficient causes of observed events. For example, how likely is it that the defendant's action was a necessary cause of the claimant's injury? How likely is it that man-made climate change is a sufficient cause of a heat wave?

Finally, Chapter 9 discusses the topic of mediation. You may have wondered, when we talked about drawing arrows in a causal diagram, whether we should draw an arrow from Drug D to Lifespan L if the drug affects lifespan only by way of its effect on blood pressure Z (a mediator). In other words, is the effect of D on L direct or indirect? And if both, how do we assess their relative importance? Such questions are not only of great scientific interest but also have practical ramifications; if we understand the mechanism through which a drug acts, we might be able to develop other drugs with the same effect that are cheaper or have fewer side effects. The reader will be pleased to discover how this age-old quest for a mediation mechanism has been reduced to an algebraic exercise and how scientists are using the new tools in the causal tool kit to solve such problems.

Chapter 10 brings the book to a close by coming back to the problem that initially led me to causation: the problem of automating human-level intelligence (sometimes called "strong AI"). I believe that causal reasoning is essential for machines to communicate with us in our own language about policies, experiments, explanations, theories, regret, responsibility, free will, and obligations—and, eventually, to make their own moral decisions.

If I could sum up the message of this book in one pithy phrase, it would be that you are smarter than your data. Data do not understand causes and effects; humans do. I hope that the new science of causal inference will enable us to better understand how we do it, because there is no better way to understand ourselves than by emulating ourselves. In the age of computers, this new understanding also brings with it the prospect of amplifying our innate abilities so that we can make better sense of data, be it big or small.

1

THE LADDER OF CAUSATION

I N the beginning . . .

I was probably six or seven years old when I first read the story of Adam and Eve in the Garden of Eden. My classmates and I were not at all surprised by God's capricious demands, forbidding them to eat from the Tree of Knowledge. Deities have their reasons, we thought. What we were more intrigued by was the idea that as soon as they ate from the Tree of Knowledge, Adam and Eve became conscious, like us, of their nakedness.

As teenagers, our interest shifted slowly to the more philosophical aspects of the story. (Israeli students read Genesis several times a year.) Of primary concern to us was the notion that the emergence of human knowledge was not a joyful process but a painful one, accompanied by disobedience, guilt, and punishment. Was it worth giving up the carefree life of Eden? some asked. Were the agricultural and scientific revolutions that followed worth the economic hardships, wars, and social injustices that modern life entails?

Don't get me wrong: we were no creationists; even our teachers were Darwinists at heart. We knew, however, that the author who choreographed the story of Genesis struggled to answer the most pressing philosophical questions of his time. We likewise suspected that this story bore the cultural footprints of the actual process by which *Homo sapiens* gained dominion over our planet. What, then, was the sequence of steps in this speedy, super-evolutionary process?

My interest in these questions waned in my early career as a professor of engineering but was reignited suddenly in the 1990s, when, while writing my book *Causality*, I confronted the Ladder of Causation.

As I reread Genesis for the hundredth time, I noticed a nuance that had somehow eluded my attention for all those years. When God finds Adam hiding in the garden, he asks, "Have you eaten from the tree which I forbade you?" And Adam answers, "The woman you gave me for a companion, she gave me fruit from the tree and I ate." "What is this you have done?" God asks Eve. She replies, "The serpent deceived me, and I ate."

As we know, this blame game did not work very well on the Almighty, who banished both of them from the garden. But here is the point I had missed before: God asked "what," and they answered "why." God asked for the facts, and they replied with explanations. Moreover, both were thoroughly convinced that naming causes would somehow paint their actions in a different light. Where did they get this idea?

For me, these nuances carried three profound implications. First, very early in our evolution, we humans realized that the world is not made up only of dry facts (what we might call data today); rather, these facts are glued together by an intricate web of cause-effect relationships. Second, causal explanations, not dry facts, make up the bulk of our knowledge, and should be the cornerstone of machine intelligence. Finally, our transition from processors of data to makers of explanations was not gradual; it was a leap that required an external push from an uncommon fruit. This matched perfectly with what I had observed theoretically in the Ladder of Causation: No machine can derive explanations from raw data. It needs a push.

If we seek confirmation of these messages from evolutionary science, we won't find the Tree of Knowledge, of course, but we still see a major unexplained transition. We understand now that humans evolved from apelike ancestors over a period of 5 million to 6 million years and that such gradual evolutionary processes are not uncommon to life on earth. But in roughly the last 50,000 years, something unique happened, which some call the Cognitive

Revolution and others (with a touch of irony) call the Great Leap Forward. Humans acquired the ability to modify their environment and their own abilities at a dramatically faster rate.

For example, over millions of years, eagles and owls have evolved truly amazing eyesight—yet they've never devised eyeglasses, microscopes, telescopes, or night-vision goggles. Humans have produced these miracles in a matter of centuries. I call this phenomenon the "super-evolutionary speedup." Some readers might object to my comparing apples and oranges, evolution to engineering, but that is exactly my point. Evolution has endowed us with the ability to engineer our lives, a gift she has not bestowed on eagles and owls, and the question, again, is "Why?" What computational facility did humans suddenly acquire that eagles did not?

Many theories have been proposed, but one is especially pertinent to the idea of causation. In his book *Sapiens*, historian Yuval Harari posits that our ancestors' capacity to imagine nonexistent things was the key to everything, for it allowed them to communicate better. Before this change, they could only trust people from their immediate family or tribe. Afterward their trust extended to larger communities, bound by common fantasies (for example, belief in invisible yet imaginable deities, in the afterlife, and in the divinity of the leader) and expectations. Whether or not you agree with Harari's theory, the connection between imagining and causal relations is almost self-evident. It is useless to ask for the causes of things unless you can imagine their consequences. Conversely, you cannot claim that Eve caused you to eat from the tree unless you can imagine a world in which, counter to facts, she did not hand you the apple.

Back to our *Homo sapiens* ancestors: their newly acquired causal imagination enabled them to do many things more efficiently through a tricky process we call "planning." Imagine a tribe preparing for a mammoth hunt. What would it take for them to succeed? My mammoth-hunting skills are rusty, I must admit, but as a student of thinking machines, I have learned one thing: a thinking entity (computer, caveman, or professor) can only accomplish a task of such magnitude by planning things in advance—by deciding how many hunters to recruit; by gauging, given wind conditions, the direction from which to approach the mammoth; in short, by imagining and comparing the consequences of several hunting strategies. To do this, the thinking entity must possess, consult, and manipulate a mental model of its reality.

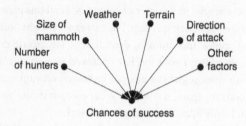

FIGURE 1.1. Perceived causes of a successful mammoth hunt.

Figure 1.1 shows how we might draw such a mental model. Each dot in Figure 1.1 represents a cause of success. Note that there are multiple causes and that none of them are deterministic. That is, we cannot be sure that having more hunters will enable success or that rain will prevent it, but these factors do change the probability of success.

The mental model is the arena where imagination takes place. It enables us to experiment with different scenarios by making local alterations to the model. Somewhere in our hunters' mental model was a subroutine that evaluated the effect of the number of hunters. When they considered adding more, they didn't have to evaluate every other factor from scratch. They could make a local change to the model, replacing "Hunters = 8" with "Hunters = 9," and reevaluate the probability of success. This modularity is a key feature of causal models.

I don't mean to imply, of course, that early humans actually drew a pictorial model like this one. But when we seek to emulate human thought on a computer, or indeed when we try to solve unfamiliar scientific problems, drawing an explicit dots-and-arrows picture is extremely useful. These causal diagrams are the computational core of the "causal inference engine" described in the Introduction.

THE THREE LEVELS OF CAUSATION

So far I may have given the impression that the ability to organize our knowledge of the world into causes and effects was monolithic and acquired all at once. In fact, my research on machine learning has taught me that a causal learner must master at least three distinct levels of cognitive ability: seeing, doing, and imagining.

The first, seeing or observing, entails detection of regularities in our environment and is shared by many animals as well as early humans before the Cognitive Revolution. The second, doing, entails predicting the effect(s) of deliberate alterations of the environment and choosing among these alterations to produce a desired outcome. Only a small handful of species have demonstrated elements of this skill. Use of tools, provided it is intentional and not just accidental or copied from ancestors, could be taken as a sign of reaching this second level. Yet even tool users do not necessarily possess a "theory" of their tool that tells them why it works and what to do when it doesn't. For that, you need to have achieved a level of understanding that permits imagining. It was primarily this third level that prepared us for further revolutions in agriculture and science and led to a sudden and drastic change in our species' impact on the planet.

I cannot prove this, but I can prove mathematically that the three levels differ fundamentally, each unleashing capabilities that the ones below it do not. The framework I use to show this goes back to Alan Turing, the pioneer of research in artificial intelligence (AI), who proposed to classify a cognitive system in terms of the queries it can answer. This approach is exceptionally fruitful when we are talking about causality because it bypasses long and unproductive discussions of what exactly causality is and focuses instead on the concrete and answerable question "What can a causal reasoner do?" Or more precisely, what can an organism possessing a causal model compute that one lacking such a model cannot?

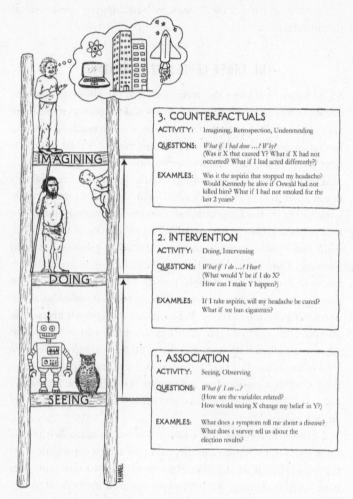

FIGURE 1.2. The Ladder of Causation, with representative organisms at each level. Most animals, as well as present-day learning machines, are on the first rung, learning from association. Tool users, such as early humans, are on the second rung if they act by planning and not merely by imitation. We can also use experiments to learn the effects of interventions, and presumably this is how babies acquire much of their causal knowledge. Counterfactual learners, on the top rung, can imagine worlds that do not exist and infer reasons for observed phenomena. (*Source:* Drawing by Maayan Harel.)

While Turing was looking for a binary classification—human or nonhuman—ours has three tiers, corresponding to progressively more powerful causal queries. Using these criteria, we can assemble the three levels of queries into one Ladder of Causation (Figure 1.2), a metaphor that we will return to again and again.

Let's take some time to consider each rung of the ladder in detail. At the first level, association, we are looking for regularities in observations. This is what an owl does when observing how a rat moves and figuring out where the rodent is likely to be a moment later, and it is what a computer Go program does when it studies a database of millions of Go games so that it can figure out which moves are associated with a higher percentage of wins. We say that one event is associated with another if observing one changes the likelihood of observing the other.

The first rung of the ladder calls for predictions based on passive observations. It is characterized by the question "What if I see . . . ?" For instance, imagine a marketing director at a department store who asks, "How likely is a customer who bought toothpaste to also buy dental floss?" Such questions are the bread and butter of statistics, and they are answered, first and foremost, by collecting and analyzing data. In our case, the question can be answered by first taking the data consisting of the shopping behavior of all customers, selecting only those who bought toothpaste, and, focusing on the latter group, computing the proportion who also bought dental floss. This proportion, also known as a "conditional probability," measures (for large data) the degree of association between "buying toothpaste" and "buying floss." Symbolically, we can write it as $P(floss \mid toothpaste)$. The "P" stands for "probability," and the vertical line means "given that you see."

Statisticians have developed many elaborate methods to reduce a large body of data and identify associations between variables. "Correlation" or "regression," a typical measure of association mentioned often in this book, involves fitting a line to a collection of data points and taking the slope of that line. Some associations might have obvious causal interpretations; others may not. But statistics alone cannot tell which is the cause and which is the effect, toothpaste or floss. From the point of view of the sales manager, it may not really matter. Good predictions need not have good explanations. The owl can be a good hunter without understanding why the rat always goes from point *A* to point *B*.

Some readers may be surprised to see that I have placed present-day learning machines squarely on rung one of the Ladder of Causation, sharing the wisdom of an owl. We hear almost every day, it seems, about rapid advances in machine learning systems— self-driving cars, speech-recognition systems, and, especially in recent years, deep-learning algorithms (or deep neural networks). How could they still be only at level one?

The successes of deep learning have been truly remarkable and have caught many of us by surprise. Nevertheless, deep learning has succeeded primarily by showing that certain questions or tasks we thought were difficult are in fact not. It has not addressed the truly difficult questions that continue to prevent us from achieving humanlike AI. As a result the public believes that "strong AI," machines that think like humans, is just around the corner or maybe even here already. In reality, nothing could be farther from the truth. I fully agree with Gary Marcus, a neuroscientist at New York University, who recently wrote in the *New York Times* that the field of artificial intelligence is "bursting with microdiscoveries"— the sort of things that make good press releases—but machines are still disappointingly far from humanlike cognition. My colleague in computer science at the University of California, Los Angeles, Adnan Darwiche, has titled a position paper "Human-Level Intelligence or Animal-Like Abilities?" which I think frames the question in just the right way. The goal of strong AI is to produce machines with humanlike intelligence, able to converse with and guide humans. Deep learning has instead given us machines with truly impressive abilities but no intelligence. The difference is profound and lies in the absence of a model of reality.

Just as they did thirty years ago, machine learning programs (including those with deep neural networks) operate almost entirely in an associational mode. They are driven by a stream of observations to which they attempt to fit a function, in much the same way that a statistician tries to fit a line to a collection of points. Deep neural networks have added many more layers to the complexity of the fitted function, but raw data still drives the fitting process. They continue to improve in accuracy as more data are fitted, but they do not benefit from the "super-evolutionary speedup." If, for example, the programmers of a driverless car want it to react differently to new situations, they have to add those new reactions explicitly. The machine will not figure out for itself that a pedestrian with a bottle of whiskey in hand is likely to respond differently to a honking horn. This lack of flexibility and adaptability is inevitable in any system that works at the first level of the Ladder of Causation.

We step up to the next level of causal queries when we begin to change the world. A typical question for this level is "What will happen to our floss sales if we double the price of toothpaste?" This already calls for a new kind of knowledge, absent from the data, which we find at rung two of the Ladder of Causation, intervention.

Intervention ranks higher than association because it involves not just seeing but changing what is. Seeing smoke tells us a totally different story about the likelihood of fire than making smoke. We cannot answer questions about interventions with passively collected data, no matter how big the data set or how deep the neural network. Many scientists have been quite traumatized to learn that none of the methods they learned in statistics is sufficient even to articulate, let alone answer, a simple question like "What happens if we double the price?" I know this because on many occasions I have helped them climb to the next rung of the ladder.

Why can't we answer our floss question just by observation? Why not just go into our vast database of previous purchases and see what happened previously when toothpaste cost twice as much? The reason is that on the previous occasions, the price may have been higher for different reasons. For example, the product may have been in short supply, and every other store also had to raise its price. But now you are considering a deliberate intervention that will set a new price regardless of market conditions. The result might be quite different from when the customer couldn't find a better deal elsewhere. If you had data on the market conditions that existed on the previous occasions, perhaps you could make a better prediction . . . but what data do you need? And then, how would you figure it out? Those are exactly the questions the science of causal inference allows us to answer.

A very direct way to predict the result of an intervention is to experiment with it under carefully controlled conditions. Big-data companies like Facebook know this and constantly perform experiments to see what happens if items on the screen are arranged differently or the customer gets a different prompt (or even a different price).

More interesting and less widely known—even in Silicon Valley—is that successful predictions of the effects of interventions can sometimes be made even without an experiment. For example, the sales manager could develop a model of consumer behavior that includes market conditions. Even if she doesn't have data on every factor, she might have data on enough key surrogates to make the prediction. A sufficiently strong and accurate causal model can allow us to use rung-one (observational) data to answer rung-two (interventional) queries. Without the causal model, we could not go from rung one to rung two. This is why deep-learning systems (as long as they use only rung-one data and do not have a causal model) will never be able to answer questions about interventions, which by definition break the rules of the environment the machine was trained in.

As these examples illustrate, the defining query of the second rung of the Ladder of Causation is "What if we do . . . ?" What will happen if we *change* the environment? We can write this kind of query as $P(floss \mid do(toothpaste))$, which asks about the probability that we will sell floss at a certain price, given that we set the price of toothpaste at another price.

Another popular question at the second level of causation is "How?," which is a cousin of "What if we do . . . ?" For instance, the manager may tell us that we have too much toothpaste in our warehouse. "How can we sell it?" he asks. That is, what price should we set for it? Again, the question refers to an intervention, which we want to perform mentally before we decide whether and how to do it in real life. That requires a causal model.

We perform interventions all the time in our daily lives, although we don't usually use such a fancy term for them. For example, when we take aspirin to cure a headache, we are intervening on one variable (the quantity of aspirin in our body) in order to affect another one (our headache status). If we are correct in our causal belief about aspirin, the "outcome" variable will respond by changing from "headache" to "no headache."

While reasoning about interventions is an important step on the causal ladder, it still does not answer all questions of interest. We might wonder, My headache is gone now, but why? Was it the aspirin I took? The food I ate? The good news I heard? These queries take us to the top rung of the Ladder of Causation, the level of counterfactuals, because to answer them we must go back in time, change history, and ask, "What would have happened if I had not taken the aspirin?" No experiment in the world can deny treatment to an already treated person and compare the two outcomes, so we must import a whole new kind of knowledge.

Counterfactuals have a particularly problematic relationship with data because data are, by definition, facts. They cannot tell us what will happen in a counterfactual or imaginary world where some observed facts are bluntly negated. Yet the human mind makes such explanation-seeking inferences reliably and repeatably. Eve did it when she identified "The serpent deceived me" as the reason for her action. This ability most distinguishes human from animal intelligence, as well as from model-blind versions of AI and machine learning.

You may be skeptical that science can make any useful statement about "would haves," worlds that do not exist and things that have not happened. But it does and always has. The laws of physics, for example, can be interpreted as counterfactual assertions, such as "Had the weight on this spring doubled, its length would have doubled as well" (Hooke's law). This statement is, of course, backed by a wealth of experimental (rung-two) evidence, derived from hundreds of springs, in dozens of laboratories, on thousands of different occasions. However, once anointed as a "law," physicists interpret it as a functional relationship that governs this very spring, at this very moment, under hypothetical values of the weight. All of these different worlds, where the weight is x pounds and the length of the spring is L_x inches, are treated as objectively knowable and simultaneously active, even though only one of them actually exists.

Going back to the toothpaste example, a top-rung question would be "What is the probability that a customer who bought toothpaste would still have bought it if we had doubled the price?" We are comparing the real world (where we know that the customer bought the toothpaste at the current price) to a fictitious world (where the price is twice as high).

The rewards of having a causal model that can answer counterfactual questions are immense. Finding out why a blunder occurred allows us to take the right corrective measures in the future. Finding out why a treatment worked on some people and not on others can lead to a new cure for a disease. Answering the question "What if things had been different?" allows us to learn from history and the experience of others, something that no other species appears to do. It is not surprising that the ancient Greek philosopher Democritus (460–370 BC) said, "I would rather discover one cause than be the King of Persia."

The position of counterfactuals at the top of the Ladder of Causation explains why I place such emphasis on them as a key moment in the evolution of human consciousness. I totally agree with Yuval Harari that the depiction of imaginary creatures was a manifestation of a new ability, which he calls the Cognitive Revolution. His prototypical example is the Lion Man sculpture, found in Stadel Cave in southwestern Germany and now held at the Ulm Museum (see Figure 1.3). The Lion Man, roughly 40,000 years old, is a mammoth tusk sculpted into the form of a chimera, half man and half lion.

We do not know who sculpted the Lion Man or what its purpose was, but we do know that anatomically modern humans made it and that it represents a break with any art or craft that had gone before. Previously, humans had fashioned tools and representational art, from beads to flutes to spear points to elegant carvings

FIGURE 1.3. The Lion Man of Stadel Cave. The earliest known representation of an imaginary creature (half man and half lion), it is emblematic of a newly developed cognitive ability, the capacity to reason about counterfactuals. (*Source:* Photo by Yvonne Mühleis, courtesy of State Office for Cultural Heritage Baden-Württemberg/Ulmer Museum, Ulm, Germany.)

of horses and other animals. The Lion Man is different: a creature of pure imagination.

As a manifestation of our newfound ability to imagine things that have never existed, the Lion Man is the precursor of every philosophical theory, scientific discovery, and technological innovation, from microscopes to airplanes to computers. Every one of these had to take shape in someone's imagination before it was realized in the physical world.

This leap forward in cognitive ability was as profound and important to our species as any of the anatomical changes that made us human. Within 10,000 years after the Lion Man's creation, all other hominids (except for the very geographically isolated Flores hominids) had become extinct. And humans have continued to change the natural world with incredible speed, using our imagination to survive, adapt, and ultimately take over. The advantage we gained from imagining counterfactuals was the same then as it is today: flexibility, the ability to reflect and improve on past actions, and, perhaps even more significant, our willingness to take responsibility for past and current actions.

As shown in Figure 1.2, the characteristic queries for the third rung of the Ladder of Causation are "What if I had done . . . ?" and "Why?" Both involve comparing the observed world to a counterfactual world. Experiments alone cannot answer such questions. While rung one deals with the seen world, and rung two deals with a brave new world that is seeable, rung three deals with a world that cannot be seen (because it contradicts what is seen). To bridge the gap, we need a model of the underlying causal process, sometimes called a "theory" or even (in cases where we are extraordinarily confident) a "law of nature." In short, we need understanding. This is, of course, a holy grail of any branch of science—the development of a theory that will enable us to predict what will happen in situations we have not even envisioned yet. But it goes even further: having such laws permits us to violate them selectively so as to create worlds that contradict ours. Our next section features such violations in action.

THE MINI-TURING TEST

In 1950, Alan Turing asked what it would mean for a computer to think like a human. He suggested a practical test, which he called "the imitation game," but every AI researcher since then has called it the "Turing test." For all practical purposes, a computer could be called a thinking machine if an ordinary human, communicating with the computer by typewriter, could not tell whether he was talking with a human or a computer. Turing was very confident that this was within the realm of feasibility. "I believe that in about fifty years' time it will be possible to program computers," he wrote, "to make them play the imitation game so well that an average interrogator will not have more than a 70 percent chance of making the right identification after five minutes of questioning."

Turing's prediction was slightly off. Every year the Loebner Prize competition identifies the most humanlike "chatbot" in the world, with a gold medal and $100,000 offered to any program that succeeds in fooling all four judges into thinking it is human. As of 2015, in twenty-five years of competition, not a single program has fooled all the judges or even half of them.

Turing didn't just suggest the "imitation game"; he also proposed a strategy to pass it. "Instead of trying to produce a program to simulate the adult mind, why not rather try to produce one which simulates the child's?" he asked. If you could do that, then you could just teach it the same way you would teach a child, and presto, twenty years later (or less, given a computer's greater speed), you would have an artificial intelligence. "Presumably the child brain is something like a notebook as one buys it from the stationer's," he wrote. "Rather little mechanism, and lots of blank sheets." He was wrong about that: the child's brain is rich in mechanisms and prestored templates.

Nonetheless, I think that Turing was on to something. We probably will not succeed in creating humanlike intelligence until we can create childlike intelligence, and a key component of this intelligence is the mastery of causation.

How can machines acquire causal knowledge? This is still a major challenge that will undoubtedly involve an intricate combination of inputs from active experimentation, passive observation, and (not least) the programmer—much the same inputs that a child receives, with evolution, parents, and peers substituted for the programmer.

However, we can answer a slightly less ambitious question: How can machines (and people) represent causal knowledge in a way that would enable them to access the necessary information swiftly, answer questions correctly, and do it with ease, as a three-year-old child can? In fact, this is the main question we address in this book.

I call this the mini-Turing test. The idea is to take a simple story, encode it on a machine in some way, and then test to see if the machine can correctly answer causal questions that a human can answer. It is "mini" for two reasons. First, it is confined to causal reasoning, excluding other aspects of human intelligence such as vision and natural language. Second, we allow the contestant to encode the story in any convenient representation, unburdening the machine of the task of acquiring the story from its own personal experience. Passing this mini-test has been my life's work—consciously for the last twenty-five years and subconsciously even before that.

Obviously, as we prepare to take the mini-Turing test, the question of representation needs to precede the question of acquisition. Without a representation, we wouldn't know how to store information for future use. Even if we could let our robot manipulate its environment at will, whatever information we learned this way would be forgotten, unless our robot were endowed with a template to encode the results of those manipulations. One major contribution of AI to the study of cognition has been the paradigm "Representation first, acquisition second." Often the quest for a good representation has led to insights into how the knowledge ought to be acquired, be it from data or a programmer.

When I describe the mini-Turing test, people commonly claim that it can easily be defeated by cheating. For example, take the list of all possible questions, store their correct answers, and then read them out from memory when asked. There is no way to distinguish (so the argument goes) between a machine that stores a dumb question-answer list and one that answers the way that you and I do—that is, by understanding the question and producing an answer using a mental causal model. So what would the mini-Turing test prove, if cheating is so easy?

The philosopher John Searle introduced this cheating possibility, known as the "Chinese Room" argument, in 1980 to challenge Turing's claim that the ability to fake intelligence amounts to having intelligence. Searle's challenge has only one flaw: cheating is not easy; in fact, it is impossible. Even with a small number of variables, the number of possible questions grows astronomically. Say that we have ten causal variables, each of which takes only two values (0 or 1). We could ask roughly 30 million possible queries, such as "What is the probability that the outcome is 1, given that we *see* variable X equals 1 and we *make* variable Y equal 0 and variable Z equal 1?" If there were more variables, or more than two states for each one, the number of possibilities would grow beyond our ability to even imagine. Searle's list would need more entries than the number of atoms in the universe. So, clearly a dumb list of questions and answers can never simulate the intelligence of a child, let alone an adult.

Humans must have some compact representation of the information needed in their brains, as well as an effective procedure to interpret each question properly and extract the right answer from the stored representation. To pass the mini-Turing test, therefore, we need to equip machines with a similarly efficient representation and answer-extraction algorithm.

Such a representation not only exists but has childlike simplicity: a causal diagram. We have already seen one example, the diagram for the mammoth hunt. Considering the extreme ease with which people can communicate their knowledge with dot-and-arrow diagrams, I believe that our brains indeed use a representation like this. But more importantly for our purposes, these models pass the mini-Turing test; no other model is known to do so. Let's look at some examples.

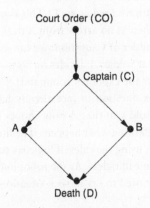

FIGURE 1.4 Causal diagram for the firing squad example.
A and *B* represent (the actions of) Soldiers *A* and *B*.

Suppose that a prisoner is about to be executed by a firing squad. A certain chain of events must occur for this to happen. First, the court orders the execution. The order goes to a captain, who signals the soldiers on the firing squad (A and B) to fire. We'll assume that they are obedient and expert marksmen, so they only fire on command, and if either one of them shoots, the prisoner dies.

Figure 1.4 shows a diagram representing the story I just told. Each of the unknowns (CO, C, A, B, D) is a true/false variable. For example, D = true means the prisoner is dead; D = false means the prisoner is alive. CO = false means the court order was not issued; CO = true means it was, and so on.

Using this diagram, we can start answering causal questions from different rungs of the ladder. First, we can answer questions of association (i.e., what one fact tells us about another). If the prisoner is dead, does that mean the court order was given? We (or a computer) can inspect the graph, trace the rules behind each of the arrows, and, using standard logic, conclude that the two soldiers wouldn't have fired without the captain's command. Likewise, the captain wouldn't have given the command if he didn't have the order in his possession. Therefore the answer to our query is yes. Alternatively, suppose we find out that A fired. What does that tell us about B? By following the arrows, the computer concludes that B must have fired too. (A would not have fired if the captain hadn't signaled, so B must have fired as well.) This is true even though A does not cause B (there is no arrow from A to B).

Going up the Ladder of Causation, we can ask questions about intervention. What if Soldier A decides on his own initiative to fire, without waiting for the captain's command? Will the prisoner be dead or alive? This question in fact already has a contradictory flavor to it. I just told you that A only shoots if commanded to, and yet now we are asking what happens if he fired without a command. If you're just using the rules of logic, as computers typically do, the question is meaningless. As the robot in the 1960s sci-fi TV series *Lost in Space* used to say in such situations, "That does not compute."

If we want our computer to understand causation, we have to teach it how to break the rules. We have to teach it the difference between merely observing an event and making it happen. "Whenever you make an event happen," we tell the computer, "remove all arrows that point to that event and continue the analysis by ordinary logic, as if the arrows had never been there." Thus, we erase all the arrows leading into the intervened variable (A). We also set that variable manually to its prescribed value (true). The rationale for this peculiar "surgery" is simple: making an event happen means that you emancipate it from all other influences and subject it to one and only one influence—that which enforces its happening.

Figure 1.5 shows the causal diagram that results from our example. This intervention leads inevitably to the prisoner's death. That is the causal function behind the arrow leading from A to D.

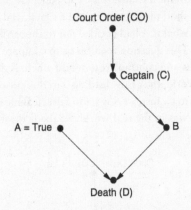

FIGURE 1.5. Reasoning about interventions. Soldier A decides to fire; arrow from C to A is deleted, and A is assigned the value true.

Note that this conclusion agrees with our intuitive judgment that A's unauthorized firing will lead to the prisoner's death, because the surgery leaves the arrow from A to D intact. Also, our judgment would be that B (in all likelihood) did *not* shoot; nothing about A's decision should affect variables in the model that are not effects of A's shot. This bears repeating. If we *see* A shoot, then we conclude that B shot too. But if A *decides* to shoot, or if we *make A* shoot, then the opposite is true. This is the difference between *see-ing* and *doing*. Only a computer capable of grasping this difference can pass the mini-Turing test.

Note also that merely collecting Big Data would not have helped us ascend the ladder and answer the above questions. Assume that you are a reporter collecting records of execution scenes day after day. Your data will consist of two kinds of events: either all five variables are true, or all of them are false. There is no way that this kind of data, in the absence of an understanding of who listens to whom, will enable you (or any machine learning algorithm) to predict the results of persuading marksman *A* not to shoot.

Finally, to illustrate the third rung of the Ladder of Causation, let's pose a counterfactual question. Suppose the prisoner is lying dead on the ground. From this we can conclude (using level one) that *A* shot, *B* shot, the captain gave the signal, and the court gave the order. But what if *A* had decided not to shoot? Would the prisoner be alive? This question requires us to compare the real world with a fictitious and contradictory world where *A* didn't shoot. In the fictitious world, the arrow leading into *A* is erased to liberate *A* from listening to *C*. Instead *A* is set to false, leaving its past history the same as it was in the real world. So the fictitious world looks like Figure 1.6.

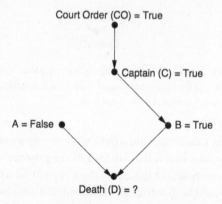

FIGURE 1.6. Counterfactual reasoning. We observe that the prisoner is dead and ask what would have happened if Soldier *A* had decided not to fire.

To pass the mini-Turing test, our computer must conclude that the prisoner would be dead in the fictitious world as well, because B's shot would have killed him. So A's courageous change of heart would not have saved his life. Undoubtedly this is one reason firing squads exist: they guarantee that the court's order will be carried out and also lift some of the burden of responsibility from the individual shooters, who can say with a (somewhat) clean conscience that their actions did not cause the prisoner's death as "he would have died anyway."

It may seem as if we are going to a lot of trouble to answer toy questions whose answer was obvious anyway. I completely agree! Causal reasoning is easy for you because you are human, and you were once a three-year-old, and you had a marvelous three-year-old brain that understood causation better than any animal or computer. The whole point of the "mini-Turing problem" is to make causal reasoning feasible for computers too. In the process, we might learn something about how humans do it. As all three examples show, we have to teach the computer how to selectively break the rules of logic. Computers are not good at breaking rules, a skill at which children excel. (Cavemen too! The Lion Man could not have been created without a breach of the rules about what head goes with what body.)

However, let's not get too complacent about human superiority. Humans may have a much harder time reaching correct causal conclusions in a great many situations. For example, there could be many more variables, and they might not be simple binary (true/false) variables. Instead of predicting whether a prisoner is alive or dead, we might want to predict how much the unemployment rate will go up if we raise the minimum wage. This kind of quantitative causal reasoning is generally beyond the power of our intuition. Also, in the firing squad example we ruled out uncertainties: maybe the captain gave his order a split second after rifleman A decided to shoot, maybe rifleman B's gun jammed, and so forth. To handle uncertainty we need information about the likelihood that the such abnormalities will occur.

Let me give you an example in which probabilities make all the difference. It echoes the public debate that erupted in Europe when the smallpox vaccine was first introduced. Unexpectedly, data showed that more people died from smallpox inoculations than from smallpox itself. Naturally, some people used this information to argue that inoculation should be banned, when in fact it was saving lives by eradicating smallpox. Let's look at some fictitious data to illustrate the effect and settle the dispute.

Suppose that out of 1 million children, 99 percent are vaccinated, and 1 percent are not. If a child is vaccinated, he or she has one chance in one hundred of developing a reaction, and the reaction has one chance in one hundred of being fatal. On the other hand, he or she has no chance of developing smallpox. Meanwhile, if a child is not vaccinated, he or she obviously has zero chance of developing a reaction to the vaccine, but he or she has one chance in fifty of developing smallpox. Finally, let's assume that smallpox is fatal in one out of five cases.

I think you would agree that vaccination looks like a good idea. The odds of having a reaction are lower than the odds of getting smallpox, and the reaction is much less dangerous than the disease. But now let's look at the data. Out of 1 million children, 990,000 get vaccinated, 9,900 have the reaction, and 99 die from it. Meanwhile, 10,000 don't get vaccinated, 200 get smallpox, and 40 die from the disease. In summary, more children die from vaccination (99) than from the disease (40).

I can empathize with the parents who might march to the health department with signs saying, "Vaccines kill!" And the data seem to be on their side; the vaccinations indeed cause more deaths than smallpox itself. But is logic on their side? Should we ban vaccination or take into account the deaths prevented? Figure 1.7 shows the causal diagram for this example.

When we began, the vaccination rate was 99 percent. We now ask the counterfactual question "What if we had set the vaccination rate to zero?" Using the probabilities I gave you above, we can conclude that out of 1 million children, 20,000 would have gotten smallpox, and 4,000 would have died. Comparing the counterfactual world with the real world, we see that not vaccinating would have cost the lives of 3,861 children (the difference between 4,000

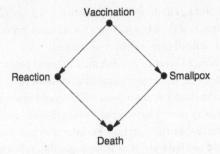

FIGURE 1.7. Causal diagram for the vaccination example.
Is vaccination beneficial or harmful?

and 139). We should thank the language of counterfactuals for helping us to avoid such costs.

The main lesson for a student of causality is that a causal model entails more than merely drawing arrows. Behind the arrows, there are probabilities. When we draw an arrow from X to Y, we are implicitly saying that some probability rule or function specifies how Y would change if X were to change. We might know what the rule is; more likely, we will have to estimate it from data. One of the most intriguing features of the Causal Revolution, though, is that in many cases we can leave those mathematical details completely unspecified. Very often the *structure of the diagram itself* enables us to estimate all sorts of causal and counterfactual relationships: simple or complicated, deterministic or probabilistic, linear or nonlinear.

From the computing perspective, our scheme for passing the mini-Turing test is also remarkable in that we used the same routine in all three examples: translate the story into a diagram, listen to the query, perform a surgery that corresponds to the given query (interventional or counterfactual; if the query is associational then no surgery is needed), and use the modified causal model to compute the answer. We did not have to train the machine in a multitude of new queries each time we changed the story. The approach is flexible enough to work whenever we can draw a causal diagram, whether it has to do with mammoths, firing squads, or vaccinations. This is exactly what we want for a causal inference engine: it is the kind of flexibility we enjoy as humans.

Of course, there is nothing inherently magical about a diagram. It succeeds because it carries causal information; that is, when we constructed the diagram, we asked, "Who could directly cause the prisoner's death?" or "What are the direct effects of vaccinations?" Had we constructed the diagram by asking about mere associations, it would not have given us these capabilities. For example, in Figure 1.7, if we reversed the arrow Vaccination → Smallpox, we would get the same associations in the data but would erroneously conclude that smallpox affects vaccination.

Decades' worth of experience with these kinds of questions has convinced me that, in both a cognitive and a philosophical sense, the idea of causes and effects is much more fundamental than the idea of probability. We begin learning causes and effects before we understand language and before we know any mathematics. (Research has shown that three-year-olds already understand the entire Ladder of Causation.) Likewise, the knowledge conveyed in a causal diagram is typically much more robust than that encoded in a probability distribution. For example, suppose that times have changed and a much safer and more effective vaccine is introduced. Suppose, further, that due to improved hygiene and socioeconomic conditions, the danger of contracting smallpox has diminished. These changes will drastically affect all the probabilities involved; yet, remarkably, the structure of the diagram will remain invariant. This is the key secret of causal modeling. Moreover, once we go through the analysis and find how to estimate the benefit of vaccination from data, we do not have to repeat the entire analysis from scratch. As discussed in the Introduction, the same estimand (i.e., recipe for answering the query) will remain valid and, as long as the diagram does not change, can be applied to the new data and produce a new estimate for our query. It is because of this robustness, I conjecture, that human intuition is organized around causal, not statistical, relations.

ON PROBABILITIES AND CAUSATION

The recognition that causation is not reducible to probabilities has been very hard-won, both for me personally and for philosophers and scientists in general. Understanding the meaning of "cause" has been the focus of a long tradition of philosophers, from David Hume and John Stuart Mill in the 1700s and 1800s, to Hans Reichenbach and Patrick Suppes in the mid-1900s, to Nancy Cartwright, Wolfgang Spohn, and Christopher Hitchcock today. In particular, beginning with Reichenbach and Suppes, philosophers have tried to define causation in terms of probability, using the notion of "probability raising": X causes Y if X *raises the probability* of Y.

This concept is solidly ensconced in intuition. We say, for example, "Reckless driving causes accidents" or "You will fail this course because of your laziness," knowing quite well that the antecedents merely tend to make the consequences more likely, not absolutely certain. One would expect, therefore, that probability raising should become the bridge between rung one and rung two of the Ladder of Causation. Alas, this intuition has led to decades of failed attempts.

What prevented the attempts from succeeding was not the idea itself but the way it was articulated formally. Almost without exception, philosophers expressed the sentence "X raises the probability of Y" using conditional probabilities and wrote $P(Y \mid X) > P(Y)$. This interpretation is wrong, as you surely noticed, because "raises" is a causal concept, connoting a causal influence of X over Y. The expression $P(Y \mid X) > P(Y)$, on the other hand, speaks only about observations and means: "If we see X, then the probability of Y increases." But this increase may come about for other reasons, including Y being a cause of X or some other variable (Z) being the cause of both of them. That's the catch! It puts the philosophers back at square one, trying to eliminate those "other reasons."

Probabilities, as given by expressions like $P(Y \mid X)$, lie on the first rung of the Ladder of Causation and cannot ever (by themselves) answer queries on the second or third rung. Any attempt to "define" causation in terms of seemingly simpler, first-rung concepts must fail. That is why I have not attempted to define causation anywhere in this book: definitions demand reduction, and reduction demands going to a lower rung. Instead, I have pursued the ultimately more constructive program of explaining how to answer causal queries and what information is needed to answer them. If this seems odd, consider that mathematicians take exactly the same approach to Euclidean geometry. Nowhere in a geometry book will you find a definition of the terms "point" and "line." Yet we can answer any and all queries about them on the basis of Euclid's axioms (or even better, the various modern versions of Euclid's axioms).

But let's look at this probability-raising criterion more carefully and see where it runs aground. The issue of a common cause, or *confounder*, of X and Y was among the most vexing for philosophers. If we take the probability-raising criterion at face value, we must conclude that high ice-cream sales cause crime because the probability of crime is higher in months when more ice cream is sold. In this particular case, we can explain the phenomenon because both ice-cream sales and crime are higher in summer, when the weather is warmer. Nevertheless, we are still left asking what general philosophical criterion could tell us that weather, not ice-cream sales, is the cause.

Philosophers tried hard to repair the definition by conditioning on what they called "background factors" (another word for confounders), yielding the criterion $P(Y \mid X, K = k) > P(Y \mid K = k)$, where K stands for some background variables. In fact, this criterion works for our ice-cream example if we treat temperature as a background variable. For example, if we look only at days when the temperature is ninety degrees ($K = 90$), we will find no residual association between ice-cream sales and crime. It's only when we compare ninety-degree days to thirty-degree days that we get the illusion of a probability raising.

Still, no philosopher has been able to give a convincingly general answer to the question "Which variables need to be included in the background set K and conditioned on?" The reason is obvious: confounding too is a causal concept and hence defies probabilistic formulation. In 1983, Nancy Cartwright broke this deadlock and enriched the description of the background context with a causal component. She proposed that we should condition on any factor that is "causally relevant" to the effect. By borrowing a concept from rung two of the Ladder of Causation, she essentially gave up on the idea of defining causes based on probability alone. This was progress, but it opened the door to the criticism that we are defining a cause in terms of itself.

Philosophical disputes over the appropriate content of K continued for more than two decades and reached an impasse. In fact, we will see a correct criterion in Chapter 4, and I will not spoil the surprise here. It suffices for the moment to say that this criterion is practically impossible to enunciate without causal diagrams.

In summary, probabilistic causality has always foundered on the rock of confounding. Every time the adherents of probabilistic causation try to patch up the ship with a new hull, the boat runs into the same rock and springs another leak. Once you misrepresent "probability raising" in the language of conditional probabilities, no amount of probabilistic patching will get you to the next rung of the ladder. As strange as it may sound, the notion of probability raising cannot be expressed in terms of probabilities.

The proper way to rescue the probability-raising idea is with the *do*-operator: we can say that X causes Y if $P(Y \mid do(X)) > P(Y)$. Since intervention is a rung-two concept, this definition can capture the causal interpretation of probability raising, and it can also be made operational through causal diagrams. In other words, if we have a causal diagram and data on hand and a researcher asks whether $P(Y \mid do(X)) > P(Y)$, we can answer his question coherently and algorithmically and thus decide if X is a cause of Y in the probability-raising sense.

I usually pay a great deal of attention to what philosophers have to say about slippery concepts such as causation, induction, and the logic of scientific inference. Philosophers have the advantage of standing apart from the hurly-burly of scientific debate and the practical realities of dealing with data. They have been less contaminated than other scientists by the anticausal biases of statistics.

They can call upon a tradition of thought about causation that goes back at least to Aristotle, and they can talk about causation without blushing or hiding it behind the label of "association."

However, in their effort to mathematize the concept of causation—itself a laudable idea—philosophers were too quick to commit to the only uncertainty-handling language they knew, the language of probability. They have for the most part gotten over this blunder in the past decade or so, but unfortunately similar ideas are being pursued in econometrics even now, under names like "Granger causality" and "vector autoregression."

Now I have a confession to make: I made the same mistake. I did not always put causality first and probability second. Quite the opposite! When I started working in artificial intelligence, in the early 1980s, I thought that uncertainty was the most important thing missing from AI. Moreover, I insisted that uncertainty be represented by probabilities. Thus, as I explain in Chapter 3, I developed an approach to reasoning under uncertainty, called Bayesian networks, that mimics how an idealized, decentralized brain might incorporate probabilities into its decisions. Given that we see certain facts, Bayesian networks can swiftly compute the likelihood that certain other facts are true or false. Not surprisingly, Bayesian networks caught on immediately in the AI community and even today are considered a leading paradigm in artificial intelligence for reasoning under uncertainty.

Though I am delighted with the ongoing success of Bayesian networks, they failed to bridge the gap between artificial and human intelligence. I'm sure you can figure out the missing ingredient: causality. True, causal ghosts were all over the place. The arrows invariably pointed from causes to effects, and practitioners often noted that diagnostic systems became unmanageable when the direction of the arrows was reversed. But for the most part we thought that this was a cultural habit, or an artifact of old thought patterns, not a central aspect of intelligent behavior.

At the time, I was so intoxicated with the power of probabilities that I considered causality a subservient concept, merely a convenience or a mental shorthand for expressing probabilistic dependencies and distinguishing relevant variables from irrelevant ones.

In my 1988 book *Probabilistic Reasoning in Intelligent Systems*, I wrote, "Causation is a language with which one can talk efficiently about certain structures of relevance relationships." The words embarrass me today, because "relevance" is so obviously a rung-one notion. Even by the time the book was published, I knew in my heart that I was wrong. To my fellow computer scientists, my book became the bible of reasoning under uncertainty, but I was already feeling like an apostate.

Bayesian networks inhabit a world where all questions are reducible to probabilities, or (in the terminology of this chapter) degrees of association between variables; they could not ascend to the second or third rungs of the Ladder of Causation. Fortunately, they required only two slight twists to climb to the top. First, in 1991, the graph-surgery idea empowered them to handle both observations and interventions. Another twist, in 1994, brought them to the third level and made them capable of handling counterfactuals. But these developments deserve a fuller discussion in a later chapter. The main point is this: while probabilities encode our beliefs about a static world, causality tells us whether and how probabilities change when the world changes, be it by intervention or by act of imagination.

Sir Francis Galton demonstrates his "Galton board" or "quincunx" at the Royal Institution. He saw this pinball-like apparatus as an analogy for the inheritance of genetic traits like stature. The pinballs accumulate in a bell-shaped curve that is similar to the distribution of human heights. The puzzle of why human heights don't spread out from one generation to the next, as the balls would, led him to the discovery of "regression to the mean." (*Source:* Drawing by Dakota Harr.)

2

FROM BUCCANEERS TO GUINEA PIGS: THE GENESIS OF CAUSAL INFERENCE

And yet it moves.

—ATTRIBUTED TO GALILEO GALILEI (1564–1642)

FOR close to two centuries, one of the most enduring rituals in British science has been the Friday Evening Discourse at the Royal Institution of Great Britain in London. Many discoveries of the nineteenth century were first announced to the public at this venue: Michael Faraday and the principles of photography in 1839; J. J. Thomson and the electron in 1897; James Dewar and the liquefaction of hydrogen in 1904.

Pageantry was an important part of the occasion; it was literally science as theater, and the audience, the cream of British society, would dress to the nines (tuxedos with black tie for men). At the appointed hour, a chime would strike, and the evening's speaker would be ushered into the auditorium. Traditionally he would begin the lecture immediately, without introduction or preamble. Experiments and live demonstrations were part of the spectacle.

On February 9, 1877, the evening's speaker was Francis Galton, FRS, first cousin of Charles Darwin, noted African explorer, inventor of fingerprinting, and the very model of a Victorian gentleman scientist. Galton's title was "Typical Laws of Heredity." His experimental apparatus for the evening was a curious contraption that he called a quincunx, now often called a "Galton board." A similar game has often appeared on the televised game show *The Price Is Right*, where it is known as Plinko. The Galton board consists of a triangular array of pins or pegs, into which small metal balls can be inserted through an opening at the top. The balls bounce downward from one row to the next, pinball style, before settling into one of a line of slots at the bottom (see frontispiece). For any individual ball, the zigs and zags to the left or right look completely random. However, if you pour a lot of balls into the Galton board, a startling regularity emerges: the accumulated balls at the bottom will always form a rough approximation to a bell-shaped curve. The slots nearest the center will be stacked high with balls, and the number of the balls in each slot gradually tapers down to zero at the edges of the quincunx.

This pattern has a mathematical explanation. The path of any individual ball is like a sequence of independent coin flips. Each time a ball hits a pin, it bounces either to the left or the right, and from a distance its choice seems completely random. The sum of the results—say, the excess of the rights over the lefts—determines which slot the ball ends up in. According to the central limit theorem, proven in 1810 by Pierre-Simon Laplace, any such random process—one that amounts to a sum of a large number of coin flips—will lead to the same probability distribution, called the normal distribution (or bell-shaped curve). The Galton board is simply a visual demonstration of Laplace's theorem.

The central limit theorem is truly a miracle of nineteenth-century mathematics. Think about it: even though the path of any individual ball is unpredictable, the path of 1,000 balls is extremely predictable—a convenient fact for the producers of *The Price Is Right*, who can estimate accurately how much money the contestants will win at Plinko over the long run. This is the same law that makes insurance companies so profitable, despite the uncertainties in human affairs.

The well-dressed audience at the Royal Institute must have wondered what all this had to do with the laws of heredity, the promised lecture topic. To illustrate the connection, Galton showed them some data collected in France on the heights of military recruits. These also follow a normal distribution: many men are of about average height, with a gradually diminishing number who are either extremely tall or extremely short. In fact, it does not matter whether you are talking about 1,000 military recruits or 1,000 balls in the Galton board: the numbers in each slot (or height category) are almost the same.

Thus, to Galton, the quincunx was a model for the inheritance of stature or, indeed, many other genetic traits. It is a causal model. In simplest terms, Galton believed the balls "inherit" their position in the quincunx in the same way that humans inherit their stature.

But if we accept this model—provisionally—it poses a puzzle, which was Galton's chief subject for the evening. The width of the bell-shaped curve depends on the number of rows of pegs placed between the top and the bottom. Suppose we doubled the number of rows. This would create a model for two generations of inheritance, with the first half of the rows representing the first generation and the second half representing the second. You would inevitably find more variation in the second generation than in the first, and in succeeding generations, the bell-shaped curve would get wider and wider still.

But this is not what happens with actual human stature. In fact, the width of the distribution of human heights stays relatively constant over time. We didn't have nine-foot humans a century ago, and we still don't. What explains the stability of the population's genetic endowment? Galton had been puzzling over this enigma for roughly eight years, since the publication of his book *Hereditary Genius* in 1869.

As the title of the book suggests, Galton's true interest was not carnival games or human stature but human intelligence. As a member of an extended family with a remarkable amount of scientific genius, Galton naturally would have liked to prove that genius runs in families. And he had set out to do exactly that in his book.

He painstakingly compiled pedigrees of 605 "eminent" Englishmen from the preceding four centuries. But he found that the sons and fathers of these eminent men were somewhat less eminent and the grandparents and grandchildren less eminent still.

It's easy enough for us now to find flaws in Galton's program. What, after all, is the definition of eminence? And isn't it possible that people in eminent families are successful because of their privilege rather than their talent? Though Galton was aware of such difficulties, he pursued this futile quest for a genetic explanation at an increasing pace and determination.

Still, Galton was on to something, which became more apparent once he started looking at features like height, which are easier to measure and more strongly linked to heredity than "eminence." Sons of tall men tend to be taller than average—but not as tall as their fathers. Sons of short men tend to be shorter than average—but not as short as their fathers. Galton first called this phenomenon "reversion" and later "regression toward mediocrity." It can be noted in many other settings. If students take two different standardized tests on the same material, the ones who scored high on the first test will usually score higher than average on the second test but not as high as they did the first time. This phenomenon of regression to the mean is ubiquitous in all facets of life, education, and business. For instance, in baseball the Rookie of the Year (a player who does unexpectedly well in his first season) often hits a "sophomore slump," in which he does not do quite as well.

Galton didn't know all of this, and he thought he had stumbled onto a law of heredity rather than a law of statistics. He believed that regression to the mean must have some cause, and in his Royal Institution lecture he illustrated his point. He showed his audience a two-layered quincunx (Figure 2.1).

After passing through the first array of pegs, the balls passed through sloping chutes that moved them closer to the center of the board. Then they would pass through a second array of pegs. Galton showed triumphantly that the chutes exactly compensated for the tendency of the normal distribution to spread out. This time, the bell-shaped probability distribution kept a constant width from generation to generation.

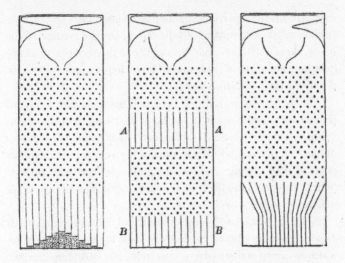

FIGURE 2.1. The Galton board, used by Francis Galton as an analogy for the inheritance of human heights. (a) When many balls are dropped through the pinball-like apparatus, their random bounces cause them to pile up in a bell-shaped curve. (b) Galton noted that on two passes, A and B, through the Galton board (the analogue of two generations) the bell-shaped curve got wider. (c) To counteract this tendency, he installed chutes to move the "second generation" back closer to the center. The chutes are Galton's causal explanation for regression to the mean. (*Source:* Francis Galton, *Natural Inheritance* [1889].)

Thus, Galton conjectured, regression toward the mean was a physical process, nature's way of ensuring that the distribution of height (or intelligence) remained the same from generation to generation. "The process of reversion cooperates with the general law of deviation," Galton told his audience. He compared it to Hooke's law, the physical law that describes the tendency of a spring to return to its equilibrium length.

Keep in mind the date. In 1877, Galton was in pursuit of a causal explanation and thought that regression to the mean was a causal process, like a law of physics. He was mistaken, but he was far from alone. Many people continue to make the same mistake to this day. For example, baseball experts always look for causal explanations for a player's sophomore slump. "He's gotten over-confident," they complain, or "the other players have figured out his weaknesses." They may be right, but the sophomore slump does not need a causal explanation. It will happen more often than not by the laws of chance alone.

The modern statistical explanation is quite simple. As Daniel Kahneman summarizes it in his book *Thinking, Fast and Slow*, "Success = talent + luck. Great success = a little more talent + a lot of luck." A player who wins Rookie of the Year is probably more talented than average, but he also (probably) had a lot of luck. Next season, he is not likely to be so lucky, and his batting average will be lower.

By 1889, Galton had figured this out, and in the process—partly disappointed but also fascinated—he took the first huge step toward divorcing statistics from causation. His reasoning is subtle but worth making the effort to understand. It is the newborn discipline of statistics uttering its first cry.

Galton had started gathering a variety of "anthropometric" statistics: height, forearm length, head length, head width, and so on. He noticed that when he plotted height against forearm length, for instance, the same phenomenon of regression to the mean took place. Tall men usually had longer-than-average forearms—but not as far above average as their height. Clearly height is not a cause of forearm length, or vice versa. If anything, both are caused by genetic inheritance. Galton started using a new word for this kind of relationship: height and forearm length were "co-related." Eventually, he opted for the more normal English word "correlated."

Later he realized an even more startling fact: in generational comparisons, the temporal order could be reversed. That is, the fathers of sons also revert to the mean. The father of a son who is taller than average is likely to be taller than average but shorter than his son (see Figure 2.2). Once Galton realized this, he had to give up any idea of a causal explanation for regression, because there is no way that the sons' heights could cause the fathers' heights.

This realization may sound paradoxical at first. "Wait!" you're saying. "You're telling me that tall dads usually have shorter sons, and tall sons usually have shorter dads. How can both of those

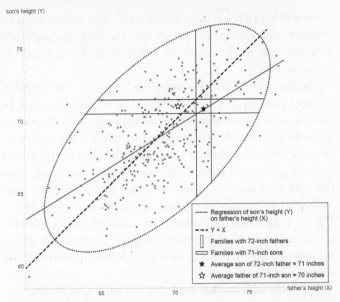

FIGURE 2.2. The scatter plot shows a data set of heights, with each dot representing the height of a father (on the x-axis) and his son (on the y-axis). The dashed line coincides with the major axis of the ellipse, while the solid line (called the regression line) connects the rightmost and leftmost points on the ellipse. The difference between them accounts for regression to the mean. For example, the black star shows that 72″ fathers have, on the average, 71″ sons. (That is, the average height of all the data points in the vertical strip is 71″.) The horizontal strip and white star show that the same loss of height occurs in the noncausal direction (backward in time). (*Source:* Figure by Maayan Harel, with a contribution from Christopher Boucher.)

statements be true? How can a son be both taller and shorter than his father?"

The answer is that we are talking not about an individual father and an individual son but about two populations. We start with the population of six-foot fathers. Because they are taller than average, their sons will regress toward the mean; let's say their sons average five feet, eleven inches. However, the population of father-son pairs with six-foot fathers is not the same as the population of father-son pairs with five-foot-eleven-inch sons. Every father in the first group is by definition six feet tall. But the second group will have a few fathers who are taller than six feet and a lot of fathers who are shorter than six feet. Their average height will be shorter than five feet, eleven inches, again displaying regression to the mean.

Another way to illustrate regression is to use a diagram called a scatter plot (Figure 2.2). Each father-son pair is represented by one dot, with the x-coordinate being the father's height and the y-coordinate being the son's height. So a father and son who are both five feet, nine inches (or sixty-nine inches) will be represented by a dot at (69, 69), right at the center of the scatter plot. A father who is six feet (or seventy-two inches) with a son who is five-foot-eleven (or seventy-one inches) will be represented by a dot at (72, 71), in the northeast corner of the scatter plot. Notice that the scatter plot has a roughly elliptical shape—a fact that was crucial to Galton's analysis and characteristic of bell-shaped distributions with two variables.

As shown in Figure 2.2, the father-son pairs with seventy-two-inch fathers lie in a vertical slice centered at 72; the father-son pairs with seventy-one-inch sons lie in a horizontal slice centered at 71. Here is visual proof that these are two different populations. If we focus only on the first population, the pairs with seventy-two-inch fathers, we can ask, "How tall are the sons on average?" It's the same as asking where the center of that vertical slice is, and by eye you can see that the center is about 71. If we focus only on the second population with seventy-one-inch sons, we can ask, "How tall are the fathers on average?" This is the same as asking for the center of the horizontal slice, and by eye you can see that its center is about 70.3.

We can go farther and think about doing the same procedure for every vertical slice. That's equivalent to asking, "For fathers of height x, what is the best prediction of the son's height (y)?" Alternatively, we can take each horizontal slice and ask where its center is: for sons of height y, what is the best "prediction" (or retrodiction) of the father's height?

As he thought about this question, Galton stumbled on an important fact: the predictions always fall on a line, which he called the regression line, which is less steep than the major axis (or axis of symmetry) of the ellipse (Figure 2.3). In fact there are two such lines, depending on which variable is being predicted and which is being used as evidence. You can predict the son's height based on the father's or the father's based on the son's. The situation is completely symmetric. Once again this shows that where regression to the mean is concerned, there is no difference between cause and effect.

The slope of the regression enables you to predict the value of one variable, given that you know the value of the other. In the context of Galton's problem, a slope of 0.5 would mean that each extra inch of height for the father would correspond, on average, to an extra half inch for the son, and vice versa. A slope of 1 would be perfect correlation, which means every extra inch for the father

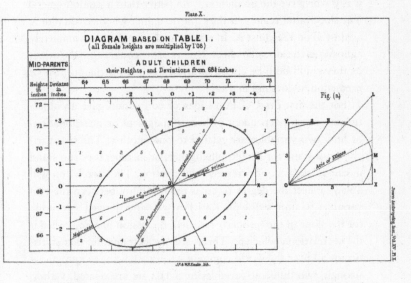

FIGURE 2.3. Galton's regression lines. Line OM gives the best prediction of a son's height if you know the height of the father; line ON gives the best prediction of a father's height if you know the height of the son. Neither is the same as the major axis (axis of symmetry) of the scatter plot. (*Source:* Francis Galton, *Journal of the Anthropological Institute of Great Britain and Ireland* [1886], 246–263, Plate X.)

is passed deterministically to the son, who would also be an inch taller. The slope can never be greater than 1; if it were, the sons of tall fathers would be taller on average, and the sons of short fathers would be shorter—and this would force the distribution of heights to become wider over time. After a few generations we would start having 9-foot people and 2-foot people, which is not observed in nature. So, provided the distribution of heights stays the same from one generation to the next, the slope of the regression line cannot exceed 1.

The law of regression applies even when we correlate two different quantities, like height and IQ. If you plot one quantity against the other in a scatter plot and rescale the two axes properly, then the slope of the best-fit line always enjoys the same properties. It equals 1 only when one quantity can predict the other precisely; it is 0 whenever the prediction is no better than a random guess. The slope (after scaling) is the same no matter whether you plot X against Y or Y against X. In other words, the slope is completely agnostic as to cause and effect. One variable could cause the other, or they could both be effects of a third cause; for the purpose of prediction, it does not matter.

For the first time, Galton's idea of correlation gave an objective measure, independent of human judgment or interpretation, of how two variables are related to one another. The two variables can stand for height, intelligence, or income; they can stand in causal, neutral, or reverse-causal relation. The correlation will always reflect the degree of cross predictability between the two variables. Galton's disciple Karl Pearson later derived a formula for the slope of the (properly rescaled) regression line and called it the correlation coefficient. This is still the first number that statisticians all over the world compute when they want to know how strongly two different variables in a data set are related. Galton and Pearson must have been thrilled to find such a universal way of describing the relationships between random variables. For Pearson, especially, the slippery old concepts of cause and effect seemed outdated and unscientific, compared to the mathematically clear and precise concept of a correlation coefficient.

GALTON AND THE ABANDONED QUEST

It is an irony of history that Galton started out in search of causation and ended up discovering correlation, a relationship that is oblivious of causation. Even so, hints of causal thinking remained in his writing. "It is easy to see that correlation [between the sizes of two organs] must be the consequence of the variations of the two organs being partly due to common causes," he wrote in 1889.

The first sacrifice on the altar of correlation was Galton's elaborate machinery to explain the stability of the population's genetic endowment. The quincunx simulated the creation of variations in height and their transmission from one generation to the next. But Galton had to invent the inclined chutes in the quincunx specifically to rein in the ever-growing diversity in the population. Having failed to find a satisfactory biological mechanism to account for this restoring force, Galton simply abandoned the effort after eight years and turned his attention to the siren song of correlation. Historian Stephen Stigler, who has written extensively about Galton, noticed this sudden shift in Galton's aims and aspirations: "What was silently missing was Darwin, the chutes, and all the 'survival of the fittest.' . . . In supreme irony, what had started out as an attempt to mathematize the framework of the *Origin of Species* ended with the essence of that great work being discarded as unnecessary!"

But to us, in the modern era of causal inference, the original problem remains. How do we explain the stability of the population, despite Darwinian variations that one generation bestows on the next?

Looking back on Galton's machine in the light of causal diagrams, the first thing I notice is that the machine was wrongly constructed. The ever-growing dispersion, which begged Galton for a counterforce, should never have been there in the first place. Indeed, if we trace a ball dropping from one level to the next in the quincunx, we see that the displacement at the next level inherits the sum total of variations bestowed upon it by all the pegs along the way. This stands in blatant contradiction to Kahneman's equations:

Success = talent + luck

Great success = A little more talent + a lot of luck.

According to these equations, success in generation 2 does not inherit the luck of generation 1. Luck, by its very definition, is a transitory occurrence; hence it has no impact on future generations. But such transitory behavior is incompatible with Galton's machine.

To compare these two conceptions side by side, let us draw their associated causal diagrams. In Figure 2.4(a) (Galton's conception), success is transmitted across generations, and luck variations accumulate indefinitely. This is perhaps natural if "success" is equated to wealth or eminence. However, for the inheritance of physical characteristics like stature, we must replace Galton's model with that in Figure 2.4(b). Here only the genetic component, shown here

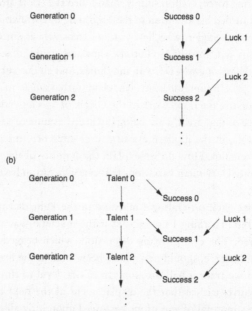

FIGURE 2.4. Two models of inheritance. (a) The Galton board model, in which luck accrues from generation to generation, leading to an ever-wider distribution of success. (b) A genetic model, in which luck does not accrue, leading to a constant distribution of success.

as talent, is passed down from one generation to the next. Luck affects each generation independently, in such a way that the chance factors in one generation have no way of affecting later generations, either directly or indirectly.

Both of these models are compatible with the bell-shaped distribution of heights. But the first model is not compatible with the stability of the distribution of heights (or success). The second model, on the other hand, shows that to explain the stability of success from one generation to the next, we only need explain the stability of the genetic endowment of the population (talent). That stability, now called the Hardy-Weinberg equilibrium, received a satisfactory mathematical explanation in the work of G. H. Hardy and Wilhelm Weinberg in 1908. And yes, they used yet another causal model—the Mendelian theory of inheritance.

In retrospect, Galton could not have anticipated the work of Mendel, Hardy, and Weinberg. In 1877, when Galton gave his lecture, Gregor Mendel's work of 1866 had been forgotten (it was only rediscovered in 1900), and the mathematics of Hardy and Weinberg's proofs would likely have been beyond him. But it is interesting to note how close he came to finding the right framework and also how the causal diagram makes it easy to zero in on his mistaken assumption: the transmission of luck from one generation to the next. Unfortunately, he was led astray by his beautiful but flawed causal model, and later, having discovered the beauty of correlation, he came to believe that causality was no longer needed.

As a final personal comment on Galton's story, I confess to committing a cardinal sin of history writing, one of many sins I will commit in this book. In the 1960s, it became unfashionable to write history from the viewpoint of modern-day science, as I have done above. "Whig history" was the epithet used to mock the hindsighted style of history writing, which focused on successful theories and experiments and gave little credit to failed theories and dead ends. The modern style of history writing became more democratic, treating chemists and alchemists with equal respect and insisting on understanding all theories in the social context of their own time.

When it comes to explaining the expulsion of causality from statistics, however, I accept the mantle of Whig historian with pride. There simply is no other way to understand how statistics became a model-blind data-reduction enterprise, except by putting on our causal lenses and retelling the stories of Galton and Pearson in the light of the new science of cause and effect. In fact, by so doing, I rectify the distortions introduced by mainstream historians who, lacking causal vocabulary, marvel at the invention of correlation and fail to note its casualty—the death of causation.

PEARSON: THE WRATH OF THE ZEALOT

It remained to Galton's disciple, Karl Pearson, to complete the task of expunging causation from statistics. Yet even he was not entirely successful.

Reading Galton's *Natural Inheritance* was one of the defining moments of Pearson's life: "I felt like a buccaneer of Drake's days—one of the order of men 'not quite pirates, but with decidedly piratical tendencies,' as the dictionary has it!" he wrote in 1934. "I interpreted . . . Galton to mean that there was a category broader than causation, namely correlation, of which causation was only the limit, and that this new conception of correlation brought psychology, anthropology, medicine and sociology in large part into the field of mathematical treatment. It was Galton who first freed me from the prejudice that sound mathematics could only be applied to natural phenomena under the category of causation."

In Pearson's eyes, Galton had enlarged the vocabulary of science. Causation was reduced to nothing more than a special case of correlation (namely, the case where the correlation coefficient is 1 or –1 and the relationship between x and y is deterministic). He expresses his view of causation with great clarity in *The Grammar of Science* (1892): "That a certain sequence has occurred and reoccurred in the past is a matter of experience to which we give expression in the concept causation. . . . Science in no case can demonstrate any inherent necessity in a sequence, nor prove with absolute certainty that it must be repeated." To summarize, causation for Pearson is only a matter of repetition and, in the deterministic sense, can never be proven. As for causality in a non-deterministic world, Pearson was even more dismissive: "the ultimate scientific statement of description of the relation between two things can always be thrown back upon . . . a contingency table." In other words, data is all there is to science. Full stop. In this view, the notions of intervention and counterfactuals discussed in Chapter 1 do not exist, and the lowest rung of the Ladder of Causation is all that is needed for doing science.

The mental leap from Galton to Pearson is breathtaking and indeed worthy of a buccaneer. Galton had proved only that one phenomenon—regression to the mean—did not require a causal explanation. Now Pearson was completely removing causation from science. What made him take this leap?

Historian Ted Porter, in his biography *Karl Pearson*, describes how Pearson's skepticism about causation predated his reading of Galton's book. Pearson had been wrestling with the philosophical foundation of physics and wrote (for example), "Force as a cause of motion is exactly on the same footing as a tree-god as a cause of growth." More generally, Pearson belonged to a philosophical school called positivism, which holds that the universe is a product of human thought and that science is only a description of those thoughts. Thus causation, construed as an objective process that happens in the world outside the human brain, could not have any scientific meaning. Meaningful thoughts can only reflect patterns of observations, and these can be completely described by correlations. Having decided that correlation was a more universal descriptor of human thought than causation, Pearson was prepared to discard causation completely.

Porter paints a vivid picture of Pearson throughout his life as a self-described *Schwärmer*, a German word that translates as "enthusiast" but can also be interpreted more strongly as "zealot." After graduating from Cambridge in 1879, Pearson spent a year abroad in Germany and fell so much in love with its culture that he promptly changed his name from Carl to Karl. He was a socialist long before it became popular, and he wrote to Karl Marx in 1881, offering to translate *Das Kapital* into English. Pearson, arguably one of England's first feminists, started the Men's and Women's

Club in London for discussions of "the woman question." He was concerned about women's subordinate position in society and advocated for them to be paid for their work. He was extremely passionate about ideas while at the same time very cerebral about his passions. It took him nearly half a year to persuade his future wife, Maria Sharpe, to marry him, and their letters suggest that she was frankly terrified of not living up to his high intellectual ideals.

When Pearson found Galton and his correlations, he at last found a focus for his passions: an idea that he believed could transform the world of science and bring mathematical rigor to fields like biology and psychology. And he moved with a buccaneer's sense of purpose toward accomplishing this mission. His first paper on statistics was published in 1893, four years after Galton's discovery of correlation. By 1901 he had founded a journal, *Biometrika*, which remains one of the most influential statistical journals (and, somewhat heretically, published my first full paper on causal diagrams in 1995). By 1903, Pearson had secured a grant from the Worshipful Company of Drapers to start a Biometrics Lab at University College London. In 1911 it officially became a department when Galton passed away and left an endowment for a professorship (with the stipulation that Pearson be its first holder). For at least two decades, Pearson's Biometrics Lab was the world center of statistics.

Once Pearson held a position of power, his zealotry came out more and more clearly. As Porter writes in his biography, "Pearson's statistical movement had aspects of a schismatic sect. He demanded the loyalty and commitment of his associates and drove dissenters from the church biometric." One of his earliest assistants, George Udny Yule, was also one of the first people to feel Pearson's wrath. Yule's obituary of Pearson, written for the Royal Society in 1936, conveys well the sting of those days, though couched in polite language.

The infection of his enthusiasm, it is true, was invaluable; but his dominance, even his very eagerness to help, could be a disadvantage.... This desire for domination, for everything to be just as he wanted it, comes out in other ways, notably the editing of *Biometrika*—surely the most personally edited journal that was ever published. . . . Those who left him and began to think for themselves were apt, as happened painfully in more instances than one, to find that after a divergence of opinion the maintenance of friendly relations became difficult, after express criticism impossible.

Even so, there were cracks in Pearson's edifice of causality-free science, perhaps even more so among the founders than among the later disciples. For instance, Pearson himself surprisingly wrote several papers about "spurious correlation," a concept impossible to make sense of without making some reference to causation.

Pearson noticed that it's relatively easy to find correlations that are just plain silly. For instance, for a fun example postdating Pearson's time, there is a strong correlation between a nation's per capita chocolate consumption and its number of Nobel Prize winners. This correlation seems silly because we cannot envision any way in which eating chocolate could cause Nobel Prizes. A more likely explanation is that more people in wealthy, Western countries eat chocolate, and the Nobel Prize winners have also been chosen preferentially from those countries. But this is a causal explanation, which, for Pearson, is not necessary for scientific thinking. To him, causation is just a "fetish amidst the inscrutable arcana of modern science." Correlation is supposed to be the goal of scientific understanding. This puts him in an awkward position when he has to explain why one correlation is meaningful and another is "spurious." He explains that a genuine correlation indicates an "organic relationship" between the variables, while a spurious correlation does not. But what is an "organic relationship"? Is it not causality by another name?

Together, Pearson and Yule compiled several examples of spurious correlations. One typical case is now called confounding, and the chocolate-Nobel story is an example. (Wealth and location are confounders, or common causes of both chocolate consumption and Nobel frequency.) Another type of "nonsense correlation" often emerges in time series data. For example, Yule found an incredibly high correlation (0.95) between England's mortality rate in a given year and the percentage of marriages conducted that year in

the Church of England. Was God punishing marriage-happy An-
glicans? No! Two separate historical trends were simply occurring
at the same time: the country's mortality rate was decreasing and
membership in the Church of England was declining. Since both
were going down at the same time, there was a positive correlation
between them, but no causal connection.

Pearson discovered possibly the most interesting kind of "spurious correlation" as early as 1899. It arises when two heterogeneous
populations are aggregated into one. Pearson, who, like Galton,
was a fanatical collector of data on the human body, had obtained
measurements of 806 male skulls and 340 female skulls from the
Paris Catacombs (Figure 2.5). He computed the correlation between skull length and skull breadth. When the computation was
done only for males or only for females, the correlations were
negligible—there was no significant association between skull
length and breadth. But when the two groups were combined, the
correlation was 0.197, which would ordinarily be considered significant. This makes sense, because a small skull length is now an
indicator that the skull likely belonged to a female and therefore
that the breadth will also be small. However, Pearson considered it
a statistical artifact. The fact that the correlation was positive had

FIGURE 2.5. Karl Pearson with a skull from the Paris Catacombs.
(*Source:* Drawing by Dakota Harr.)

no biological or "organic" meaning; it was just a result of combining two distinct populations inappropriately.

This example is a case of a more general phenomenon called Simpson's paradox. Chapter 6 will discuss when it is appropriate to segregate data into separate groups and will explain why spurious correlations can emerge from aggregation. But let's take a look at what Pearson wrote: "To those who persist in looking upon all correlations as cause and effect, the fact that correlation can be produced between two quite uncorrelated characters A and B by taking an artificial mixture of two closely allied races, must come rather as a shock." As Stephen Stigler comments, "I cannot resist the speculation that he himself was the first one shocked." In essence, Pearson was scolding himself for the tendency to think causally.

Looking at the same example through the lens of causality, we can only say, What a missed opportunity! In an ideal world, such examples might have spurred a talented scientist to think about the reason for his shock and develop a science to predict when spurious correlations appear. At the very least, he should explain when to aggregate the data and when not to. But Pearson's only guidance to his followers is that an "artificial" mixture (whatever that means) is bad. Ironically, using our causal lens, we now know that in some cases the aggregated data, not the partitioned data, give the correct result. The logic of causal inference can actually tell us which one to trust. I wish that Pearson were here to enjoy it!

Pearson's students did not all follow in lockstep behind him. Yule, who broke with Pearson for other reasons, broke with him over this too. Initially he was in the hard-line camp holding that correlations say everything we could ever wish to understand about science. However, he changed his mind to some extent when he needed to explain poverty conditions in London. In 1899, he studied the question of whether "out-relief" (that is, welfare delivered to a pauper's home versus a poorhouse) increased the rate of poverty. The data showed that districts with more out-relief had a higher poverty rate, but Yule realized that the correlation was possibly spurious: these districts might also have more elderly people, who tend to be poorer. However, he then showed that even in comparisons of districts with equal proportions of elderly people, the correlation remained. This emboldened him to say that the increased poverty rate was due to out-relief. But after stepping out of line to make this assertion, he fell back into line again, writing in a footnote, "Strictly speaking, for 'due to' read 'associated with.'" This set the pattern for generations of scientists after him. They would think "due to" and say "associated with."

With Pearson and his followers actively hostile toward causation, and with halfhearted dissidents such as Yule fearful of antagonizing their leader, the stage was set for another scientist from across the ocean to issue the first direct challenge to the causality-avoiding culture.

SEWALL WRIGHT, GUINEA PIGS, AND PATH DIAGRAMS

When Sewall Wright arrived at Harvard University in 1912, his academic background scarcely suggested the kind of lasting effect he would have on science. He had attended a small (and now defunct) college in Illinois, Lombard College, graduating in a class of only seven students. One of his teachers had been his own father, Philip Wright, an academic jack-of-all-trades who even ran the college's printing press. Sewall and his brother Quincy helped out with the press, and among other things they published the first poetry by a not-yet-famous Lombard student, Carl Sandburg.

Sewall Wright's ties with his father remained very close long after he graduated from college. Papa Philip moved to Massachusetts when Sewall did. Later, when Sewall worked in Washington, DC, Philip did likewise, first at the US Tariff Commission and then at the Brookings Institution as an economist. Although their academic interests diverged, they nevertheless found ways to collaborate, and Philip was the first economist to make use of his son's invention of path diagrams.

Wright came to Harvard to study genetics, at the time one of the hottest topics in science because Gregor Mendel's theory of dominant and recessive genes had just been rediscovered. Wright's advisor, William Castle, had identified eight different hereditary factors (or genes, as we would call them today) that affected fur color in rabbits. Castle assigned Wright to do the same thing for guinea pigs. After earning his doctorate in 1915, Wright got an offer for which he was uniquely qualified: taking care of guinea pigs at the US Department of Agriculture (USDA).

One wonders if the USDA knew what it was getting when it hired Wright. Perhaps it expected a diligent animal caretaker who could straighten out the chaos of twenty years of poorly kept records. Wright did all that and much, much more. Wright's guinea pigs were the springboard to his whole career and his whole theory of evolution, much like the finches on the Galapagos islands that had inspired Charles Darwin. Wright was an early advocate of the view that evolution is not gradual, as Darwin had posited, but takes place in relatively sudden bursts.

In 1925, Wright moved on to a faculty position at the University of Chicago that was probably better suited to someone with his wide-ranging theoretical interests. Even so, he remained very devoted to his guinea pigs. An often told anecdote says that he was once holding an unruly guinea pig under his arm while lecturing, and absentmindedly began using it to erase the blackboard (see Figure 2.6). While his biographers agree that this story is likely apocryphal, such stories often contain more truth than dry biographies do.

Wright's early work at the USDA interests us most here. The inheritance of coat color in guinea pigs stubbornly refused to play by Mendelian rules. It proved virtually impossible to breed an all-white or all-colored guinea pig, and even the most inbred families (after multiple generations of brother-sister mating) still had pronounced variation, from mostly white to mostly colored. This contradicted the prediction of Mendelian genetics that a particular trait should become "fixed" by multiple generations of inbreeding.

Wright began to doubt that genetics alone governed the amount of white and postulated that "developmental factors" in the womb were causing some of the variations. With hindsight, we know that he was correct. Different color genes are expressed in different places on the body, and the patterns of color depend not only on what genes the animal has inherited but where and in what combinations they happen to be expressed or suppressed.

FIGURE 2.6. Sewall Wright was the first person to develop a mathematical method for answering causal questions from data, known as path diagrams. His love of mathematics surrendered only to his passion for guinea pigs. (*Source:* Drawing by Dakota Harr.)

As it often happens (at least to the ingenious!), a pressing research problem leads to new methods of analysis, which vastly transcended their origins in guinea pig genetics. Yet, for Sewall Wright, estimating the developmental factors probably seemed like a college-level problem that he could have solved in his father's math class at Lombard. When looking for the magnitude of some unknown quantity, you first assign a symbol to that quantity, next you express what you know about this and other quantities in the form of mathematical equations, and finally, if you have enough patience and enough equations, you can solve them and find your quantity of interest.

In Wright's case, the desired and unknown quantity (shown in Figure 2.7) was d, the effect of "developmental factors" on white fur. Other causal quantities that entered into his equations included h, for "hereditary" factors, also unknown. Finally—and here comes Wright's ingenuity—he showed that if we knew the causal quantities in Figure 2.7, we could predict correlations in the

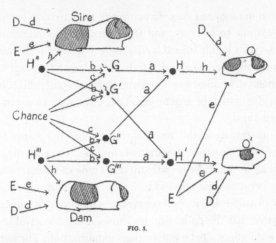

FIG. 5.

FIGURE 2.7. Sewall Wright's first path diagram, illustrating the factors leading to coat color in guinea pigs. D = developmental factors (after conception, before birth), E = environmental factors (after birth), G = genetic factors from each individual parent, H = combined hereditary factors from both parents, O, O' = offspring. The objective of analysis was to estimate the strength of the effects of D, E, H (written as d, e, h in the diagram). (*Source:* Sewall Wright, *Proceedings of the National Academy of Sciences* [1920], 320–332.)

data (not shown in the diagram) by a simple graphical rule. This rule sets up a bridge from the deep, hidden world of causation to the surface world of correlations. It was the first bridge ever built between causality and probability, the first crossing of the barrier between rung two and rung one on the Ladder of Causation. Having built this bridge, Wright could travel backward over it, *from* the correlations measured in the data (rung one) *to* the hidden causal quantities, d and h (rung two). He did this by solving algebraic equations. This idea must have seemed simple to Wright but turned out to be revolutionary because it was the first proof that the mantra "Correlation does not imply causation" should give way to "Some correlations do imply causation."

In the end, Wright showed that the hypothesized developmental factors were more important than heredity. In a randomly bred population of guinea pigs, 42 percent of the variation in coat pattern was due to heredity, and 58 percent was developmental. By contrast, in a highly inbred family, only 3 percent of the variation in white fur coverage was due to heredity, and 92 percent was developmental. In other words, twenty generations of inbreeding had all but eliminated the genetic variation, but the developmental factors remained.

As interesting as this result is, the crux of the matter for our history is the way that Wright made his case. The path diagram in Figure 2.7 is the street map that tells us how to navigate over this bridge between rung one and rung two. It is a scientific revolution in one picture—and it comes complete with adorable guinea pigs!

Notice that the path diagram shows every conceivable factor that could affect a baby guinea pig's pigmentation. The letters D, E, and H refer to developmental, environmental, and hereditary factors, respectively. Each parent (the sire and the dam) and each child (offspring O and O') has its own set of D, E, and H factors. The two offspring share environmental factors but have different developmental histories. The diagram incorporates the then novel insights of Mendelian genetics: a child's heredity (H) is determined by its parents' sperm and egg cells (G and G''), and these in turn are determined from the parents' heredity (H'' and H''') via a mixing process that was not yet understood (because DNA had not been discovered). It was understood, though, that the mixing process included an element of randomness (labeled "Chance" in the diagram).

One thing the diagram does not show explicitly is the difference between an inbred family and a normal family. In an inbred family there would be a strong correlation between the heredity of the sire and the dam, which Wright indicated with a two-headed arrow between H'' and H'''. Aside from that, every arrow in the diagram is one-way and leads from a cause to an effect. For example, the arrow from G to H indicates that the sire's sperm cell may have a direct causal effect on the offspring's heredity. The absence of an arrow from G to H' indicates that the sperm cell that gave rise to offspring O has no causal effect on the heredity of offspring O'.

When you take apart the diagram arrow by arrow in this way, I think you will find that every one of them makes perfect sense. Note also that each arrow is accompanied by a small letter (a, b, c, etc.). These letters, called path coefficients, represent the strength of the causal effects that Wright wanted to solve for. Roughly speaking, a path coefficient represents the amount of variability in the *target* variable that is accounted for by the *source* variable. For instance, it is fairly evident that 50 percent of each child's hereditary makeup should come from each parent, so that a should be 1/2. (For technical reasons, Wright preferred to take the square root, so that $a = 1/\sqrt{2}$ and $a^2 = 1/2$.)

This interpretation of path coefficients, in terms of the amount of variation explained by a variable, was reasonable at the time. The modern causal interpretation is different: the path coefficients represent the results of a hypothetical intervention on the source variable. However, the notion of an intervention would have to wait until the 1940s, and Wright could not have anticipated it when he wrote his paper in 1920. Fortunately, in the simple models he analyzed then, the two interpretations yield the same result.

I want to emphasize that the path diagram is not just a pretty picture; it is a powerful computational device because the rule for computing correlations (the bridge from rung two to rung one) involves tracing the paths that connect two variables to each other and multiplying the coefficients encountered along the way. Also, notice that the omitted arrows actually convey more significant assumptions than those that are present. An omitted arrow restricts the causal effect to zero, while a present arrow remains totally agnostic about the magnitude of the effect (unless we a priori impose some value on the path coefficient).

Wright's paper was a tour de force and deserves to be considered one of the landmark results of twentieth-century biology. Certainly it is a landmark for the history of causality. Figure 2.7 is the first causal diagram ever published, the first step of twentieth-century science onto the second rung of the Ladder of Causation. And not a tentative step but a bold and decisive one! The following year Wright published a much more general paper called "Correlation and Causation" that explained how path analysis worked in other settings than guinea pig breeding.

I don't know what kind of reaction the thirty-year-old scientist expected, but the reaction he got surely must have stunned him. It came in the form of a rebuttal published in 1921 by one Henry Niles, a student of American statistician Raymond Pearl (no relation), who in turn was a student of Karl Pearson, the godfather of statistics.

Academia is full of genteel savagery, which I have had the honor to weather at times in my own otherwise placid career, but even so I have seldom seen a criticism as savage as Niles's. He begins with a long series of quotes from his heroes, Karl Pearson and Francis Galton, attesting to the redundancy or even meaninglessness of the word "cause." He concludes, "To contrast 'causation' and 'correlation' is unwarranted because causation is simply perfect correlation." In this sentence he is directly echoing what Pearson wrote in *Grammar of Science*.

Niles further disparages Wright's entire methodology. He writes, "The basic fallacy of the method appears to be the assumption that it is possible to set up *a priori* a comparatively simple graphic system which will truly represent the lines of action of several variables upon each other, and upon a common result." Finally, Niles works through some examples and, bungling the computations because he has not taken the trouble to understand Wright's rules, he arrives at opposite conclusions. In summary, he declares, "We therefore conclude that philosophically the basis of the method of path coefficients is faulty, while practically the results of applying it where it can be checked prove it to be wholly unreliable."

From the scientific point of view a detailed discussion of Niles's criticism is perhaps not worth the time, but his paper is very important to us as historians of causation. First, it faithfully reflects the attitude of his generation toward causation and the total grip that his mentor, Karl Pearson, had on the scientific thinking of his time. Second, we continue to hear Niles's objections today.

Of course, at times scientists do not know the entire web of relationships between their variables. In that case, Wright argued, we can use the diagram in exploratory mode; we can postulate certain causal relationships and work out the predicted correlations between variables. If these contradict the data, then we have evidence that the relationships we assumed were false. This way of using path diagrams, rediscovered in 1953 by Herbert Simon (a 1978 Nobel laureate in economics), inspired much work in the social sciences.

Although we don't need to know every causal relation between the variables of interest and might be able to draw some conclusions with only partial information, Wright makes one point with absolute clarity: you cannot draw causal conclusions without some causal hypotheses. This echoes what we concluded in Chapter 1: you cannot answer a question on rung two of the Ladder of Causation using only data collected from rung one.

Sometimes people ask me, "Doesn't that make causal reasoning circular? Aren't you just assuming what you want to prove?" The answer is no. By combining very mild, qualitative, and obvious assumptions (e.g., coat color of the son does not influence that of the parents) with his twenty years of guinea pig data, he obtained a quantitative and by no means obvious result: that 42 percent of the variation in coat color is due to heredity. Extracting the nonobvious from the obvious is not circular—it is a scientific triumph and deserves to be hailed as such.

Wright's contribution is unique because the information leading to the conclusion (of 42 percent heritability) resided in two distinct, almost incompatible mathematical languages: the language of diagrams on one side and that of data on the other. This heretical idea of marrying qualitative "arrow-information" to quantitative "data-information" (two foreign languages!) was one of the miracles that first attracted me, as a computer scientist, to this enterprise.

Many people still make Niles's mistake of thinking that the goal of causal analysis is to prove that X is a cause of Y or else to find the cause of Y from scratch. That is the problem of causal discovery, which was my ambitious dream when I first plunged into graphical modeling and is still an area of vigorous research. In contrast, the focus of Wright's research, as well as this book, is representing plausible causal knowledge in some mathematical language, combining it with empirical data, and answering causal queries that are of practical value. Wright understood from the very beginning that causal discovery was much more difficult and perhaps impossible. In his response to Niles, he writes, "The writer [i.e., Wright himself] has never made the preposterous claim that the theory of path coefficients provides a general formula for the deduction of causal relations. He wishes to submit that the combination of knowledge of correlations with knowledge of causal relations to obtain certain results, is a different thing from the deduction of causal relations from correlations implied by Niles' statement."

E PUR SI MUOVE (AND YET IT MOVES)

If I were a professional historian, I would probably stop here. But as the "Whig historian" that I promised to be, I cannot contain myself from expressing my sheer admiration for the precision of Wright's words in the quote ending the previous section, which have not gone stale in the ninety years since he first articulated them and which essentially defined the new paradigm of modern causal analysis.

My admiration for Wright's precision is second only to my admiration for his courage and determination. Imagine the situation in 1921. A self-taught mathematician faces the hegemony of the statistical establishment alone. They tell him, "Your method is based on a complete misapprehension of the nature of causality in the scientific sense." And he retorts, "Not so! My method generates something that is important and goes beyond anything that you can generate." They say, "Our gurus looked into these problems already, two decades ago, and concluded that what you have done is nonsense. You have only combined correlations with correlations and gotten correlations. When you grow up, you will understand." And he continues, "I am not dismissing your gurus, but a spade is a spade. My path coefficients are *not* correlations. They are something totally different: causal effects."

Imagine that you are in kindergarten, and your friends mock you for believing that 3 + 4 = 7, when everybody knows that 3 + 4 = 8. Then imagine going to your teacher for help and hearing her say, too, that 3 + 4 = 8. Would you not go home and ask yourself if perhaps there was something wrong with the way you were thinking? Even the strongest man would start to waver in his convictions. I have been in that kindergarten, and I know.

But Wright did not blink. And this was not just a matter of arithmetic, where there can be some sort of independent verification. Only philosophers had dared to express an opinion on the nature of causation. Where did Wright get this inner conviction that he was on the right track and the rest of the kindergarten class was just plain wrong? Maybe his Midwestern upbringing and the tiny college he went to encouraged his self-reliance and taught him that the surest kind of knowledge is what you construct yourself.

One of the earliest science books I read in school told of how the Inquisition forced Galileo to recant his teaching that Earth revolves around the sun and how he whispered under his breath, "And yet it moves" (*E pur si muove*). I don't think that there is a child in the world who has read this legend and not been inspired by Galileo's courage in defending his convictions. Yet as much as we admire him for his stand, I can't help but think that he at least had his astronomical observations to fall back on. Wright had only untested conclusions—say, that developmental factors account for 58 percent, not 3 percent, of variation. With nothing to lean on except his internal conviction that path coefficients tell you what correlations do not, he still declared, "And yet it moves!"

Colleagues tell me that when Bayesian networks fought against the artificial intelligence establishment (see Chapter 3), I acted stubbornly, single-mindedly, and uncompromisingly. Indeed, I recall being totally convinced of my approach, with not an iota of hesitation. But I had probability theory on my side. Wright didn't have even one theorem to lean on. Scientists had abandoned causation, so Wright could not fall back on any theoretical framework. Nor could he rely on authorities, as Niles did, because there was no one for him to quote; the gurus had already pronounced their verdicts three decades earlier.

But one solace to Wright, and one sign that he was on the right path, must have been his understanding that he could answer questions that cannot be answered in any other way. Determining the relative importance of several factors was one such question.

Another beautiful example of this can be found in his "Correlation and Causation" paper, from 1921, which asks how much a guinea pig's birth weight will be affected if it spends one more day in the womb. I would like to examine Wright's answer in some detail to enjoy the beauty of his method and to satisfy readers who would like to see how the mathematics of path analysis works.

Notice that we cannot answer Wright's question directly, because we can't weigh a guinea pig in the womb. What we can do, though, is compare the birth weights of guinea pigs that spend (say) sixty-six days gestating with those that spend sixty-seven days. Wright noted that the guinea pigs that spent a day longer in the womb weighed an average of 5.66 grams more at birth. So, one might naively suppose that a guinea pig embryo grows at 5.66 grams per day just before it is born.

"Wrong!" says Wright. The pups born later are usually born later for a reason: they have fewer litter mates. This means that they have had a more favorable environment for growth through-out the pregnancy. A pup with only two siblings, for instance, will already weigh more on day sixty-six than a pup with four siblings. Thus the difference in birth weights has two causes, and we want to disentangle them. How much of the 5.66 grams is due to spend-ing an additional day in utero and how much is due to having fewer siblings to compete with?

Wright answered this question by setting up a path diagram (Figure 2.8). X represents the pup's birth weight. Q and P represent the two known causes of the birth weight: the length of gestation (P) and rate of growth in utero (Q). L represents litter size, which

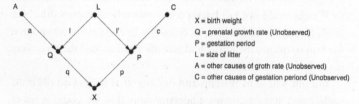

X = birth weight
Q = prenatal growth rate (Unobserved)
P = gestation period
L = size of litter
A = other causes of groth rate (Unobserved)
C = other causes of gestation periond (Unobserved)

FIGURE 2.8. Causal (path) diagram for birth-weight example.

affects both P and Q (a larger litter causes the pup to grow slower and also have fewer days in utero). It's very important to realize that X, P, and L can be measured, for each guinea pig, but Q cannot. Finally, A and C are exogenous causes that we don't have any data about (e.g., hereditary and environmental factors that control growth rate and gestation time independently of litter size). The important assumption that these factors are independent of each other is conveyed by the absence of any arrow between them, as well as of any common ancestor.

Now the question facing Wright was, "What is the direct effect of the gestation period P on the birth weight X?" The data (5.66 grams per day) don't tell you the direct effect; they give you a correlation, biased by the litter size L. To get the direct effect, we need to remove this bias.

In Figure 2.8, the direct effect is represented by the path coefficient p, corresponding to the path $P \to X$. The bias due to litter size corresponds to the path $P \leftarrow L \to Q \to X$. And now the algebraic magic: the amount of bias is equal to the product of the path coefficients along that path (in other words, l times l' times q). The total correlation, then, is just the sum of the path coefficients along the two paths: algebraically, $p + (l \times l' \times q) = 5.66$ grams per day.

If we knew the path coefficients l, l', and q, then we could just work out the second term and subtract it from 5.66 to get the desired quantity p. But we don't know them, because Q (for example) is not measured. But here's where the ingenuity of path coefficients really shines. Wright's methods tell us how to express each of the measured correlations in terms of the path coefficients. After doing this for each of the measured pairs (P, X), (L, X), and (L, P), we obtain three equations that can be solved algebraically for the unknown path coefficients, p, l', and $l \times q$. Then we are done, because the desired quantity p has been obtained.

Today we can skip the mathematics altogether and calculate p by cursory inspection of the diagram. But in 1920, this was the first time that mathematics was summoned to connect causation and correlation. And it worked! Wright calculated p to be 3.34 grams per day. In other words, had all the other variables (A, L, C, Q)

been held constant and only the time of gestation increased by a day, the average increase in birth weight would be 3.34 grams per day. Note that this result is biologically meaningful. It tells us how rapidly the pups are growing per day before birth. By contrast, the number 5.66 grams per day has no biological significance, because it conflates two separate processes, one of which is not causal but anticausal (or diagnostic) in the link $P \leftarrow L$. Lesson one from this example: causal analysis allows us to quantify processes in the real world, not just patterns in the data. The pups are growing at 3.34 grams per day, not 5.66 grams per day. Lesson two, whether you followed the mathematics or not: in path analysis you draw conclusions about individual causal relationships by examining the diagram as a whole. The entire structure of the diagram may be needed to estimate each individual parameter.

In a world where science progresses logically, Wright's response to Niles should have produced a scientific excitement followed by an enthusiastic adoption of his methods by other scientists and statisticians. But that is not what happened. "One of the mysteries of the history of science from 1920 to 1960 is the virtual absence of any appreciable use of path analysis, except by Wright himself and by students of animal breeding," wrote one of Wright's geneticist colleagues, James Crow. "Although Wright had illustrated many diverse problems to which the method was applicable, none of these leads was followed."

Crow didn't know it, but the mystery extended to social sciences as well. In 1972, economist Arthur Goldberger lamented the "scandalous neglect" of Wright's work during that period and noted, with the enthusiasm of a convert, that "[Wright's] approach . . . sparked the recent upsurge of causal modeling in sociology."

If only we could go back and ask Wright's contemporaries, "Why didn't you pay attention?" Crow suggests one reason: path analysis "doesn't lend itself to 'canned' programs. The user has to have a hypothesis and must devise an appropriate diagram of multiple causal sequences." Indeed, Crow put his finger on an essential point: path analysis requires scientific thinking, as does every exercise in causal inference. Statistics, as frequently practiced, discourages it and encourages "canned" procedures instead. Scientists will always prefer routine calculations on data to methods that challenge their scientific knowledge.

R. A. Fisher, the undisputed high priest of statistics in the generation after Galton and Pearson, described this difference succinctly. In 1925, he wrote, "Statistics may be regarded as . . . the study of methods of the reduction of data." Pay attention to the words "methods," "reduction," and "data." Wright abhorred the idea of statistics as merely a collection of methods; Fisher embraced it. Causal analysis is emphatically not just about data; in causal analysis we must incorporate some understanding of the process that produces the data, and then we get something that was not in the data to begin with. But Fisher was right about one point: once you remove causation from statistics, reduction of data is the only thing left.

Although Crow did not mention it, Wright's biographer William Provine points out another factor that may have affected the lack of support for path analysis. From the mid-1930s onward, Fisher considered Wright his enemy. I previously quoted Yule on how relations with Pearson became strained if you disagreed with him and impossible if you criticized him. Exactly the same thing could be said about Fisher. The latter carried out nasty feuds with anyone he disagreed with, including Pearson, Pearson's son Egon, Jerzy Neyman (more will be said on these two in Chapter 8), and of course Wright.

The real focus of the Fisher-Wright rivalry was not path analysis but evolutionary biology. Fisher disagreed with Wright's theory (called "genetic drift") that a species can evolve rapidly when it undergoes a population bottleneck. The details of the dispute are beyond the scope of this book, and the interested reader should consult Provine. Relevant here is this: from the 1920s to the 1950s, the scientific world for the most part turned to Fisher as its oracle for statistical knowledge. And you can be certain that Fisher never said one kind word to anyone about path analysis.

In the 1960s, things began to change. A group of social scientists, including Otis Duncan, Hubert Blalock, and the economist Arthur Goldberger (mentioned earlier), rediscovered path analysis as a method of predicting the effect of social and educational policies. In yet another irony of history, Wright had actually been asked to speak to an influential group of econometricians called the Cowles Commission in 1947, but he utterly failed to communicate to them what path diagrams were about. Only when economists arrived at similar ideas themselves was a short-lived connection forged.

The fates of path analysis in economics and sociology followed different trajectories, each leading to a betrayal of Wright's ideas. Sociologists renamed path analysis as structural equation modeling (SEM), embraced diagrams, and used them extensively until 1970, when a computer package called LISREL automated the calculation of path coefficients (in some cases). Wright would have predicted what followed: path analysis turned into a rote method, and researchers became software users with little interest in what was going on under the hood. In the late 1980s, a public challenge (by statistician David Freedman) to explain the assumptions behind SEM went unanswered, and some leading SEM experts even disavowed that SEMs had anything to do with causality.

In economics, the algebraic part of path analysis became known as simultaneous equation models (no acronym). Economists essentially never used path diagrams and continue not to use them to this day, relying instead on numerical equations and matrix algebra. A dire consequence of this is that, because algebraic equations are nondirectional (that is, $x = y$ is the same as $y = x$), economists had no notational means to distinguish causal from regression equations and thus were unable to answer policy-related questions, even after solving the equations. As late as 1995, most economists refrained from explicitly attributing causal or counterfactual meaning to their equations. Even those who used structural equations for policy decisions remained incurably suspicious of diagrams, which could have saved them pages and pages of computation. Not surprisingly, some economists continue to claim that "it's all in the data" to this very day.

For all these reasons, the promise of path diagrams remained only partially realized, at best, until the 1990s. In 1983, Wright himself was called back into the ring one more time to defend them, this time in the *American Journal of Human Genetics*. At the time he wrote this article, Wright was past ninety years old. It is both wonderful and tragic to read his essay, written in 1983, on the very same topic he had written about in 1923. How many times in the history of science have we had the privilege of hearing from a theory's creator sixty years after he first set it down on paper? It would be like Charles Darwin coming back from the grave to testify at the Scopes Monkey Trial in 1925. But it is also tragic, because in the intervening sixty years his theory should have developed, grown, and flourished; instead it had advanced little since the 1920s.

The motivation for Wright's paper was a critique of path analysis, published in the same journal, by Samuel Karlin (a Stanford mathematician and recipient of the 1989 National Medal of Science, who made fundamental contributions to economics and population genetics) and two coauthors. Of interest to us are two of Karlin's arguments.

First, Karlin objects to path analysis for a reason that Niles did not raise: it assumes that all the relationships between any two variables in the path diagram are linear. This assumption allows Wright to describe the causal relationships with a single number, the path coefficient. If the equations were not linear, then the effect on Y of a one-unit change in X might depend on the current value of X. Neither Karlin nor Wright realized that a general nonlinear theory was just around the corner. (It would be developed three years later by a star student in my lab, Thomas Verma.)

But Karlin's most interesting criticism was also the one that he considered the most important: "Finally, and we think most fruitfully, one can adopt an essentially model-free approach, seeking to understand the data interactively by using a battery of displays, indices, and contrasts. This approach emphasizes the concept of robustness in interpreting results." In this one sentence Karlin articulates how little had changed from the days of Pearson and how much influence Pearson's ideology still had in 1983. He is saying that the data themselves already contain all scientific wisdom; they need only be cajoled and massaged (by "displays, indices, and contrasts") into dispensing those pearls of wisdom. There is no need for our analysis to take into account the process that generated the data. We would do just as well, if not better, with a "model-free approach." If Pearson were alive today, living in the era of Big Data, he would say exactly this: the answers are all in the data.

Of course, Karlin's statement violates everything we learned in Chapter 1. To speak of causality, we must have a mental model of the real world. A "model-free approach" may take us to the first rung of the Ladder of Causation, but no farther.

Wright, to his great credit, understood the enormous stakes and stated in no uncertain terms, "In treating the model-free approach (3) as preferred alternative . . . Karlin et al. are urging not merely a change in method, but an abandonment of the purpose of path analysis and evaluation of the relative importance of varying causes. There can be no such analysis without a model. Their advice to anyone with an urge to make such an evaluation is to repress it and do something else."

Wright understood that he was defending the very essence of the scientific method and the interpretation of data. I would give the same advice today to big-data, model-free enthusiasts. Of course, it is okay to tease out all the information that the data can provide, but let's ask how far this will get us. It will never get us beyond the first rung of the Ladder of Causation, and it will never answer even as simple a question as "What is the relative importance of various causes?" *E pur si muove!*

FROM OBJECTIVITY TO SUBJECTIVITY— THE BAYESIAN CONNECTION

One other theme in Wright's rebuttal may hint at another reason for the resistance of statisticians to causality. He repeatedly states that he did not want path analysis to become "stereotyped." According to Wright, "The unstereotyped approach of path analysis differs profoundly from the stereotyped modes of description designed to avoid any departures from complete objectivity."

What does he mean? First, he means that path analysis should be based on the user's personal understanding of causal processes, reflected in the causal diagram. It cannot be reduced to mechanical routines, such as those laid out in statistics manuals. For Wright, drawing a path diagram is not a statistical exercise; it is an exercise in genetics, economics, psychology, or whatever the scientist's own field of expertise is.

Second, Wright traces the allure of "model-free" methods to their objectivity. This has indeed been a holy grail for statisticians since day one—or since March 15, 1834, when the Statistical Society of London was founded. Its founding charter said that data were to receive priority in all cases over opinions and interpretations. Data are objective; opinions are subjective. This paradigm long predates Pearson. The struggle for objectivity—the idea of reasoning exclusively from data and experiment—has been part of the way that science has defined itself ever since Galileo.

Unlike correlation and most of the other tools of mainstream statistics, causal analysis requires the user to make a subjective commitment. She must draw a causal diagram that reflects her qualitative belief—or, better yet, the consensus belief of researchers in her field of expertise—about the topology of the causal processes at work. She must abandon the centuries-old dogma of objectivity for objectivity's sake. Where causation is concerned, a grain of wise subjectivity tells us more about the real world than any amount of objectivity.

In the above paragraph, I said that "most of" the tools of statistics strive for complete objectivity. There is one important exception to this rule, though. A branch of statistics called Bayesian statistics has achieved growing popularity over the last fifty years or so. Once considered almost anathema, it has now gone completely mainstream, and you can attend an entire statistics conference without hearing any of the great debates between "Bayesians" and "frequentists" that used to thunder in the 1960s and 1970s.

The prototype of Bayesian analysis goes like this: Prior Belief + New Evidence → Revised Belief. For instance, suppose you toss a coin ten times and find that in nine of those tosses the coin came up heads. Your belief that the coin is fair is probably shaken, but how much? An orthodox statistician would say, "In the absence of any additional evidence, I would believe that this coin is loaded, so I would bet nine to one that the next toss turns up heads."

A Bayesian statistician, on the other hand, would say, "Wait a minute. We also need to take into account our prior knowledge about the coin." Did it come from the neighborhood grocery or a shady gambler? If it's just an ordinary quarter, most of us would not let the coincidence of nine heads sway our belief so dramatically. On the other hand, if we already suspected the coin was weighted, we would conclude more willingly that the nine heads provided serious evidence of bias.

Bayesian statistics give us an objective way of combining the observed evidence with our prior knowledge (or subjective belief) to obtain a revised belief and hence a revised prediction of the outcome of the coin's next toss. Still, what frequentists could not abide was that Bayesians were allowing opinion, in the form of subjective probabilities, to intrude into the pristine kingdom of statistics. Mainstream statisticians were won over only grudgingly, when Bayesian analysis proved a superior tool for a variety of applications, such as weather prediction and tracking enemy submarines. In addition, in many cases it can be proven that the influence of prior beliefs vanishes as the size of the data increases, leaving a single objective conclusion in the end.

Unfortunately, the acceptance of Bayesian subjectivity in mainstream statistics did nothing to help the acceptance of causal subjectivity, the kind needed to specify a path diagram. Why? The answer rests on a grand linguistic barrier. To articulate subjective assumptions, Bayesian statisticians still use the language of probability, the native language of Galton and Pearson. The assumptions entering causal inference, on the other hand, require a richer language (e.g., diagrams) that is foreign to Bayesians and frequentists alike. The reconciliation between Bayesians and frequentists shows that philosophical barriers can be bridged with goodwill and a common language. Linguistic barriers are not surmounted so easily.

Moreover, the subjective component in causal information does not necessarily diminish over time, even as the amount of data increases. Two people who believe in two different causal diagrams can analyze the same data and may never come to the same conclusion, regardless of how "big" the data are. This is a terrifying prospect for advocates of scientific objectivity, which explains their refusal to accept the inevitability of relying on subjective causal information.

On the positive side, causal inference is objective in one critically important sense: once two people agree on their assumptions, it provides a 100 percent objective way of interpreting any new evidence (or data). It shares this property with Bayesian inference. So the savvy reader will probably not be surprised to find out that I arrived at the theory of causality through a circuitous route that started with Bayesian probability and then took a huge detour through Bayesian networks. I will tell that story in the next chapter.

Sherlock Holmes meets his modern counterpart, a robot equipped with a Bayesian network. In different ways both are tackling the question of how to infer causes from observations. The formula on the computer screen is Bayes's rule. (*Source:* Drawing by Maayan Harel.)

3

FROM EVIDENCE TO CAUSES:
REVEREND BAYES MEETS MR. HOLMES

Do two men travel together unless they have agreed?
Does the lion roar in the forest if he has no prey?

—AMOS 3:3

"IT'S elementary, my dear Watson."

So spoke Sherlock Holmes (at least in the movies) just before dazzling his faithful assistant with one of his famously nonelementary deductions. But in fact, Holmes performed not just deduction, which works from a hypothesis to a conclusion. His great skill was induction, which works in the opposite direction, from evidence to hypothesis.

Another of his famous quotes suggests his modus operandi: "When you have eliminated the impossible, whatever remains, however improbable, must be the truth." Having *induced* several hypotheses, Holmes eliminated them one by one in order to *deduce* (by elimination) the correct one. Although induction and deduction go hand in hand, the former is by far the more mysterious. This fact kept detectives like Sherlock Holmes in business.

However, in recent years experts in artificial intelligence (AI) have made considerable progress toward automating the process of reasoning from evidence to hypothesis and likewise from effect to cause. I was fortunate enough to participate in the very earliest stages of this progress by developing one of its basic tools, called Bayesian networks. This chapter explains what these are, looks at some of their current-day applications, and discusses the circuitous route by which they led me to study causation.

BONAPARTE, THE COMPUTER DETECTIVE

On July 17, 2014, Malaysia Airlines Flight 17 took off from Amsterdam's Schiphol Airport, bound for Kuala Lumpur. Alas, the airplane never reached its destination. Three hours into the flight, as the jet flew over eastern Ukraine, it was shot down by a Russian-made surface-to-air missile. All 298 people on board, 283 passengers and 15 crew members, were killed.

July 23, the day the first bodies arrived in the Netherlands, was declared a national day of mourning. But for investigators at the Netherlands Forensic Institute (NFI) in The Hague, July 23 was the day when the clock started ticking. Their job was to identify the remains of the deceased as quickly as possible and return them to their loved ones for burial. Time was of the essence, because every day of uncertainty would bring fresh anguish to the grieving families.

The investigators faced many obstacles. The bodies had been badly burned, and many were stored in formaldehyde, which breaks down DNA. Also, because eastern Ukraine was a war zone, forensics experts had only sporadic access to the crash site. Newly recovered remains continued to arrive for ten more months. Finally, the investigators did not have previous records of the victims' DNA, for the simple reason that the victims were not criminals. They would have to rely instead on partial matches with family members.

Fortunately, the scientists at NFI had a powerful tool working in their favor, a state-of-the-art disaster victim identification program called Bonaparte. This software, developed in the mid-2000s by a team from Radboud University in Nijmegen, uses Bayesian networks to combine DNA information taken from several different family members of the victims.

Thanks in part to Bonaparte's accuracy and speed, the NFI managed to identify remains from 294 of the 298 victims by December 2014. As of 2016, only two victims of the crash (both Dutch citizens) have vanished without a trace.

Bayesian networks, the machine-reasoning tool that underlies the Bonaparte software, affect our lives in many ways that most people are not aware of. They are used in speech-recognition software, in spam filters, in weather forecasting, in the evaluation of potential oil wells, and in the Food and Drug Administration's approval process for medical devices. If you play video games on a Microsoft Xbox, a Bayesian network ranks your skill. If you own a cell phone, the codes that your phone uses to pick your call out of thousands of others are decoded by belief propagation, an algorithm devised for Bayesian networks. Vint Cerf, the chief Internet evangelist at another company you might have heard of, Google, puts it this way: "We're huge consumers of Bayesian methods."

In this chapter I will tell the story of Bayesian networks from their roots in the eighteenth century to their development in the 1980s, and I will give some more examples of how they are used today. They are related to causal diagrams in a simple way: a causal diagram is a Bayesian network in which every arrow signifies a direct causal relation, or at least the possibility of one, in the direction of that arrow. Not all Bayesian networks are causal, and in many applications it does not matter. However, if you ever want to ask a rung-two or rung-three query about your Bayesian network, you must draw it with scrupulous attention to causality.

REVEREND BAYES AND THE PROBLEM OF INVERSE PROBABILITY

Thomas Bayes, after whom I named the networks in 1985, never dreamed that a formula he derived in the 1750s would one day be used to identify disaster victims. He was concerned only with the probabilities of two events, one (the hypothesis) occurring before the other (the evidence). Nevertheless, causality was very much on his mind. In fact, causal aspirations were the driving force behind his analysis of "inverse probability."

A Presbyterian minister who lived from 1702 to 1761, the Reverend Thomas Bayes appears to have been a mathematics geek. As a dissenter from the Church of England, he could not study at Oxford or Cambridge and was educated instead at the University of Scotland, where he likely picked up quite a bit of math. He continued to dabble in it and organize math discussion circles after he returned to England.

In an article published after his death (see Figure 3.1), Bayes tackled a problem that was the perfect match for him, pitting math against theology. To set the context, in 1748, the Scottish philosopher David Hume had written an essay titled "On Miracles," in which he argued that eyewitness testimony could never prove that a miracle had happened. The miracle Hume had in mind was, of course, the resurrection of Christ, although he was smart enough

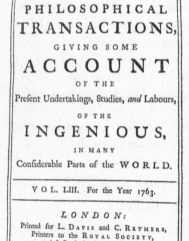

FIGURE 3.1. Title page of the journal where Thomas Bayes's posthumous article on inverse probability was published and the first page of Richard Price's introduction.

not to say so. (Twenty years earlier, theologian Thomas Woolston had gone to prison for blasphemy for writing such things.) Hume's main point was that inherently fallible evidence cannot overrule a proposition with the force of natural law, such as "Dead people stay dead."

For Bayes, this assertion provoked a natural, one might say Holmesian question: How much evidence would it take to convince us that something we consider improbable has actually happened? When does a hypothesis cross the line from impossibility to improbability and even to probability or virtual certainty? Although the question was phrased in the language of probability, the implications were intentionally theological. Richard Price, a fellow minister who found the essay among Bayes's possessions after his death and sent it for publication with a glowing introduction that he wrote himself, made this point abundantly clear:

> The purpose I mean is, to shew what reason we have for believing that there are in the constitution of things fixt laws according to which things happen, and that, therefore, the frame of the world must be the effect of the wisdom and power of an intelligent cause; and thus to confirm the argument taken from final causes for the existence of the Deity. It will be easy to see that the converse problem solved in this essay is more directly applicable to this purpose; for it shews us, with distinctness and precision, in every case of any particular order or recurrency of events, what reason there is to think that such recurrency or order is derived from stable causes or regulations in nature, and not from any irregularities of chance.

Bayes himself did not discuss any of this in his paper; Price highlighted these theological implications, perhaps to make the impact of his friend's paper more far-reaching. But it turned out that Bayes didn't need the help. His paper is remembered and argued about 250 years later, not for its theology but because it shows that you can deduce the probability of a cause from an effect. If we know the cause, it is easy to estimate the probability of the effect, which is a forward probability. Going the other direction—a problem known in Bayes's time as "inverse probability"—is harder. Bayes did not explain why it is harder; he took that as self-evident, proved that it is doable, and showed us how.

To appreciate the nature of the problem, let's look at a slightly simplified version of an example he suggested in his posthumous paper of 1763. Imagine that we shoot a billiard ball on a table, making sure that it bounces many times so that we have no idea where it will end up. What is the probability that it will stop within x feet of the left-hand end of the table? If we know the length of the table and it is perfectly smooth and flat, this is a very easy question (Figure 3.2, top). For example, on a twelve-foot snooker table, the probability of the ball stopping within a foot of the end would be 1/12. On an eight-foot billiard table, the probability would be 1/8.

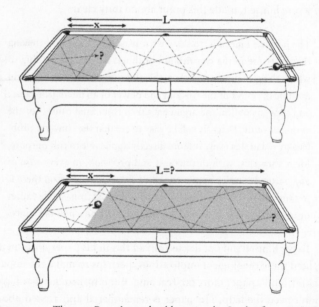

FIGURE 3.2. Thomas Bayes's pool table example. In the first version, a forward-probability question, we know the length of the table and want to calculate the probability of the ball stopping within x feet of the end. In the second, an inverse-probability question, we observe that the ball stopped x feet from the end and want to estimate the likelihood that the table's length is L. (*Source:* Drawing by Maayan Harel.)

Our intuitive understanding of the physics tells us that, in general, if the length of the table is L feet, the probability of the ball's stopping within x feet of the end is x/L. The longer the table length (L), the lower the probability, because there are more positions competing for the honor of being the ball's resting place. On the other hand, the larger x is, the higher the probability, because it includes a larger set of stopping positions.

Now consider the inverse-probability problem. We observe the final position of the ball to be $x = 1$ foot from the end, but we are not given the length L (Figure 3.2, bottom). Reverend Bayes asked, What is the probability that the length was, say, one hundred feet? Common sense tells us that L is more likely to be fifty feet than one hundred feet, because the longer table makes it harder to explain why the ball ended up so close to the end. But how much more likely is it? "Intuition" or "common sense" gives us no clear guidance.

Why was the forward probability (of x given L) so much easier to assess mentally than the probability of L given x? In this example, the asymmetry comes from the fact that L acts as the cause and x is the effect. If we observe a cause—for example, Bobby throws a ball toward a window—most of us can predict the effect (the ball will probably break the window). Human cognition works in this direction. But given the effect (the window is broken), we need much more information to deduce the cause (which boy threw the ball that broke it or even the fact that it was broken by a ball in the first place). It takes the mind of a Sherlock Holmes to keep track of all the possible causes. Bayes set out to break this cognitive asymmetry and explain how even ordinary humans can assess inverse probabilities.

To see how Bayes's method works, let's start with a simple example about customers in a teahouse, for whom we have data documenting their preferences. Data, as we know from Chapter 1, are totally oblivious to cause-effect asymmetries and hence should offer us a way to resolve the inverse-probability puzzle.

Suppose two-thirds of the customers who come to the shop order tea, and half of the tea drinkers also order scones. What fraction of the clientele orders both tea and scones? There's no trick to this question, and I hope that the answer is almost obvious.

Because half of two-thirds is one-third, it follows that one-third of the customers order both tea and scones.

For a numerical illustration, suppose that we tabulate the orders of the next twelve customers who come in the door. As Table 3.1 shows, two-thirds of the customers (1, 5, 6, 7, 8, 9, 10, 12) ordered tea, and one-half of those people ordered scones (1, 5, 8, 12). So the proportion of customers who ordered both tea and scones is indeed $(1/2) \times (2/3) = 1/3$, just as we predicted prior to seeing the specific data.

TABLE 3.1. Fictitious data for the tea-scones example.

Customer	Tea	Scones	Customer	Tea	Scones
1	Yes	Yes	7	Yes	No
2	No	Yes	8	Yes	Yes
3	No	No	9	Yes	No
4	No	No	10	Yes	No
5	Yes	Yes	11	No	No
6	Yes	No	12	Yes	Yes

The starting point for Bayes's rule is to notice that we could have analyzed the data in the reverse order. That is, we could have observed that five-twelfths of the customers (1, 2, 5, 8, 12) ordered scones, and four-fifths of these (1, 5, 8, 12) ordered tea. So the proportion of customers who ordered both tea and scones is $(4/5) \times (5/12) = 1/3$. Of course it's no coincidence that it came out the same; we were merely computing the same quantity in two different ways. The temporal order in which the customers announce their order makes no difference.

To make this a general rule, we can let $P(T)$ denote the probability that a customer orders tea and $P(S)$ denote the probability he orders scones. If we already know a customer has ordered tea, then $P(S \mid T)$ denotes the probability that he orders scones. (Remember that the vertical line stands for "given that.") Likewise, $P(T \mid S)$ denotes the probability that he orders tea, given that we already know he ordered scones. Then the first calculation we did says,

$$P(S \text{ AND } T) = P(S \mid T) \, P(T).$$

The second calculation says,

$$P(S \text{ AND } T) = P(T \mid S) \, P(S).$$

Now, as Euclid said 2,300 years ago, two things that each equal a third thing also equal one another. That means it must be the case that

$$P(S \mid T) \, P(T) = P(T \mid S) \, P(S) \tag{3.1}$$

This innocent-looking equation came to be known as "Bayes's rule." If we look carefully at what it says, we find that it offers a general solution to the inverse-probability problem. It tells us that if we know the probability of S given T, $P(S \mid T)$, we ought to be able to figure out the probability of T given S, $P(T \mid S)$, assuming of course that we know $P(T)$ and $P(S)$. This is perhaps the most important role of Bayes's rule in statistics: we can estimate the conditional probability directly in one direction, for which our judgment is more reliable, and use mathematics to derive the conditional probability in the other direction, for which our judgment is rather hazy. The equation also plays this role in Bayesian networks; we tell the computer the forward probabilities, and the computer tells us the inverse probabilities when needed.

To see how Bayes's rule works in the teahouse example, suppose you didn't bother to calculate $P(T \mid S)$ and left your spreadsheet containing the data at home. However, you happen to remember that half of those who order tea also order scones, and two-thirds of the customers order tea and five-twelfths order scones. Unexpectedly, your boss asks you, "But what proportion of scone eaters order tea?" There's no need to panic, because you can work it out from the other probabilities. Bayes's rule says that $P(T \mid S)(5/12) = (1/2)(2/3)$, so your answer is $P(T \mid S) = 4/5$, because 4/5 is the only value for $P(T \mid S)$ that will make this equation true.

We can also look at Bayes's rule as a way to update our belief in a particular hypothesis. This is extremely important to understand, because a large part of human belief about future events rests on the frequency with which they or similar events have occurred in the past. Indeed, when a customer walks in the door of the restaurant, we believe, based on our past encounters with similar customers, that she probably wants tea. But if she first orders scones, we become even more certain. In fact, we might even suggest it: "I presume you want tea with that?" Bayes's rule simply lets us attach numbers to this reasoning process. From Table 3.1, we see that the prior probability that the customer wants tea (meaning when she walks in the door, before she orders anything) is two-thirds. But if the customer orders scones, now we have additional information about her that we didn't have before. The updated probability that she wants tea, given that she has ordered scones, is $P(T \mid S) = 4/5$.

Mathematically, that's all there is to Bayes's rule. It seems almost trivial. It involves nothing more than the concept of conditional probability, plus a little dose of ancient Greek logic. You might justifiably ask how such a simple gimmick could make Bayes famous and why people have argued over his rule for 250 years. After all, mathematical facts are supposed to settle controversies, not create them.

Here I must confess that in the teahouse example, by deriving Bayes's rule from data, I have glossed over two profound objections, one philosophical and the other practical. The philosophical one stems from the interpretation of probabilities as a degree of belief, which we used implicitly in the teahouse example. Who ever said that beliefs act, or should act, like proportions in the data?

The crux of the philosophical debate is whether we can legitimately translate the expression "given that I know" into the language of probabilities. Even if we agree that the unconditional probabilities $P(S)$, $P(T)$, and $P(S \text{ AND } T)$ reflect my degree of belief in those propositions, who says that my revised degree of belief in T should equal the ratio $P(S \text{ AND } T)/P(T)$, as dictated by Bayes's rule? Is "given that I know T" the same as "among cases where T occurred"? The language of probability, expressed in symbols like $P(S)$, was intended to capture the concept of frequencies in games of chance. But the expression "given that I know" is epistemological and should be governed by the logic of knowledge, not that of frequencies and proportions.

From the philosophical perspective, Thomas Bayes's accomplishment lies in his proposing the first formal definition of conditional probability as the ratio $P(S \mid T) = P(S \text{ AND } T)/P(T)$. His essay was admittedly hazy; he has no term "conditional probability" and instead uses the cumbersome language "the probability of the 2nd [event] on supposition that the 1st happens." The recognition that the relation "given that" deserves its own symbol evolved only in the 1880s, and it was not until 1931 that Harold Jeffreys (known more as a geophysicist than a probability theorist) introduced the now standard vertical bar in $P(S \mid T)$.

As we saw, Bayes's rule is formally an elementary consequence of his definition of conditional probability. But epistemologically, it is far from elementary. It acts, in fact, as a normative rule for updating beliefs in response to evidence. In other words, we should view Bayes's rule not just as a convenient definition of the new concept of "conditional probability" but as an empirical claim to faithfully represent the English expression "given that I know." It asserts, among other things, that the belief a person attributes to S after discovering T is never lower than the degree of belief that person attributes to S AND T before discovering T. Also, it implies that the more surprising the evidence T—that is, the smaller $P(T)$ is—the more convinced one should become of its cause S. No wonder Bayes and his friend Price, as Episcopal ministers, saw this as an effective rejoinder to Hume. If T is a miracle ("Christ rose from the dead"), and S is a closely related hypothesis ("Christ is the son of God"), our degree of belief in S is very dramatically increased if we know for a fact that T is true. The more miraculous the miracle, the more credible the hypothesis that explains its occurrence. This explains why the writers of the New Testament were so impressed by their eyewitness evidence.

Now let me discuss the practical objection to Bayes's rule—which may be even more consequential when we exit the realm of theology and enter the realm of science. If we try to apply the rule to the billiard-ball puzzle, in order to find $P(L \mid x)$ we need a quantity that is not available to us from the physics of billiard balls: we need the prior probability of the length L, which is every bit as tough to estimate as our desired $P(L \mid x)$. Moreover, this probability will vary significantly from person to person, depending on a given individual's previous experience with tables of different lengths. A person who has never in his life seen a snooker table would be very doubtful that L could be longer than ten feet. A person who has only seen snooker tables and never seen a billiard table would, on the other hand, give a very low prior probability to L being less than ten feet. This variability, also known as "subjectivity," is sometimes seen as a deficiency of Bayesian inference. Others regard it as a powerful advantage; it permits us to express our personal experience mathematically and combine it with data in a principled and transparent way. Bayes's rule informs our reasoning in cases where ordinary intuition fails us or where emotion might lead us astray. We will demonstrate this power in a situation familiar to all of us.

Suppose you take a medical test to see if you have a disease, and it comes back positive. How likely is it that you have the disease? For specificity, let's say the disease is breast cancer, and the test is a mammogram. In this example the *forward* probability is the probability of a positive test, given that you have the disease: $P(test \mid disease)$. This is what a doctor would call the "sensitivity" of the test, or its ability to correctly detect an illness. Generally it is the same for all types of patients, because it depends only on the technical capability of the testing instrument to detect the abnormalities associated with the disease. The *inverse* probability is the one you surely care more about: What is the probability that I have the disease, given that the test came out positive? This is $P(disease \mid test)$, and it represents a flow of information in the noncausal direction, from the result of the test to the probability of disease. This probability is not necessarily the same for all types of patients; we would certainly view the positive test with more alarm in a patient with a family history of the disease than in one with no such history.

Notice that we have started to talk about causal and noncausal directions. We didn't do that in the teahouse example because it did not matter which came first, ordering tea or ordering scones. It only mattered which conditional probability we felt more capable of assessing. But the causal setting clarifies why we feel less comfortable assessing the "inverse probability," and Bayes's essay makes clear that this is exactly the sort of problem that interested him.

Suppose a forty-year-old woman gets a mammogram to check for breast cancer, and it comes back positive. The hypothesis, D (for "disease"), is that she has cancer. The evidence, T (for "test"), is the result of the mammogram. How strongly should she believe the hypothesis? Should she have surgery?

We can answer these questions by rewriting Bayes's rule as follows:

$$\text{(Updated probability of } D) = P(D \mid T) =$$
$$\text{(likelihood ratio)} \times \text{(prior probability of } D) \qquad (3.2)$$

where the new term "likelihood ratio" is given by $P(T \mid D)/P(T)$. It measures how much more likely the positive test is in people with the disease than in the general population. Equation 3.2 therefore tells us that the new evidence T augments the probability of D by a fixed ratio, no matter what the prior probability was.

Let's do an example to see how this important concept works. For a typical forty-year-old woman, the probability of getting breast cancer in the next year is about one in seven hundred, so we'll use that as our prior probability.

To compute the likelihood ratio, we need to know $P(T \mid D)$ and $P(T)$. In the medical context, $P(T \mid D)$ is the sensitivity of the mammogram—the probability that it will come back positive if you have cancer. According to the Breast Cancer Surveillance Consortium (BCSC), the sensitivity of mammograms for forty-year-old women is 73 percent.

The denominator, $P(T)$, is a bit trickier. A positive test, T, can come both from patients who have the disease and from patients who don't. Thus, $P(T)$ should be a weighted average of $P(T \mid D)$ (the probability of a positive test among those who have the disease) and $P(T \mid \sim D)$ (the probability of a positive test among those who don't). The second is known as the false positive rate. According to the BCSC, the false positive rate for forty-year-old women is about 12 percent.

Why a weighted average? Because there are many more healthy women (~D) than women with cancer (D). In fact, only 1 in 700 women has cancer, and the other 699 do not, so the probability of a positive test for a randomly chosen woman should be much more strongly influenced by the 699 women who don't have cancer than by the one woman who does.

Mathematically, we compute the weighted average as follows: $P(T) = (1/700) \times (73 \text{ percent}) + (699/700) \times (12 \text{ percent}) \approx 12.1$ percent. The weights come about because only 1 in 700 women has a 73 percent chance of a positive test, and the other 699 have a 12 percent chance. Just as you might expect, $P(T)$ came out very close to the false positive rate.

Now that we know $P(T)$, we finally can compute the updated probability—the woman's chances of having breast cancer after the test comes back positive. The likelihood ratio is 73 percent/12.1 percent ≈ 6. As I said before, this is the factor by which we augment her prior probability to compute her updated probability of having cancer. Since her prior probability was one in seven hundred, her updated probability is $6 \times 1/700 \approx 1/116$. In other words, she still has less than a 1 percent chance of having cancer.

The conclusion is startling. I think that most forty-year-old women who have a positive mammogram would be astounded to learn that they still have less than a 1 percent chance of having breast cancer. Figure 3.3 might make the reason easier to understand: the tiny number of true positives (i.e., women with breast cancer) is overwhelmed by the number of false positives. Our sense of surprise at this result comes from the common cognitive confusion between the forward probability, which is well studied and thoroughly documented, and the inverse probability, which is needed for personal decision making.

The conflict between our perception and reality partially explains the outcry when the US Preventive Services Task Force, in 2009, recommended that forty-year-old women should not get annual mammograms. The task force understood what many women

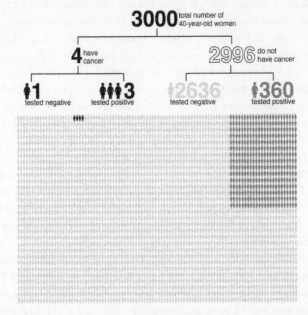

FIGURE 3.3. In this example, based on false-positive and false-negative rates provided by the Breast Cancer Surveillance Consortium, only 3 out of 363 forty-year-old women who test positive for breast cancer actually have the disease. (Proportions do not exactly match the text because of rounding.) (*Source:* Infographic by Maayan Harel.)

did not: a positive test at that age is way more likely to be a false alarm than to detect cancer, and many women were unnecessarily terrified (and getting unnecessary treatment) as a result.

However, the story would be very different if our patient had a gene that put her at high risk for breast cancer—say, a one-in-twenty chance within the next year. Then a positive test would increase the probability to almost one in three. For a woman in this situation, the chances that the test provides lifesaving information are much higher. That is why the task force continued to recommend annual mammograms for high-risk women.

This example shows that P(*disease* | *test*) is not the same for everyone; it is context dependent. If you know that you are at high risk for a disease to begin with, Bayes's rule allows you to factor that information in. Or if you know that you are immune, you need not even bother with the test! In contrast, $P(test \mid disease)$ does not depend on whether you are at high risk or not. It is "robust" to such variations, which explains to some degree why physicians organize their knowledge and communicate with forward probabilities. The former are properties of the disease itself, its stage of progression, or the sensitivity of the detecting instruments; hence they remain relatively invariant to the reasons for the disease (epidemic, diet, hygiene, socioeconomic status, family history). The inverse probability, $P(disease \mid test)$, is sensitive to these conditions.

The history-minded reader will surely wonder how Bayes handled the subjectivity of $P(L)$, where L is the length of a billiard table. The answer has two parts. First, Bayes was interested not in the length of the table per se but in its future consequences (i.e., the probability that the next ball would end up at some specified location on the table). Second, Bayes assumed that L is determined mechanically by shooting a billiard ball from a greater distance, say L^*. In this way he bestowed objectivity onto $P(L)$ and transformed the problem into one where prior probabilities are estimable from data, as we see in the teahouse and cancer test examples.

In many ways, Bayes's rule is a distillation of the scientific method. The textbook description of the scientific method goes something like this: (1) formulate a hypothesis, (2) deduce a testable consequence of the hypothesis, (3) perform an experiment and collect evidence, and (4) update your belief in the hypothesis. Usually the textbooks deal with simple yes-or-no tests and updates; the evidence either confirms or refutes the hypothesis. But life and science are never so simple! All evidence comes with a certain amount of uncertainty. Bayes's rule tells us how to perform step (4) in the real world.

FROM BAYES'S RULE TO BAYESIAN NETWORKS

In the early 1980s, the field of artificial intelligence had worked itself into a cul-de-sac. Ever since Alan Turing first laid out the challenge in his 1950 paper "Computing Machinery and Intelligence," the leading approach to AI had been so-called rule-based systems or expert systems, which organize human knowledge as a collection of specific and general facts, along with inference rules to connect them. For example: Socrates is a man (specific fact). All men are mortals (general fact). From this knowledge base we (or an intelligent machine) can derive the fact that Socrates is a mortal, using the universal rule of inference: if all A's are B's, and x is an A, then x is a B.

The approach was fine in theory, but hard-and-fast rules can rarely capture real-life knowledge. Perhaps without realizing it, we deal with exceptions to rules and uncertainties in evidence all the time. By 1980, it was clear that expert systems struggled with making correct inferences from uncertain knowledge. The computer could not replicate the inferential process of a human expert because the experts themselves were not able to articulate their thinking process within the language provided by the system.

The late 1970s, then, were a time of ferment in the AI community over the question of how to deal with uncertainty. There was no shortage of ideas. Lotfi Zadeh of Berkeley offered "fuzzy logic," in which statements are neither true nor false but instead take a range of possible truth values. Glen Shafer of the University of Kansas proposed "belief functions," which assign two probabilities to each fact, one indicating how likely it is to be "possible," the other, how likely it is to be "provable." Edward Feigenbaum and his colleagues at Stanford University tried "certainty factors," which inserted numerical measures of uncertainty into their deterministic rules for inference.

Unfortunately, although ingenious, these approaches suffered a common flaw: they modeled the expert, not the world, and therefore tended to produce unintended results. For example, they could not operate in both diagnostic and predictive modes, the uncontested specialty of Bayes's rule. In the certainty factor approach, the rule "If fire, then smoke (with certainty c_1)" could not combine coherently with "If smoke, then fire (with certainty c_2)" without triggering a runaway buildup of belief.

Probability was also considered at the time but immediately fell into ill repute, since the demands on storage space and processing time became formidable. I entered the arena rather late, in 1982, with an obvious yet radical proposal: instead of reinventing a new uncertainty theory from scratch, let's keep probability as a guardian of common sense and merely repair its computational deficiencies. More specifically, instead of representing probability in huge tables, as was previously done, let's represent it with a network of loosely coupled variables. If we only allow each variable to interact with a few neighboring variables, then we might overcome the computational hurdles that had caused other probabilists to stumble.

The idea did not come to me in a dream; it came from an article by David Rumelhart, a cognitive scientist at University of California, San Diego, and a pioneer of neural networks. His article about children's reading, published in 1976, made clear that reading is a complex process in which neurons on many different levels are active at the same time (see Figure 3.4). Some of the neurons are simply recognizing individual features—circles or lines. Above them, another layer of neurons is combining these shapes and forming conjectures about what the letter might be. In Figure 3.4, the network is struggling with a great deal of ambiguity about the second word. At the letter level, it could be "FHP," but that doesn't make much sense at the word level. At the word level it could be "FAR" or "CAR" or "FAT." The neurons pass this information up to the syntactic level, which decides that after the word "THE," it's expecting a noun. Finally this information gets passed all the way up to the semantic level, which realizes that the previous sentence mentioned a Volkswagen, so the phrase is likely to be "THE CAR," referring to that same Volkswagen. The key point is that all the neurons are passing information back and forth, from the top down and from the bottom up and from side to side. It's a highly parallel system, and one that is quite different from our self-perception of the brain as a monolithic, centrally controlled system.

Reading Rumelhart's paper, I felt convinced that any artificial intelligence would have to model itself on what we know about

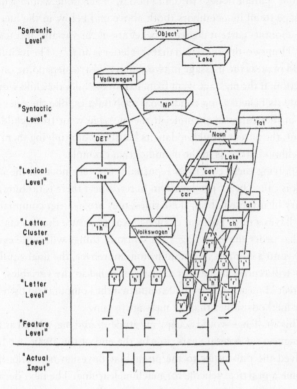

FIGURE 3.4. David Rumelhart's sketch of how a message-passing network would learn to read the phrase "THE CAR." (*Source:* Courtesy of Center for Brain and Cognition, University of California, San Diego.)

human neural information processing and that machine reasoning under uncertainty would have to be constructed with a similar message-passing architecture. But what are the messages? This took me quite a few months to figure out. I finally realized that the messages were conditional probabilities in one direction and likelihood ratios in the other.

More precisely, I assumed that the network would be hierarchical, with arrows pointing from higher neurons to lower ones, or from "parent nodes" to "child nodes." Each node would send a message to all its neighbors (both above and below in the hierarchy) about its current degree of belief about the variable it tracked (e.g., "I'm two-thirds certain that this letter is an *R*"). The recipient would process the message in two different ways, depending on its direction. If the message went from parent to child, the child would update its beliefs using conditional probabilities, like the ones we saw in the teahouse example. If the message went from child to parent, the parent would update its beliefs by multiplying them by a likelihood ratio, as in the mammogram example.

Applying these two rules repeatedly to every node in the network is called belief propagation. In retrospect there is nothing arbitrary or invented about these rules; they are in strict compliance with Bayes's rule. The real challenge was to ensure that no matter in what order these messages are sent out, things will settle eventually into a comfortable equilibrium; moreover, the final equilibrium will represent the correct state of belief in the variables. By "correct" I mean, as if we had conducted the computation by textbook methods rather than by message passing.

This challenge would occupy my students and me, as well as my colleagues, for several years. But by the end of the 1980s, we had resolved the difficulties to the point that Bayesian networks had become a practical scheme for machine learning. The next decade saw a continual increase in real-world applications, such as spam filtering and voice recognition. However, by then I was already trying to climb the Ladder of Causation, while entrusting the probabilistic side of Bayesian networks to the safekeeping of others.

BAYESIAN NETWORKS: WHAT CAUSES SAY ABOUT DATA

Although Bayes didn't know it, his rule for inverse probability represents the simplest Bayesian network. We have seen this network in several guises now: Tea → Scones, Disease → Test, or, more generally, Hypothesis → Evidence. Unlike the causal diagrams we will deal with throughout the book, a Bayesian network carries no assumption that the arrow has any causal meaning. The arrow merely signifies that we know the "forward" probability, $P(scones \mid tea)$ or $P(test \mid disease)$. Bayes's rule tells us how to reverse the procedure, specifically by multiplying the prior probability by a likelihood ratio.

Belief propagation formally works in exactly the same way whether the arrows are noncausal or causal. Nevertheless, you may have the intuitive feeling that we have done something more meaningful in the latter case than in the former. That is because our brains are endowed with special machinery for comprehending cause-effect relationships (such as cancer and mammograms). Not so for mere associations (such as tea and scones).

The next step after a two-node network with one link is, of course, a three-node network with two links, which I will call a "junction." These are the building blocks of all Bayesian networks (and causal networks as well). There are three basic types of junctions, with the help of which we can characterize any pattern of arrows in the network.

1. $A \rightarrow B \rightarrow C$. This junction is the simplest example of a "chain," or of mediation. In science, one often thinks of B as the mechanism, or "mediator," that transmits the effect of A to C. A familiar example is Fire $\rightarrow$ Smoke $\rightarrow$ Alarm. Although we call them "fire alarms," they are really smoke alarms. The fire by itself does not set off an alarm, so there is no direct arrow from Fire to Alarm. Nor does the fire set off the alarm through any other variable, such as heat. It works only by releasing smoke molecules in the air. If we disable that link in the chain, for instance by sucking all the smoke molecules away with a fume hood, then there will be no alarm.

   This observation leads to an important conceptual point about chains: the mediator B "screens off" information about A from C, and vice versa. (This was first pointed out by Hans Reichenbach, a German-American philosopher of science.) For example, once we know the value of Smoke, learning about Fire does not give us any reason to raise or lower our belief in Alarm. This stability of belief is a rung-one concept; hence it should also be seen in the data, when it is available. Suppose we had a database of all the instances when there was fire, when there was smoke, or when the alarm went off. If we looked at only the rows where Smoke = 1, we would expect Alarm = 1 every time, regardless of whether Fire = 0 or Fire = 1. This screening-off pattern still holds if the effect is not deterministic. For example, imagine a faulty alarm system that fails to respond correctly 5 percent of the time. If we look only at the rows where Smoke = 1, we will find that the probability of Alarm = 1 is the same (95 percent), regardless of whether Fire = 0 or Fire = 1.

The process of looking only at rows in the table where Smoke = 1 is called conditioning on a variable. Likewise, we say that Fire and Alarm are conditionally independent, given the value of Smoke. This is important to know if you are programming a machine to update its beliefs; conditional independence gives the machine a license to focus on the relevant information and disregard the rest. We all need this kind of license in our everyday thinking, or else we will spend all our time chasing false signals. But how do we decide which information to disregard, when every new piece of information changes the boundary between the relevant and the irrelevant? For humans, this understanding comes naturally. Even three-year-old toddlers understand the screening-off effect, though they don't have a name for it. Their instinct must have come from some mental representation, possibly resembling a causal diagram. But machines do not have this instinct, which is one reason that we equip them with causal diagrams.

2. $A \leftarrow B \rightarrow C$. This kind of junction is called a "fork," and B is often called a common cause or confounder of A and C. A confounder will make A and C statistically correlated even though there is no direct causal link between them. A good example (due to David Freedman) is Shoe Size ← Age of Child → Reading Ability. Children with larger shoes tend to read at a higher level. But the relationship is not one of

cause and effect. Giving a child larger shoes won't make him read better! Instead, both variables are explained by a third, which is the child's age. Older children have larger shoes, and they also are more advanced readers.

We can eliminate this spurious correlation, as Karl Pearson and George Udny Yule called it, by conditioning on the child's age. For instance, if we look only at seven-year-olds, we expect to see no relationship between shoe size and reading ability. As in the case of chain junctions, A and C are conditionally independent, given B.

Before we go on to our third junction, we need to add a word of clarification. The conditional independences I have just mentioned are exhibited whenever we look at these junctions in isolation. If additional causal paths surround them, these paths need also be taken into account. The miracle of Bayesian networks lies in the fact that the three kinds of junctions we are now describing in isolation are sufficient for reading off all the independencies implied by a Bayesian network, regardless of how complicated.

3. $A \to B \leftarrow C$. This is the most fascinating junction, called a "collider." Felix Elwert and Chris Winship have illustrated this junction using three features of Hollywood actors: Talent $\to$ Celebrity $\leftarrow$ Beauty. Here we are asserting that both talent and beauty contribute to an actor's success, but beauty and talent are completely unrelated to one another in the general population.

We will now see that this collider pattern works in exactly the opposite way from chains or forks when we condition on the variable in the middle. If A and C are independent to begin with, conditioning on B will make them dependent. For example, if we look only at famous actors (in other words, we observe the variable Celebrity = 1), we will see a negative correlation between talent and beauty: finding out that a celebrity is unattractive increases our belief that he or she is talented.

This negative correlation is sometimes called collider bias or the "explain-away" effect. For simplicity, suppose that you don't need both talent and beauty to be a celebrity; one is sufficient. Then if Celebrity A is a particularly good actor, that "explains away" his success, and he doesn't need to be any more beautiful than the average person. On the other hand, if Celebrity B is a really bad actor, then the only way to explain his success is his good looks. So, given the outcome Celebrity = 1, talent and beauty are inversely related—even though they are not related in the population as a whole. Even in a more realistic situation, where success is a complicated function of beauty and talent, the explain-away effect will still be present. This example is admittedly somewhat apocryphal, because beauty and talent are hard to measure objectively; nevertheless, collider bias is quite real, and we will see lots of examples in this book.

These three junctions—chains, forks, and colliders—are like keyholes through the door that separates the first and second levels of the Ladder of Causation. If we peek through them, we can see the secrets of the causal process that generated the data we observe; each stands for a distinct pattern of causal flow and leaves its mark in the form of conditional dependences and independences in the data. In my public lectures I often call them "gifts from the gods" because they enable us to test a causal model, discover new models, evaluate effects of interventions, and much more. Still, standing in isolation, they give us only a glimpse. We need a key that will completely open the door and let us step out onto the second rung. That key, which we will learn about in Chapter 7, involves all three junctions, and is called d-separation. This concept tells us, for any given pattern of paths in the model, what patterns of dependencies we should expect in the data. This fundamental connection between causes and probabilities constitutes the main contribution of Bayesian networks to the science of causal inference.

WHERE IS MY BAG? FROM AACHEN TO ZANZIBAR

So far I have emphasized only one aspect of Bayesian networks—namely, the diagram and its arrows that preferably point from cause to effect. Indeed, the diagram is like the engine of the Bayesian network. But like any engine, a Bayesian network runs on fuel. The fuel is called a *conditional probability table*.

Another way to put this is that the diagram describes the relation of the variables in a qualitative way, but if you want quantitative answers, you also need quantitative inputs. In a Bayesian network, we have to specify the conditional probability of each node given its "parents." (Remember that the parents of a node are all the nodes that feed into it.) These are the forward probabilities, $P(evidence \mid hypotheses)$.

In the case where A is a root node, with no arrows pointing into it, we need only specify the prior probability for each state of A. In our second network, Disease → Test, Disease is a root node. Therefore we specified the prior probability that a person has the disease (1/700 in our example) and that she does not have the disease (699/700 in our example).

By depicting A as a root node, we do not really mean that A has no prior causes. Hardly any variable is entitled to such a status. We really mean that any prior causes of A can be adequately summarized in the prior probability $P(A)$ that A is true. For example, in the Disease → Test example, family history might be a cause of Disease. But as long as we are sure that this family history will not affect the variable Test (once we know the status of Disease), we need not represent it as a node in the graph. However, if there is a cause of Disease that also directly affects Test, then that cause must be represented explicitly in the diagram.

In the case where the node A has a parent, A has to "listen" to its parent before deciding on its own state. In our mammogram example, the parent of Test was Disease. We can show this "listening" process with a 2 × 2 table (see Table 3.2). For example, if Test

TABLE 3.2. A simple conditional probability table.

Probability of →, given ↓	T = 0	T = 1
D = 0	88	12
D = 1	27	73

"hears" that $D = 0$, then 88 percent of the time it will take the value $T = 0$, and 12 percent of the time it will take the value $T = 1$. Notice that the second column of this table contains the same information we saw earlier from the Breast Cancer Surveillance Consortium: the false positive rate (upper right corner) is 12 percent, and the sensitivity (lower right corner) is 73 percent. The remaining two entries are filled in to make each row sum to 100 percent.

As we move to more complicated networks, the conditional probability table likewise gets more complicated. For example, if we have a node with two parents, the conditional probability table has to take into account the four possible states of both parents. Let's look at a concrete example, suggested by Stefan Conrady and Lionel Jouffe of BayesiaLab, Inc. It's a scenario familiar to all travelers: we can call it "Where Is My Bag?"

Suppose you've just landed in Zanzibar after making a tight connection in Aachen, and you're waiting for your suitcase to appear on the carousel. Other passengers have started to get their bags, but you keep waiting . . . and waiting . . . and waiting. What are the chances that your suitcase did not actually make the connection from Aachen to Zanzibar? The answer depends, of course, on how long you have been waiting. If the bags have just started to show up on the carousel, perhaps you should be patient and wait a little bit longer. If you've been waiting a long time, then things are looking bad. We can quantify these anxieties by setting up a causal diagram (Figure 3.5).

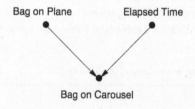

FIGURE 3.5. Causal diagram for airport/bag example.

This diagram reflects the intuitive idea that there are two causes for the appearance of any bag on the carousel. First, it had to be on the plane to begin with; otherwise, it will certainly never appear on the carousel. Second, the presence of the bag on the carousel becomes more likely as time passes . . . provided it was actually on the plane.

To turn the causal diagram into a Bayesian network, we have to specify the conditional probability tables. Let's say that all the bags at Zanzibar airport get unloaded within ten minutes. (They are very efficient in Zanzibar!) Let's also suppose that the probability your bag made the connection, $P(bag\ on\ plane$ = true) is 50 percent. (I apologize if this offends anybody who works at the Aachen airport. I am only following Conrady and Jouffe's example. Personally, I would prefer to assume a higher prior probability, like 95 percent.)

The real workhorse of this Bayesian network is the conditional probability table for "Bag on Carousel" (see Table 3.3).

This table, though large, should be easy to understand. The first eleven rows say that if your bag didn't make it onto the plane (*bag on plane* = false) then, no matter how much time has elapsed, it won't be on the carousel (*carousel* = false). That is, $P(carousel$ = false | *bag on plane* = false) is 100 percent. That is the meaning of the 100s in the first eleven rows.

The other eleven rows say that the bags are unloaded from the plane at a steady rate. If your bag is indeed on the plane, there is a 10 percent probability it will be unloaded in the first minute, a 10 percent probability in the second minute, and so forth. For example, after 5 minutes there is a 50 percent probability it has been unloaded, so we see a 50 for $P(carousel$ = true | *bag on plane* = true, *time* = 5). After ten minutes, all the bags have been unloaded, so $P(carousel$ = true | *bag on plane* = true, *time* = 10) is 100 percent. Thus we see a 100 in the last entry of the table.

The most interesting thing to do with this Bayesian network, as with most Bayesian networks, is to solve the inverse-probability problem: if x minutes have passed and I still haven't gotten my bag, what is the probability that it was on the plane? Bayes's rule

TABLE 3.3. A more complicated conditional probability table.

Probability of →, Given ↓		*carousel* = false	*carousel* = true
bag on plane	*time elapsed*		
False	0	100	0
False	1	100	0
False	2	100	0
False	3	100	0
False	4	100	0
False	5	100	0
False	6	100	0
False	7	100	0
False	8	100	0
False	9	100	0
False	10	100	0
True	0	100	0
True	1	90	10
True	2	80	20
True	3	70	30
True	4	60	40
True	5	50	50
True	6	40	60
True	7	30	70
True	8	20	80
True	9	10	90
True	10	0	100

automates this computation and reveals an interesting pattern. After one minute, there is still a 47 percent chance that it was on the plane. (Remember that our prior assumption was a 50 percent probability.) After five minutes, the probability drops to 33 percent. After ten minutes, of course, it drops to zero. Figure 3.6 shows a plot of the probability over time, which one might call the "Curve of Abandoning Hope." To me the interesting thing is that it *is* a

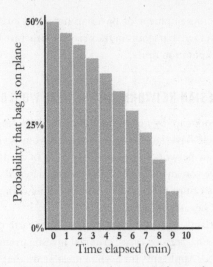

FIGURE 3.6. The probability of seeing your bag on the carousel decreases slowly at first, then more rapidly. (*Source:* Graph by Maayan Harel, data from Stefan Conrady and Lionel Jouffe.)

curve: I think that most people would expect it to be a straight line. It actually sends us a pretty optimistic message: don't give up hope too soon! According to this curve, you should abandon only one-third of your hope in the first half of the allotted time.

Besides a life lesson, we've learned that you don't want to do this by hand. Even with this tiny network of three nodes, there were 2 × 11 = 22 parent states, each contributing to the probability of the child state. For a computer, though, such computations are elementary . . . up to a point. If they aren't done in an organized fashion, the sheer number of computations can overwhelm even the fastest supercomputer. If a node has ten parents, each of which has two states, the conditional probability table will have more than 1,000 rows. And if each of the ten parents has ten states, the table will have 10 billion rows! For this reason one usually has to winnow the connections in the network so that only the most important ones remain and the network is "sparse." One technical

advance in the development of Bayesian networks entailed finding ways to leverage sparseness in the network structure to achieve
reasonable computation times.

BAYESIAN NETWORKS IN THE REAL WORLD

Bayesian networks are by now a mature technology, and you can
buy off-the-shelf Bayesian network software from several companies. Bayesian networks are also embedded in many "smart"
devices. To give you an idea of how they are used in real-world applications, let's return to the Bonaparte DNA-matching software
with which we began this chapter.

The Netherlands Forensic Institute uses Bonaparte every day,
mostly for missing-persons cases, criminal investigations, and immigration cases. (Applicants for asylum must prove that they have
fifteen family members in the Netherlands.) However, the Bayesian
network does its most impressive work after a massive disaster,
such as the crash of Malaysia Airlines Flight 17.

Few, if any, of the victims of the plane crash could be identified by comparing DNA from the wreckage to DNA in a central
database. The next best thing to do was to ask family members
to provide DNA swabs and look for partial matches to the DNA
of the victims. Conventional (non-Bayesian) methods can do this
and have been instrumental in solving a number of cold cases in
the Netherlands, the United States, and elsewhere. For example, a
simple formula called the "Paternity Index" or the "Sibling Index"
can estimate the likelihood that the unidentified DNA comes from
the father or the brother of the person whose DNA was tested.

However, these indices are inherently limited because they
work for only one specified relation and only for close relations.
The idea behind Bonaparte is to make it possible to use DNA in-
formation from more distant relatives or from multiple relatives.
Bonaparte does this by converting the pedigree of the family (see
Figure 3.7) into a Bayesian network.

In Figure 3.8, we see how Bonaparte converts one small piece of
a pedigree to a (causal) Bayesian network. The central problem is
that the genotype of an individual, detected in a DNA test, contains a

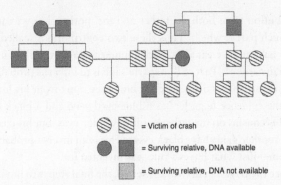

= Victim of crash

= Surviving relative, DNA available

= Surviving relative, DNA not available

FIGURE 3.7. Actual pedigree of a family with multiple victims in the Malaysia Airlines crash. (*Source:* Data provided by Willem Burgers.)

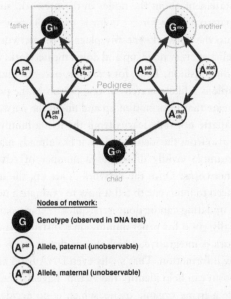

Nodes of network:

G Genotype (observed in DNA test)

A^pat Allele, paternal (unobservable)

A^mat Allele, maternal (unobservable)

FIGURE 3.8. From DNA tests to Bayesian networks. In Bayesian network, unshaded nodes represent alleles, and shaded nodes represent genotypes. Data are only available on shaded nodes because genotypes cannot indicate which allele came from the father and which from the mother. The Bayesian network enables inference on the unobserved nodes and also allows us to estimate the likelihood that a given DNA sample came from the child. (*Source:* Infographic by Maayan Harel.)

contribution from both the father and the mother, but we cannot tell which part is which. Thus these two contributions (called "alleles") have to be treated as hidden, unmeasurable variables in the Bayesian network. Part of Bonaparte's job is to infer the probability of the cause (the victim's gene for blue eyes came from his father) from the evidence (e.g., he has a blue-eyed gene and a black-eyed gene; his cousins on the father's side have blue eyes, but his cousins on the mother's side have black eyes). This is an inverse-probability problem—just what Bayes's rule was invented for.

Once the Bayesian network is set up, the final step is to input the victim's DNA and compute the likelihood that it fits into a specific slot in the pedigree. This is done by belief propagation with Bayes's rule. The network begins with a particular degree of belief in each possible statement about the nodes in the network, such as "this person's paternal allele for eye color is blue." As new evidence is entered into the network—at any place in the network—the degrees of belief at every node, up and down the network, will change in a cascading fashion. Thus, for example, once we find out that a given sample is a likely match for one person in the pedigree, we can propagate that information up and down the network. In this way, Bonaparte not only learns from the living family members' DNA but also from the identifications it has already made.

This example vividly illustrates a number of advantages of Bayesian networks. Once the network is set up, the investigator does not need to intervene to tell it how to evaluate a new piece of data. The updating can be done very quickly. (Bayesian networks are especially good for programming on a distributed computer.) The network is integrative, which means that it reacts as a whole to any new information. That's why even DNA from an aunt or a second cousin can help identify the victim. Bayesian networks are almost like a living organic tissue, which is no accident because this is precisely the picture I had in mind when I was struggling to make them work. I wanted Bayesian networks to operate like the neurons of a human brain; you touch one neuron, and the entire network responds by propagating the information to every other neuron in the system.

The transparency of Bayesian networks distinguishes them from most other approaches to machine learning, which tend to produce inscrutable "black boxes." In a Bayesian network you can follow every step and understand how and why each piece of evidence changed the network's beliefs.

As elegant as Bonaparte is, it's worth noting one feature it does not (yet) incorporate: human intuition. Once it has finished the analysis, it provides the NFI's experts with a ranking of the most likely identifications for each DNA sample and a likelihood ratio for each. The investigators are then free to combine the DNA evidence with other physical evidence recovered from the crash site, as well as their intuition, to make their final determinations. At present, no identifications are made by the computer acting alone. One goal of causal inference is to create a smoother human-machine interface, which might allow the investigators' intuition to join the belief propagation dance.

This example of DNA identification with Bonaparte only scratches the surface of the applications of Bayesian networks to genomics. However, I would like to move on to a second application that has become ubiquitous in today's society. In fact, there is a very good chance that you have a Bayesian network in your pocket right now. It's called a cell phone, every one of which uses error-correction algorithms based on belief propagation.

To begin at the beginning, when you talk into a phone, it converts your beautiful voice into a string of ones and zeros (called bits) and transmits these using a radio signal. Unfortunately, no radio signal is received with perfect fidelity. As the signal makes its way to the cell tower and then to your friend's phone, some random bits will flip from zero to one or vice versa.

To correct these errors, we can add redundant information. An ultrasimple scheme for error correction is simply to repeat each information bit three times: encode a one as "111" and a zero as "000." The valid strings "111" and "000" are called codewords. If the receiver hears an invalid string, such as "101," it will search for the most likely valid codeword to explain it. The zero is more likely to be wrong than both ones, so the decoder will interpret this message as "111" and therefore conclude that the information bit was a one.

Alas, this code is highly inefficient, because it makes all our messages three times longer. However, communication engineers have worked for seventy years on finding better and better error-correcting codes.

The problem of decoding is identical to the other inverse-probability problems we have discussed, because we once again want to infer the probability of a hypothesis (the message sent was "Hello world!") from evidence (the message received was "Hxllo wovld!"). The situation seems ripe for an application of belief propagation.

In 1993, an engineer for France Telecom named Claude Berrou stunned the coding world with an error-correcting code that achieved near-optimal performance. (In other words, the amount of redundant information required is close to the theoretical minimum.) His idea, called a "turbo code," can be best illustrated by representing it with a Bayesian network.

Figure 3.9(a) shows how a traditional code works. The information bits, which you speak into the phone, are shown in the first row. They are encoded, using any code you like—call it code A—into codewords (second row), which are then received with some errors (third row). This diagram is a Bayesian network, and we can use belief propagation to infer from the received bits what the information bits were. However, this would not in any way improve on code A.

Berrou's brilliant idea was to encode each message twice, once directly and once after scrambling the message. This results in the creation of two separate codewords and the receipt of two noisy messages (Figure 3.9b). There is no known formula for directly decoding such a dual message. But Berrou showed empirically that if you apply the belief propagation formulas on Bayesian networks repeatedly, two amazing things happen. Most of the time (and by this I mean something like 99.999 percent of the time) you get the correct information bits. Not only that, you can use much shorter codewords. To put it simply, two copies of code A are way better than one.

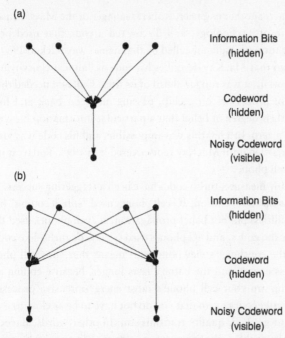

FIGURE 3.9. (a) Bayesian network representation of ordinary coding process. Information bits are transformed into codewords; these are transmitted and received at the destination with noise (errors). (b) Bayesian network representation of turbo code. Information bits are scrambled and encoded twice. Decoding proceeds by belief propagation on this network. Each processor at the bottom uses information from the other processor to improve its guess of the hidden codeword, in an iterative process.

This capsule history is correct except for one thing: Berrou did not know that he was working with Bayesian networks! He had simply discovered the belief propagation algorithm himself. It wasn't until five years later that David MacKay of Cambridge realized that it was the same algorithm that he had been enjoying in the late 1980s while playing with Bayesian networks. This placed Berrou's algorithm in a familiar theoretical context and allowed information theorists to sharpen their understanding of its performance.

In fact, another engineer, Robert Gallager of the Massachusetts Institute of Technology, had discovered a code that used belief propagation (though not called by that name) way back in 1960, so long ago that MacKay describes his code as "almost clairvoyant." In any event, it was too far ahead of its time. Gallager needed thousands of processors on a chip, passing messages back and forth about their degree of belief that a particular information bit was a one or a zero. In 1960 this was impossible, and his code was virtually forgotten until MacKay rediscovered it in 1998. Today, it is in every cell phone.

By any measure, turbo codes have been a staggering success. Before the turbo revolution, 2G cell phones used "soft decoding" (i.e., probabilities) but not belief propagation. 3G cell phones used Berrou's turbo codes, and 4G phones used Gallager's turbo-like codes. From the consumer's viewpoint, this means that your cell phone uses less energy and the battery lasts longer, because coding and decoding are your cell phone's most energy-intensive processes. Also, better codes mean that you do not have to be as close to a cell tower to get high-quality transmission. In other words, Bayesian networks enabled phone manufacturers to deliver on their promise: more bars in more places.

FROM BAYESIAN NETWORKS TO CAUSAL DIAGRAMS

After a chapter devoted to Bayesian networks, you might wonder how they relate to the rest of this book and in particular to causal diagrams, the kind we met in Chapter 1. Of course, I have discussed them in such detail in part because they were my personal route into causality. But more importantly from both a theoretical and practical point of view, Bayesian networks hold the key that enables causal diagrams to interface with data. All the probabilistic properties of Bayesian networks (including the junctions we discussed earlier in this chapter) and the belief propagation algorithms that were developed for them remain valid in causal diagrams. They are in fact indispensable for understanding causal inference.

The main differences between Bayesian networks and causal diagrams lie in how they are constructed and the uses to which they are put. A Bayesian network is literally nothing more than a compact representation of a huge probability table. The arrows mean only that the probabilities of child nodes are related to the values of parent nodes by a certain formula (the conditional probability tables) and that this relation is sufficient. That is, knowing additional ancestors of the child will not change the formula. Likewise, a missing arrow between any two nodes means that they are independent, once we know the values of their parents. We saw a simple version of this statement earlier, when we discussed the screening-off effect in chains and links. In a chain $A \to B \to C$, the missing arrow between A and C means that A and C are independent once we know the values of their parents. Because A has no parents, and the only parent of C is B, it follows that A and C are independent once we know the value of B, which agrees with what we said before.

If, however, the same diagram has been constructed as a causal diagram, then both the thinking that goes into the construction and the interpretation of the final diagram change. In the construction phase, we need to examine each variable, say C, and ask ourselves which other variables it "listens" to before choosing its value. The chain structure $A \to B \to C$ means that B listens to A only, C listens to B only, and A listens to no one; that is, it is determined by external forces that are not part of our model.

This listening metaphor encapsulates the entire knowledge that a causal network conveys; the rest can be derived, sometimes by leveraging data. Note that if we reverse the order of arrows in the chain, thus obtaining $A \gets B \gets C$, the causal reading of the structure will change drastically, but the independence conditions will remain the same. The missing arrow between A and C will still mean that A and C are independent once we know the value of B, as in the original chain. This has two enormously important implications. First, it tells us that causal assumptions cannot be invented at our whim; they are subject to the scrutiny of data and can be falsified. For instance, if the observed data do not show A and C

to be independent, conditional on B, then we can safely conclude that the chain model is incompatible with the data and needs to be discarded (or repaired). Second, the graphical properties of the diagram dictate which causal models can be distinguished by data and which will forever remain indistinguishable, no matter how large the data. For example, we cannot distinguish the fork $A \leftarrow B \rightarrow C$ from the chain $A \rightarrow B \rightarrow C$ by data alone because, with C listening to B only, the two imply the same independence conditions.

Another convenient way of thinking about the causal model is in terms of hypothetical experiments. Each arrow can be thought of as a statement about the outcome of a hypothetical experiment. An arrow from A to C means that if we could wiggle only A, then we would expect to see a change in the probability of C. A missing arrow from A to C means that in the same experiment we would not see any change in C, once we held constant the parents of C (in other words, B in the example above). Note that the probabilistic expression "once we know the value of B" has given way to the causal expression "once we hold B constant," which implies that we are physically preventing B from varying and disabling the arrow from A to B.

The causal thinking that goes into the construction of the causal network will pay off, of course, in the type of questions the network can answer. Whereas a Bayesian network can only tell us how likely one event is, given that we observed another (rung-one information), causal diagrams can answer interventional and counterfactual questions. For example, the causal fork $A \leftarrow B \rightarrow C$ tells us in no uncertain terms that wiggling A would have no effect on C, no matter how intense the wiggle. On the other hand, a Bayesian network is not equipped to handle a "wiggle," or to tell the difference between seeing and doing, or indeed to distinguish a fork from a chain. In other words, both a chain and a fork would predict that observed changes in A are associated with changes in C, making no prediction about the effect of "wiggling" A.

Now we come to the second, and perhaps more important, impact of Bayesian networks on causal inference. The relationships that were discovered between the graphical structure of the diagram and the data that it represents now permit us to emulate wiggling without physically doing so. Specifically, applying a smart sequence of conditioning operations enables us to predict the effect of actions or interventions without actually conducting an experiment. To demonstrate, consider again the causal fork $A \leftarrow B \rightarrow C$, in which we proclaimed the correlation between A and C to be spurious. We can verify this by an experiment in which we wiggle A and find no correlation between A and C. But we can do better. We can ask the diagram to emulate the experiment and tell us if any conditioning operation can reproduce the correlation that would prevail in the experiment. The answer would come out affirmative: "The correlation between A and C that would be measured after conditioning on B would equal the correlation seen in the experiment." This correlation can be estimated from the data, and in our case it would be zero, faithfully confirming our intuition that wiggling A would have no effect on C.

This ability to emulate interventions by smart observations could not have been acquired had the statistical properties of Bayesian networks not been unveiled between 1980 and 1988. We can now decide which set of variables we must measure in order to predict the effects of interventions from observational studies. We can also answer "Why?" questions. For example, someone may ask why wiggling A makes C vary. Is it really the direct effect of A, or is it the effect of a mediating variable B? If both, can we assess what portion of the effect is mediated by B?

To answer such mediation questions, we have to envision two simultaneous interventions: wiggling A and holding B constant (to be distinguished from conditioning on B). If we can perform this intervention physically, we obtain the answer to our question. But if we are at the mercy of observational studies, we need to emulate the two actions with a clever set of observations. Again, the graphical structure of the diagram will tell us whether this is possible.

All these capabilities were still in the future in 1988, when I started thinking about how to marry causation to diagrams. I only knew that Bayesian networks, as then conceived, could not answer the questions I was asking. The realization that you cannot even tell $A \leftarrow B \rightarrow C$ apart from $A \rightarrow B \rightarrow C$ from data alone was a painful frustration.

I know that you, the reader, are eager now to learn how causal diagrams enable us to do calculations like the ones I have just described. And we will get there—in Chapters 7 through 9. But we are not ready yet, because the moment we start talking about observational versus experimental studies, we leave the relatively friendly waters of the AI community for the much stormier waters of statistics, which have been stirred up by its unhappy divorce from causality. In retrospect, fighting for the acceptance of Bayesian networks in AI was a picnic—no, a luxury cruise!—compared with the fight I had to wage for causal diagrams. That battle is still ongoing, with a few remaining islands of resistance.

To navigate these new waters, we will have to understand the ways in which orthodox statisticians have learned to address causation and the limitations of those methods. The questions we raised above, concerning the effect of interventions, including direct and indirect effects, are not part of mainstream statistics, primarily because the field's founding fathers purged it of the language of cause and effect. But statisticians nevertheless consider it permissible to talk about causes and effects in one situation: a randomized controlled trial (RCT) in which a treatment A is randomly assigned to some individuals and not to others and the observed changes in B are then compared. Here, both orthodox statistics and causal inference agree on the meaning of the sentence "A causes B."

Before we turn to the new science of cause and effect— illuminated by causal models—we should first try to understand the strengths and limitations of the old, model-blind science: why randomization is needed to conclude that A causes B and the nature of the threat (called "confounding") that RCTs are intended to disarm. The next chapter takes up these topics. In my experience, most statisticians as well as modern data analysts are not comfortable with any of these questions, since they cannot articulate them using a data-centric vocabulary. In fact, they often disagree on what "confounding" means!

After we examine these issues in the light of causal diagrams, we can place randomized controlled trials into their proper context. Either we can view them as a special case of our inference engine, or we can view causal inference as a vast extension of RCTs. Either viewpoint is fine, and perhaps people trained to see RCTs as the arbiter of causation will find the latter more congenial.

Intervention
(Vegetarian Diet)

Control
(King's Diet)

The biblical story of Daniel, often cited as the first controlled experiment. Daniel (third from left?) realized that a proper comparison of two diets could only be made when they were given to two groups of similar individuals, chosen in advance. King Nebuchadnezzar (rear) was impressed with the results. (*Source:* Drawing by Dakota Harr.)

4

CONFOUNDING AND DECONFOUNDING: OR, SLAYING THE LURKING VARIABLE

If our conception of causal effects had anything to do with randomized experiments, the latter would have been invented 500 years before Fisher.

—THE AUTHOR (2016)

A SHPENAZ, the overseer of King Nebuchadnezzar's court, had a major problem. In 597 BC, the king of Babylon had sacked the kingdom of Judah and brought back thousands of captives, many of them the nobility of Jerusalem. As was customary in his kingdom, Nebuchadnezzar wanted some of them to serve in his court, so he commanded Ashpenaz to seek out "children in whom was no blemish, but well favoured, and skilful in all wisdom, and cunning in knowledge, and understanding science." These lucky children were to be educated in the language and culture of Babylon so that they could serve in the administration of the empire, which stretched from the Persian Gulf to the Mediterranean Sea. As part of their education, they would get to eat royal meat and drink royal wine.

And therein lay the problem. One of his favorites, a boy named Daniel, refused to touch the food. For religious reasons, he could not eat meat not prepared according to Jewish laws, and he asked that he and his friends be given a diet of vegetables instead. Ashpenaz would have liked to comply with the boy's wishes, but he was afraid that the king would notice: "Once he sees your frowning faces, different from the other children your age, it will cost me my head."

Daniel tried to assure Ashpenaz that the vegetarian diet would not diminish their capacity to serve the king. As befits a person "cunning in knowledge, and understanding science," he proposed an experiment. Try us for ten days, he said. Take four of us and feed us only vegetables; take another group of children and feed them the king's meat and wine. After ten days, compare the two groups. Said Daniel, "And as thou seest, deal with thy servants."

Even if you haven't read the story, you can probably guess what happened next. Daniel and his three companions prospered on the vegetarian diet. The king was so impressed with their wisdom and learning—not to mention their healthy appearance—that he gave them a favored place in his court, where "he found them ten times better than all the magicians and astrologers that were in all his realm." Later Daniel became an interpreter of the king's dreams and survived a memorable encounter in a lion's den.

Believe it or not, the biblical story of Daniel encapsulates in a profound way the conduct of experimental science today. Ashpenaz asks a question about causation: Will a vegetarian diet cause my servants to lose weight? Daniel proposes a methodology to deal with any such questions: Set up two groups of people, identical in all relevant ways. Give one group a new treatment (a diet, a drug, etc.), while the other group (called the control group) either gets the old treatment or no special treatment at all. If, after a suitable amount of time, you see a measurable difference between the two supposedly identical groups of people, then the new treatment must be the cause of the difference.

Nowadays we call this a controlled experiment. The principle is simple. To understand the causal effect of the diet, we would like to compare what happens to Daniel on one diet with what would have happened if he had stayed on the other. But we can't go back in time and rewrite history, so instead we do the next best thing: we compare a group of people who get the treatment with a group of similar people who don't. It's obvious, but nevertheless crucial, that the groups be comparable and representative of some population. If these conditions are met, then the results should be transferable to the population at large. To Daniel's credit, he seems to understand this. He isn't just asking for vegetables on his own behalf: if the trial shows the vegetarian diet is better, then all the Israelite servants should be allowed that diet in the future. That, at least, is how I interpret the phrase, "As thou seest, deal with thy servants."

Daniel also understood that it was important to compare groups. In this respect he was already more sophisticated than many people today, who choose a fad diet (for example) just because a friend went on that diet and lost weight. If you choose a diet based only on one friend's experience, you are essentially saying that you believe you are similar to your friend in all relevant details: age, heredity, home environment, previous diet, and so forth. That is a lot to assume.

Another key point of Daniel's experiment is that it was prospective: the groups were chosen in advance. By contrast, suppose that you see twenty people in an infomercial who all say they lost weight on a diet. That seems like a pretty large sample size, so some viewers might consider it convincing evidence. But that would amount to basing their decision on the experience of people who already had a good response. For all you know, for every person who lost weight, ten others just like him or her tried the diet and had no success. But of course, they weren't chosen to appear on the infomercial.

Daniel's experiment was strikingly modern in all these ways. Prospective controlled trials are still a hallmark of sound science. However, Daniel didn't think of one thing: confounding bias. Suppose that Daniel and his friends are healthier than the control group to start with. In that case, their robust appearance after ten days on the diet may have nothing to do with the diet itself; it may reflect their overall health. Maybe they would have prospered even more if they had eaten the king's meat!

Confounding bias occurs when a variable influences both who is selected for the treatment and the outcome of the experiment.

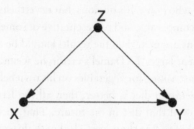

FIGURE 4.1. The most basic version of confounding: Z is a confounder of the proposed causal relationship between X and Y.

Sometimes the confounders are known; other times they are merely suspected and act as a "lurking third variable." In a causal diagram, confounders are extremely easy to recognize: in Figure 4.1, the variable Z at the center of the fork is a confounder of X and Y. (We will see a more universal definition later, but this triangle is the most recognizable and common situation.)

The term "confounding" originally meant "mixing" in English, and we can understand from the diagram why this name was chosen. The true causal effect X → Y is "mixed" with the spurious correlation between X and Y induced by the fork X ← Z → Y. For example, if we are testing a drug and give it to patients who are younger on average than the people in the control group, then age becomes a confounder—a lurking third variable. If we don't have any data on the ages, we will not be able to disentangle the true effect from the spurious effect.

However, the converse is also true. If we do have measurements of the third variable, then it is very easy to deconfound the true and spurious effects. For instance, if the confounding variable Z is age, we compare the treatment and control groups in every age group separately. We can then take an average of the effects, weighting each age group according to its percentage in the target population.

This method of compensation is familiar to all statisticians; it is called "adjusting for Z" or "controlling for Z."

Oddly, statisticians both over- and underrate the importance of adjusting for possible confounders. They overrate it in the sense that they often control for many more variables than they need to and even for variables that they should *not* control for. I recently came across a quote from a political blogger named Ezra Klein who expresses this phenomenon of "overcontrolling" very clearly: "You see it all the time in studies. 'We controlled for . . . ' And then the list starts. The longer the better. Income. Age. Race. Religion. Height. Hair color. Sexual preference. Crossfit attendance. Love of parents. Coke or Pepsi. The more things you can control for, the stronger your study is—or, at least, the stronger your study seems. Controls give the feeling of specificity, of precision. . . . But sometimes, you can control for too much. Sometimes you end up controlling for the thing you're trying to measure." Klein raises a valid concern. Statisticians have been immensely confused about what variables should and should not be controlled for, so the default practice has been to control for everything one can measure. The vast majority of studies conducted in this day and age subscribe to this practice. It is a convenient, simple procedure to follow, but it is both wasteful and ridden with errors. A key achievement of the Causal Revolution has been to bring an end to this confusion.

At the same time, statisticians greatly underrate controlling in the sense that they are loath to talk about causality at all, even if the controlling has been done correctly. This too stands contrary to the message of this chapter: if you have identified a sufficient set of deconfounders in your diagram, gathered data on them, and properly adjusted for them, then you have every right to say that you have computed the causal effect $X \to Y$ (provided, of course, that you can defend your causal diagram on scientific grounds).

The textbook approach of statisticians to confounding is quite different and rests on an idea most effectively advocated by R. A. Fisher: the randomized controlled trial (RCT). Fisher was exactly right, but not for exactly the right reasons. The randomized controlled trial is indeed a wonderful invention—but until recently the generations of statisticians who followed Fisher could not prove that what they got from the RCT was indeed what they sought to obtain. They did not have a language to write down what they were looking for—namely, the causal effect of X on Y. One of my goals in this chapter is to explain, from the point of view of causal diagrams, precisely why RCTs allow us to estimate the causal effect $X \rightarrow Y$ without falling prey to confounder bias. Once we have understood why RCTs work, there is no need to put them on a pedestal and treat them as the gold standard of causal analysis, which all other methods should emulate. Quite the opposite: we will see that the so-called gold standard in fact derives its legitimacy from more basic principles.

This chapter will also show that causal diagrams make possible a shift of emphasis from confounders to deconfounders. The former cause the problem; the latter cure it. The two sets may overlap, but they don't have to. If we have data on a sufficient set of deconfounders, it does not matter if we ignore some or even all of the confounders.

This shift of emphasis is a main way in which the Causal Revolution allows us to go beyond Fisherian experiments and infer causal effects from nonexperimental studies. It enables us to determine which variables should be controlled for to serve as deconfounders. This question has bedeviled both theoretical and practical statisticians; it has been an Achilles' heel of the field for decades. That is because it has nothing to do with data or statistics. Confounding is a causal concept—it belongs on rung two of the Ladder of Causation.

Graphical methods, beginning in the 1990s, have totally deconfounded the confounding problem. In particular, we will soon meet a method called the back-door criterion, which unambiguously identifies which variables in a causal diagram are deconfounders. If the researcher can gather data on those variables, she can adjust for them and thereby make predictions about the result of an intervention even without performing it.

In fact, the Causal Revolution has gone even farther than this. In some cases we can control for confounding even when we do not have data on a sufficient set of deconfounders. In these cases we can use different adjustment formulas—not the conventional one,

which is only appropriate for use with the back-door criterion—
and still eradicate all confounding. We will save these exciting developments for Chapter 7.

Although confounding has a long history in all areas of science, the recognition that the problem requires causal, not statistical, solutions is very recent. Even as recently as 2001, reviewers
rebuked a paper of mine while insisting, "Confounding is solidly
founded in standard statistics." Fortunately, the number of such
reviewers has shrunk dramatically in the past decade. There is now
an almost universal consensus, at least among epidemiologists,
philosophers, and social scientists, that (1) confounding needs and
has a causal solution, and (2) causal diagrams provide a complete
and systematic way of finding that solution. The age of confusion
over confounding has come to an end!

THE CHILLING FEAR OF CONFOUNDING

In 1998, a study in the *New England Journal of Medicine* revealed
an association between regular walking and reduced death rates
among retired men. The researchers used data from the Honolulu
Heart Program, which has followed the health of 8,000 men of Japanese ancestry since 1965.

The researchers, led by Robert Abbott, a biostatistician at the
University of Virginia, wanted to know whether the men who exer-
cised more lived longer. They chose a sample of 707 men from the
larger group of 8,000, all of whom were physically healthy enough
to walk. Abbott's team found that the death rate over a twelve-
year period was two times higher among men who walked less
than a mile a day (I'll call them "casual walkers") than among men
who walked more than two miles a day ("intense walkers"). To be
precise, 43 percent of the casual walkers had died, while only 21.5
percent of the intense walkers had died.

However, because the experimenters did not prescribe who
would be a casual walker and who would be an intense walker,
we have to take into consideration the possibility of confounding
bias. An obvious confounder might be age: younger men might
be more willing to do a vigorous workout and also would be

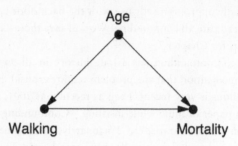

FIGURE 4.2. Causal diagram for walking example.

less likely to die. So we would have a causal diagram like that in Figure 4.2.

The classic forking pattern at the "Age" node tells us that age is a confounder of walking and mortality. I'm sure you can think of other possible confounders. Perhaps the casual walkers were slacking off for a reason; maybe they couldn't walk as much. Thus, physical condition could be a confounder. We could go on and on like this. What if the light walkers were alcohol drinkers? What if they ate more?

The good news is, the researchers thought about all these factors. The study has accounted and adjusted for every reasonable factor—age, physical condition, alcohol consumption, diet, and several others. For example, it's true that the intense walkers tended to be slightly younger. So the researchers adjusted the death rate for age and found that the difference between casual and intense walkers was still very large. (The age-adjusted death rate for the casual walkers was 41 percent, compared to 24 percent for the intense walkers.)

Even so, the researchers were very circumspect in their conclusions. At the end of the article, they wrote, "Of course, the effects on longevity of intentional efforts to increase the distance walked per day by physically capable older men cannot be addressed in our study." To use the language of Chapter 1, they decline to say anything about your probability of surviving twelve years given that you *do(exercise)*.

In fairness to Abbott and the rest of his team, they may have had good reasons for caution. This was a first study, and the sample was relatively small and homogeneous. Nevertheless, this caution reflects a more general attitude, transcending issues of homogeneity and sample size. Researchers have been taught to believe that an observational study (one where subjects choose their own treatment) can never illuminate a causal claim. I assert that this caution is overexaggerated. Why else would one bother adjusting for all these confounders, if not to get rid of the spurious part of the association and thereby get a better view of the causal part?

Instead of saying "Of course we can't," as they did, we should proclaim that of course we can say something about an intentional intervention. If we believe that Abbott's team identified all the important confounders, we must also believe that intentional walking tends to prolong life (at least in Japanese males).

This provisional conclusion, predicated on the assumption that no other confounders could play a major role in the relationships found, is an extremely valuable piece of information. It tells a potential walker precisely what kind of uncertainty remains in taking the claim at face value. It tells him that the remaining uncertainty is not higher than the possibility that additional confounders exist that were not taken into account. It is also valuable as a guide to future studies, which should focus on those other factors (if they exist), not the ones neutralized in the current study. In short, knowing the set of assumptions that stand behind a given conclusion is not less valuable than attempting to circumvent those assumptions with an RCT, which, as we shall see, has complications of its own.

THE SKILLFUL INTERROGATION OF NATURE: WHY RCTS WORK

As I have mentioned already, the one circumstance under which scientists will abandon some of their reticence to talk about causality is when they have conducted a randomized controlled trial. You can read it on Wikipedia or in a thousand other places: "The RCT is often considered the gold standard of a clinical trial." We have one person to thank for this, R. A. Fisher, so it is very interesting to read what a person very close to him wrote about his reasons. The passage is lengthy, but worth quoting in full:

> The whole art and practice of scientific experimentation is comprised in the skillful interrogation of Nature. Observation has provided the scientist with a picture of Nature in some aspect, which has all the imperfections of a voluntary statement. He wishes to check his interpretation of this statement by asking specific questions aimed at establishing causal relationships. His questions, in the form of experimental operations, are necessarily particular, and he must rely on the consistency of Nature in making general deductions from her response in a particular instance or in predicting the outcome to be anticipated from similar operations on other occasions. His aim is to draw valid conclusions of determinate precision and generality from the evidence he elicits.
>
> Far from behaving consistently, however, Nature appears vacillating, coy, and ambiguous in her answers. She responds to the form of the question as it is set out in the field and not necessarily to the question in the experimenter's mind; she does not interpret for him; she gives no gratuitous information; and she is a stickler for accuracy. In consequence, the experimenter who wants to compare two manurial treatments wastes his labor if, dividing his field into two equal parts, he dresses each half with one of his manures, grows a crop, and compares the yields from the two halves. The form of his question was: what is the difference between the yield of plot A under the first treatment and that of plot B under the second? He has not asked whether plot A would yield the same as plot B under uniform treatment, and he cannot distinguish plot effects from treatment effects, for Nature has recorded, as requested, not only the contribution of the manurial differences to the plot yields but also the contributions of differences in soil fertility, texture, drainage, aspect, microflora, and innumerable other variables.

The author of this passage is Joan Fisher Box, the daughter of Ronald Aylmer Fisher, and it is taken from her biography of her illustrious father. Though not a statistician herself, she has clearly absorbed very deeply the central challenge statisticians face. She states in no uncertain terms that the questions they ask are "aimed at establishing causal relationships." And what gets in their way is confounding, although she does not use that word. They want to know the effect of a fertilizer (or "manurial treatment," as fertilizers were called in that era)—that is, the expected yield under one fertilizer compared with the yield under an alternative. Nature, however, tells them about the effect of the fertilizer mixed (remember, this is the original meaning of "confounded") with a variety of other causes.

I like the image that Fisher Box provides in the above passage: Nature is like a genie that answers exactly the question we pose, not necessarily the one we intend to ask. But we have to believe, as Fisher Box clearly does, that the answer to the question we wish to ask does exist in nature. Our experiments are a sloppy means of uncovering the answer, but they do not by any means define the answer. If we follow her analogy exactly, then $do(X = x)$ must come first, because it is a property of nature that represents the answer we seek: What is the effect of using the first fertilizer on the whole field? Randomization comes second, because it is only a man-made means to elicit the answer to that question. One might compare it to the gauge on a thermometer, which is a means to elicit the temperature but is not the temperature itself.

In his early years at Rothamsted Experimental Station, Fisher usually took a very elaborate, systematic approach to disentangling the effects of fertilizer from other variables. He would divide his fields into a grid of subplots and plan carefully so that each fertilizer was tried with each combination of soil type and plant (see Figure 4.3). He did this to ensure the comparability of each sample; in reality, he could never anticipate all the confounders that might determine the fertility of a given plot. A clever enough genie could defeat any structured layout of the field.

Around 1923 or 1924, Fisher began to realize that the only experimental design that the genie could not defeat was a random one. Imagine performing the same experiment one hundred times on a field with an unknown distribution of fertility. Each time you assign fertilizers to subplots randomly. Sometimes you may be very unlucky and use Fertilizer 1 in all the least fertile subplots. Other

FIGURE 4.3. R. A. Fisher with one of his many innovations: a Latin square experimental design, intended to ensure that one plot of each plant type appears in each row (fertilizer type) and column (soil type). Such designs are still used in practice, but Fisher would later argue convincingly that a randomized design is even more effective. (*Source:* Drawing by Dakota Harr.)

times you may get lucky and apply it to the most fertile subplots. But by generating a new random assignment each time you perform the experiment, you can guarantee that the great majority of the time you will be neither lucky nor unlucky. In those cases, Fertilizer 1 will be applied to a selection of subplots that is representative of the field as a whole. This is exactly what you want for a controlled trial. Because the distribution of fertility in the field is fixed throughout your series of experiments—the genie can't change it—he is tricked into answering (most of the time) the causal question you wanted to ask.

From our perspective, in an era when randomized trials are the gold standard, all of this may appear obvious. But at the time, the idea of a randomly designed experiment horrified Fisher's statistical colleagues. Fisher's literally drawing from a deck of cards to assign subplots to each fertilizer may have contributed to their dismay. Science subjected to the whims of chance?

But Fisher realized that an uncertain answer to the right question is much better than a highly certain answer to the wrong question.

If you ask the genie the wrong question, you will never find out what you want to know. If you ask the right question, getting an answer that is occasionally wrong is much less of a problem. You can still estimate the amount of uncertainty in your answer, because the uncertainty comes from the randomization procedure (which is known) rather than the characteristics of the soil (which are unknown).

Thus, randomization actually brings two benefits. First, it eliminates confounder bias (it asks Nature the right question). Second, it enables the researcher to quantify his uncertainty. However, according to historian Stephen Stigler, the second benefit was really Fisher's main reason for advocating randomization. He was the world's master of quantifying uncertainty, having developed many new mathematical procedures for doing so. By comparison, his understanding of deconfounding was purely intuitive, for he lacked a mathematical notation for articulating what he sought.

Now, ninety years later, we can use the *do*-operator to fill in what Fisher wanted to but couldn't ask. Let's see, from a causal point of view, how randomization enables us to ask the genie the right question.

Let's start, as usual, by drawing a causal diagram. Model 1, shown in Figure 4.4, describes how the yield of each plot is determined under normal conditions, where the farmer decides by whim or bias which fertilizer is best for each plot. The query he wants to pose to the genie Nature is "What is the yield under a uniform application of Fertilizer 1 (versus Fertilizer 2) to the entire field?" Or, in *do*-operator notation, what is $P(yield \mid do(fertilizer = 1))$?

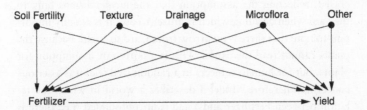

FIGURE 4.4. Model 1: an improperly controlled experiment.

If the farmer performs the experiment naively, for example applying Fertilizer 1 to the high end of his field and Fertilizer 2 to the low end, he is probably introducing Drainage as a confounder. If he uses Fertilizer 1 one year and Fertilizer 2 the next year, he is probably introducing Weather as a confounder. In either case, he will get a biased comparison.

The world that the farmer wants to know about is described by Model 2, where all plots receive the same fertilizer (see Figure 4.5). As explained in Chapter 1, the effect of the *do*-operator is to erase all the arrows pointing to Fertilizer and force this variable to a particular value—say, Fertilizer = 1.

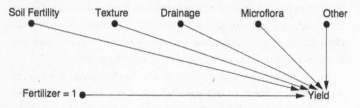

FIGURE 4.5. Model 2: the world we would like to know about.

Finally, let's see what the world looks like when we apply randomization. Now some plots will be subjected to *do(fertilizer* = 1) and others to *do(fertilizer* = 2), but the choice of which treatment goes to which plot is random. The world created by such a model is shown by Model 3 in Figure 4.6, showing the variable Fertilizer obtaining its assignment by a random device—say, Fisher's deck of cards.

Notice that all the arrows pointing toward Fertilizer have been erased, reflecting the assumption that the farmer listens only to the card when deciding which fertilizer to use. It is equally important to note that there is no arrow from Card to Yield, because the plants cannot read the cards. (This is a fairly safe assumption for plants, but for human subjects in a randomized trial it is a serious concern.) Therefore Model 3 describes a world in which the relation between Fertilizer and Yield is unconfounded (i.e., there is no common cause of Fertilizer and Yield). This means that in the

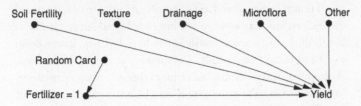

FIGURE 4.6. Model 3: the world simulated by a randomized controlled trial.

world described by Figure 4.6, there is no difference between seeing Fertilizer = 1 and doing Fertilizer = 1.

That brings us to the punch line: randomization is a way of simulating Model 2. It disables all the old confounders without introducing any new confounders. That is the source of its power; there is nothing mysterious or mystical about it. It is nothing more or less than, as Joan Fisher Box said, "the skillful interrogation of Nature."

The experiment would, however, fail in its objective of simulating Model 2 if either the experimenter were allowed to use his own judgment to choose a fertilizer or the experimental subjects, in this case the plants, "knew" which card they had drawn. This is why clinical trials with human subjects go to great lengths to conceal this information from both the patients and the experimenters (a procedure known as double blinding).

I will add to this a second punch line: there are other ways of simulating Model 2. One way, if you know what all the possible confounders are, is to measure and adjust for them. However, randomization does have one great advantage: it severs every incoming link to the randomized variable, including the ones we don't know about or cannot measure (e.g., "Other" factors in Figures 4.4 to 4.6).

By contrast, in a nonrandomized study, the experimenter must rely on her knowledge of the subject matter. If she is confident that her causal model accounts for a sufficient number of deconfounders and she has gathered data on them, then she can estimate the effect of Fertilizer on Yield in an unbiased way. But the danger is that she may have missed a confounding factor, and her estimate may therefore be biased.

All things being equal, RCTs are still preferred to observational studies, just as safety nets are recommended for tightrope walkers. But all things are not necessarily equal. In some cases, intervention may be physically impossible (for instance, in a study of the effect of obesity on heart disease, we cannot randomly assign patients to be obese or not). Or intervention may be unethical (in a study of the effects of smoking, we can't ask randomly selected people to smoke for ten years). Or we may encounter difficulties recruiting subjects for inconvenient experimental procedures and end up with volunteers who do not represent the intended population.

Fortunately, the *do*-operator gives us scientifically sound ways of determining causal effects from nonexperimental studies, which challenge the traditional supremacy of RCTs. As discussed in the walking example, such causal estimates produced by observational studies may be labeled "provisional causality," that is, causality contingent upon the set of assumptions that our causal diagram advertises. It is important that we not treat these studies as second-class citizens: they have the advantage of being conducted in the natural habitat of the target population, not in the artificial setting of a laboratory, and they can be "pure" in the sense of not being contaminated by issues of ethics or feasibility.

Now that we understand that the principal objective of an RCT is to eliminate confounding, let's look at the other methods that the Causal Revolution has given us. The story begins with a 1986 paper by two of my longtime colleagues, which started a reevaluation of what confounding means.

THE NEW PARADIGM OF CONFOUNDING

"While confounding is widely recognized as one of the central problems in epidemiological research, a review of the literature will reveal little consistency among the definitions of confounding or confounder." With this one sentence, Sander Greenland of the University of California, Los Angeles, and Jamie Robins of Harvard University put their finger on the central reason why the control of confounding had not advanced one bit since Fisher. Lacking a principled understanding of confounding, scientists could not say anything meaningful in observational studies where physical control over treatments is infeasible.

How was confounding defined then, and how should it be defined? Armed with what we now know about the logic of causality, the answer to the second question is easier. The quantity we observe is the conditional probability of the outcome given the treatment, $P(Y \mid X)$. The question we want to ask of Nature has to do with the causal relationship between X and Y, which is captured by the interventional probability $P(Y \mid do(X))$. Confounding, then, should simply be defined as anything that leads to a discrepancy between the two: $P(Y \mid X) \neq P(Y \mid do(X))$. Why all the fuss?

Unfortunately, things were not as easy as that before the 1990s because the *do*-operator had yet to be formalized. Even today, if you stop a statistician in the street and ask, "What does 'confounding' mean to you?" you will probably get one of the most convoluted and confounded answers you ever heard from a scientist. One recent book, coauthored by leading statisticians, spends literally two pages trying to explain it, and I have yet to find a reader who understood the explanation.

The reason for the difficulty is that confounding is not a statistical notion. It stands for the discrepancy between what we want to assess (the causal effect) and what we actually do assess using statistical methods. If you can't articulate mathematically what you want to assess, you can't expect to define what constitutes a discrepancy.

Historically, the concept of "confounding" has evolved around two related conceptions: incomparability and a lurking third variable. Both of these concepts have resisted formalization. When we talked about comparability, in the context of Daniel's experiment, we said that the treatment and control groups should be identical in all relevant ways. But this begs us to distinguish relevant from irrelevant attributes. How do we know that age is relevant in the Honolulu walking study? How do we know that the alphabetical order of a participant's name is not relevant? You might say it's obvious or common sense, but generations of scientists have struggled to articulate that common sense formally, and a robot cannot rely on our common sense when asked to act properly.

The same ambiguity plagues the third-variable definition. Should a confounder be a common cause of both X and Y or merely correlated with each? Today we can answer such questions by referring to the causal diagram and checking which variables produce a discrepancy between $P(X \mid Y)$ and $P(X \mid do(Y))$. Lacking a diagram or a *do*-operator, five generations of statisticians and health scientists had to struggle with surrogates, none of which were satisfactory. Considering that the drugs in your medicine cabinet may have been developed on the basis of a dubious definition of "confounders," you should be somewhat concerned.

Let's take a look at some of the surrogate definitions of confounding. These fall into two main categories, declarative and procedural. A typical (and wrong) declarative definition would be "A confounder is any variable that is correlated with both X and Y." On the other hand, a procedural definition would attempt to characterize a confounder in terms of a statistical test. This appeals to statisticians, who love any test that can be performed on the data directly without appealing to a model.

Here is a procedural definition that goes by the scary name of "noncollapsibility." It comes from a 1996 paper by the Norwegian epidemiologist Sven Hernberg: "Formally one can compare the crude relative risk and the relative risk resulting after adjustment for the potential confounder. A difference indicates confounding, and in that case one should use the adjusted risk estimate. If there is no or a negligible difference, confounding is not an issue and the crude estimate is to be preferred." In other words, if you suspect a confounder, try adjusting for it and try not adjusting for it. If there is a difference, it is a confounder, and you should trust the adjusted value. If there is no difference, you are off the hook. Hernberg was by no means the first person to advocate such an approach; it has misguided a century of epidemiologists, economists, and social scientists, and it still reigns in certain quarters of applied statistics. I have picked on Hernberg only because he was unusually explicit about it and because he wrote this in 1996, well after the Causal Revolution was already underway.

The most popular of the declarative definitions evolved over a period of time. Alfredo Morabia, author of *A History of*

Epidemiologic Methods and Concepts, calls it "the classic epidemiological definition of confounding," and it consists of three parts. A confounder of X (the treatment) and Y (the outcome) is a variable Z that is (1) associated with X in the population at large, and (2) associated with Y among people who have not been exposed to the treatment X. In recent years, this has been supplemented by a third condition: (3) Z should not be on the causal path between X and Y.

Observe that all the terms in the "classic" version (1 and 2) are statistical. In particular, Z is only assumed to be associated with—not a cause of—X and Y. Edward Simpson proposed the rather convoluted condition "Y is associated with Z among the unexposed" in 1951. From the causal point of view, it seems that Simpson's idea was to discount the part of the correlation of Z with Y that is due to the causal effect of X on Y; in other words, he wanted to say that Z has an effect on Y independent of its effect on X. The only way he could think to express this discounting was to condition on X by focusing on the control group ($X = 0$). Statistical vocabulary, deprived of the word "effect," gave him no other way of saying it.

If this is a bit confusing, it should be! How much easier it would have been if he could have simply written a causal diagram, like Figure 4.1, and said, "Y is associated with Z via paths not going through X." But he didn't have this tool, and he couldn't talk about paths, which were a forbidden concept.

The "classical epidemiological definition" of a confounder has other flaws, as the following two examples show:

$$(i)\ X \to Z \to Y$$

and

$$(ii)\ X \to M \to Y$$
$$\downarrow$$
$$Z$$

In example (i), Z satisfies conditions (1) and (2) but is not a confounder. It is known as a mediator: it is the variable that explains the causal effect of X on Y. It is a disaster to control for Z if you are trying to find the causal effect of X on Y. If you look only at those individuals in the treatment and control groups for whom $Z = 0$, then you have completely blocked the effect of X, because it works by changing Z. So you will conclude that X has no effect on Y. This is exactly what Ezra Klein meant when he said, "Sometimes you end up controlling for the thing you're trying to measure."

In example (ii), Z is a proxy for the mediator M. Statisticians very often control for proxies when the actual causal variable can't be measured; for instance, party affiliation might be used as a proxy for political beliefs. Because Z isn't a perfect measure of M, some of the influence of X on Y might "leak through" if you control for Z. Nevertheless, controlling for Z is still a mistake. While the bias might be less than if you controlled for M, it is still there.

For this reason later statisticians, notably David Cox in his textbook *The Design of Experiments* (1958), warned that you should only control for Z if you have a strong prior reason to believe that it is "quite unaffected" by X. This "quite unaffected" condition is of course a causal assumption. He adds, "Such hypotheses may be perfectly in order, but the scientist should always be aware when they are being appealed to." Remember that it's 1958, in the midst of the great prohibition on causality. Cox is saying that you can go ahead and take a swig of causal moonshine when adjusting for confounders, but don't tell the preacher. A daring suggestion! I never fail to commend him for his bravery.

By 1980, Simpson's and Cox's conditions had been combined into the three-part test for confounding that I mentioned above. It is about as trustworthy as a canoe with only three leaks. Even though it does make a halfhearted appeal to causality in part (3), each of the first two parts can be shown to be both unnecessary and insufficient.

Greenland and Robins drew that conclusion in their landmark 1986 paper. The two took a completely new approach to confounding, which they called "exchangeability." They went back to the original idea that the control group ($X = 0$) should be comparable to the treatment group ($X = 1$). But they added a counterfactual twist. (Remember from Chapter 1 that counterfactuals are at rung three of the Ladder of Causation and therefore powerful enough to detect confounding.) Exchangeability requires the researcher to consider the treatment group, imagine what would have happened to its constituents if they had not gotten treatment, and then judge whether the outcome would be the same as for those who (in reality) did not receive treatment. Only then can we say that no confounding exists in the study.

In 1986, talking counterfactuals to an audience of epidemiologists took some courage, because they were still very much under the influence of classical statistics, which holds that all the answers are in the data—not in what might have been, which will remain forever unobserved. However, the statistical community was somewhat prepared to listen to such heresy, thanks to the pioneering work of another Harvard statistician, Donald Rubin. In Rubin's "potential outcomes" framework, proposed in 1974, counterfactual variables like "Blood Pressure of Person X had he received Drug D" and "Blood Pressure of Person X had he not received Drug D" are just as legitimate as a traditional variable like Blood Pressure—despite the fact that one of those two variables will remain forever unobserved.

Robins and Greenland set out to express their conception of confounding in terms of potential outcomes. They partitioned the population into four types of individuals: doomed, causative, preventive, and immune. The language is suggestive, so let's think of the treatment X as a flu vaccination and the outcome Y as coming down with flu. The doomed people are those for whom the vaccine doesn't work; they will get flu whether they get the vaccine or not. The causative group (which may be nonexistent) includes those for whom the vaccine actually causes the disease. The preventive group consists of people for whom the vaccine prevents the disease: they will get flu if they are not vaccinated, and they will not get flu if they are vaccinated. Finally, the immune group consists of people who will not get flu in either case. Table 4.1 sums up these considerations.

Ideally, each person would have a sticker on his forehead identifying which group he belonged to. Exchangeability simply means that the percentage of people with each kind of sticker (d percent, c percent, p percent, and i percent, respectively) should be the same in

TABLE 4.1. Classification of individuals according to response type.

Group	Percentage in Group	Outcome If Vaccinated	Outcome If Not Vaccinated
Doomed	d	Flu	Flu
Causative	c	Flu	No flu
Preventive	p	No flu	Flu
Immune	i	No flu	No flu

both the treatment and control groups. Equality among these proportions guarantees that the outcome would be just the same if we switched the treatments and controls. Otherwise, the treatment and control groups are not alike, and our estimate of the effect of the vaccine will be confounded. Note that the two groups may be different in many ways. They can differ in age, sex, health conditions, and a variety of other characteristics. Only equality among d, c, p, and i determines whether they are exchangeable or not. So exchangeability amounts to equality between two sets of four proportions, a vast reduction in complexity from the alternative of assessing the innumerable factors by which the two groups may differ.

Using this commonsense definition of confounding, Greenland and Robins showed that the "statistical" definitions, both declarative and procedural, give incorrect answers. A variable can satisfy the three-part test of epidemiologists and still increase bias, if adjusted for.

Greenland and Robins's definition was a great achievement, because it enabled them to give explicit examples showing that the previous definitions of confounding were inadequate. However, the definition could not be translated into practice. To put it simply, those stickers on the forehead don't exist. We do not even have a count of the proportions d, c, p, and i. In fact, this is precisely the kind of information that the genie of Nature keeps locked inside her magic lantern and doesn't show to anybody. Lacking this information, the researcher is left to intuit whether the treatment and control groups are exchangeable or not.

By now, I hope that your curiosity is well piqued. How can causal diagrams turn this massive headache of confounding into a fun game? The trick lies in an operational test for confounding, called the back-door criterion. This criterion turns the problem of defining confounding, identifying confounders, and adjusting for them into a routine puzzle that is no more challenging than solving a maze. It has thus brought the thorny, age-old problem to a happy conclusion.

THE *DO*-OPERATOR AND THE BACK-DOOR CRITERION

To understand the back-door criterion, it helps first to have an intuitive sense of how information flows in a causal diagram. I like to think of the links as pipes that convey information from a starting point X to a finish Y. Keep in mind that the conveying of information goes in both directions, causal and noncausal, as we saw in Chapter 3.

In fact, the noncausal paths are precisely the source of confounding. Remember that I define confounding as anything that makes $P(Y \mid do(X))$ differ from $P(Y \mid X)$. The *do*-operator erases all the arrows that come into X, and in this way it prevents any information about X from flowing in the noncausal direction. Randomization has the same effect. So does statistical adjustment, if we pick the right variables to adjust.

In the last chapter, we looked at three rules that tell us how to stop the flow of information through any individual junction. I will repeat them for emphasis:

(a) In a chain junction, $A \rightarrow B \rightarrow C$, controlling for B prevents information about A from getting to C or vice versa.

(b) Likewise, in a fork or confounding junction, $A \leftarrow B \rightarrow C$, controlling for B prevents information about A from getting to C or vice versa.

(c) Finally, in a collider, $A \rightarrow B \leftarrow C$, exactly the opposite rules hold. The variables A and C start out independent, so that information about A tells you nothing about C. But if you control for *B*, then information starts flowing through the "pipe," due to the explain-away effect.

We must also keep in mind another fundamental rule:

(d) Controlling for descendants (or proxies) of a variable is like "partially" controlling for the variable itself. Controlling for a descendant of a mediator partly closes the pipe; controlling for a descendant of a collider partly opens the pipe.

Now, what if we have longer pipes with more junctions, like this:

$$A \leftarrow B \leftarrow C \rightarrow D \leftarrow E \rightarrow F \rightarrow G \leftarrow H \rightarrow I \rightarrow J?$$

The answer is very simple: if a single junction is blocked, then *J* cannot "find out" about *A* through this path. So we have many options to block communication between *A* and *J*: control for *B*, control for *C*, don't control for *D* (because it's a collider), control for *E*, and so forth. Any one of these is sufficient. This is why the usual statistical procedure of controlling for everything that we can measure is so misguided. In fact, this particular path is blocked if we don't control for anything! The colliders at *D* and *G* block the path without any outside help. Controlling for *D* and *G* would open this path and enable *J* to listen to *A*.

Finally, to deconfound two variables *X* and *Y*, we need only block every noncausal path between them without blocking or perturbing any causal paths. More precisely, a *back-door path* is any path from *X* to *Y* that starts with an arrow pointing into *X*. *X* and *Y* will be deconfounded if we block every back-door path (because such paths allow spurious correlation between *X* and *Y*). If we do this by controlling for some set of variables *Z*, we also need to make sure that no member of *Z* is a descendant of *X* on a causal path; otherwise we might partly or completely close off that path.

That's all there is to it! With these rules, deconfounding becomes so simple and fun that you can treat it like a game. I urge you to try a few examples just to get the hang of it and see how easy it is. If you still find it hard, be assured that algorithms exist that can crack all such problems in a matter of nanoseconds. In each case, the goal of the game is to specify a set of variables that will deconfound X and Y. In other words, they should not be descended from X, and they should block all the back-door paths.

GAME 1.

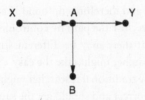

This one is easy! There are no arrows leading into X, therefore no back-door paths. We don't need to control for anything.

Nevertheless, some researchers would consider B a confounder. It is associated with X because of the chain $X \to A \to B$. It is associated with Y among individuals with $X = 0$ because there is an open path $B \leftarrow A \to Y$ that does not pass through X. And B is not on the causal path $X \to A \to Y$. It therefore passes the three-step "classical epidemiological definition" for confounding, but it does not pass the back-door criterion and will lead to disaster if controlled for.

GAME 2.

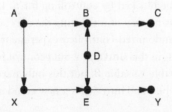

In this example you should think of *A*, *B*, *C*, and *D* as "pretreatment" variables. (The treatment, as usual, is *X*.) Now there is one back-door path $X \leftarrow A \rightarrow B \leftarrow D \rightarrow E \rightarrow Y$. This path is already blocked by the collider at *B*, so we don't need to control for anything. Many statisticians would control for *B* or *C*, thinking there is no harm in doing so as long as they occur before the treatment. A leading statistician even recently wrote, "To avoid conditioning on some observed covariates . . . is nonscientific ad hockery." He is wrong; conditioning on *B* or *C* is a poor idea because it would open the noncausal path and therefore confound *X* and *Y*. Note that in this case we could reclose the path by controlling for *A* or *D*. This example shows that there may be different strategies for deconfounding. One researcher might take the easy way and not control for anything; a more traditional researcher might control for *C* and *D*. Both would be correct and should get the same result (provided that the model is correct, and we have a large enough sample).

GAME 3.

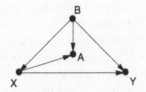

In Games 1 and 2 you didn't have to do anything, but this time you do. There is one back-door path from *X* to *Y*, $X \leftarrow B \rightarrow Y$, which can only be blocked by controlling for *B*. If *B* is unobservable, then there is no way of estimating the effect of *X* on *Y* without running a randomized controlled experiment. Some (in fact, most) statisticians in this situation would control for *A*, as a proxy for the unobservable variable *B*, but this only partially eliminates the confounding bias and introduces a new collider bias.

GAME 4.

This one introduces a new kind of bias, called "M-bias" (named for the shape of the graph). Once again there is only one back-door path, and it is already blocked by a collider at B. So we don't need to control for anything. Nevertheless, all statisticians before 1986 and many today would consider B a confounder. It is associated with X (via $X \leftarrow A \rightarrow B$) and associated with Y via a path that doesn't go through X ($B \leftarrow C \rightarrow Y$). It does not lie on a causal path and is not a descendant of anything on a causal path, because there is no causal path from X to Y. Therefore B passes the traditional three-step test for a confounder.

M-bias puts a finger on what is wrong with the traditional approach. It is incorrect to call a variable, like B, a confounder merely because it is associated with both X and Y. To reiterate, X and Y are unconfounded if we do not control for B. B only becomes a confounder when you control for it!

When I started showing this diagram to statisticians in the 1990s, some of them laughed it off and said that such a diagram was extremely unlikely to occur in practice. I disagree! For example, seatbelt usage (B) has no causal effect on smoking (X) or lung disease (Y); it is merely an indicator of a person's attitudes toward societal norms (A) as well as safety and health-related measures (C). Some of these attitudes may affect susceptibility to lung disease (Y). In practice, seatbelt usage was found to be correlated with both X and Y; indeed, in a study conducted in 2006 as part of a tobacco litigation, seat-belt usage was listed as one of the first variables to be controlled for. If you accept the above model, then controlling for *B* alone would be a mistake.

Note that it's all right to control for *B* if you also control for *A* or *C*. Controlling for the collider *B* opens the "pipe," but controlling for *A* or *C* closes it again. Unfortunately, in the seat-belt example, *A* and *C* are variables relating to people's attitudes and not likely to be observable. If you can't observe it, you can't adjust for it.

GAME 5.

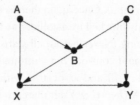

Game 5 is just Game 4 with a little extra wrinkle. Now a second back-door path $X \leftarrow B \leftarrow C \rightarrow Y$ needs to be closed. If we close this path by controlling for *B*, then we open up the *M*-shaped path $X \leftarrow A \rightarrow B \leftarrow C \rightarrow Y$. To close that path, we must control for *A* or *C* as well. However, notice that we could just control for *C* alone; that would close the path $X \leftarrow B \leftarrow C \rightarrow Y$ and not affect the other path.

Games 1 through 3 come from a 1993 paper by Clarice Weinberg, a deputy chief at the National Institutes of Health, called "Toward a Clearer Definition of Confounding." It came out during the transitional period between 1986 and 1995, when Greenland and Robins's paper was available but causal diagrams were still not widely known. Weinberg therefore went through the considerable arithmetic exercise of verifying exchangeability in each of the cases shown. Although she used graphical displays to communicate the scenarios involved, she did not use the logic of diagrams to assist in distinguishing confounders from deconfounders. She is the only person I know of who managed this feat. Later, in 2012, she collaborated on an updated version that analyzes the same examples with causal diagrams and verifies that all her conclusions from 1993 were correct.

In both of Weinberg's papers, the medical application was to estimate the effect of smoking (X) on miscarriages, or "spontaneous abortions" (Y). In Game 1, A represents an underlying abnormality that is induced by smoking; this is not an observable variable because we don't know what the abnormality is. B represents a history of previous miscarriages. It is very, very tempting for an epidemiologist to take previous miscarriages into account and adjust for them when estimating the probability of future miscarriages. But that is the wrong thing to do here! By doing so we are partially inactivating the mechanism through which smoking acts, and we will thus underestimate the true effect of smoking.

Game 2 is a more complicated version where there are two different smoking variables: X represents whether the mother smokes now (at the beginning of the second pregnancy), while A represents whether she smoked during the first pregnancy. B and E are underlying abnormalities caused by smoking, which are unobservable, and D represents other physiological causes of those abnormalities. Note that this diagram allows for the fact that the mother could have changed her smoking behavior between pregnancies, but the other physiological causes would not change. Again, many epidemiologists would adjust for prior miscarriages (C), but this is a bad idea unless you also adjust for smoking behavior in the first pregnancy (A).

Games 4 and 5 come from a paper published in 2014 by Andrew Forbes of Monash University and Elizabeth Williamson, now at the London School of Hygiene and Tropical Medicine. They are interested in the effect of smoking on adult asthma. In Game 4, X represents an individual's smoking behavior, and Y represents whether the person has asthma as an adult. B represents childhood asthma, which is a collider because it is affected by both A, parental smoking, and C, an underlying (and unobservable) predisposition toward asthma. In Game 5 the variables have the same meanings, but they added two arrows for greater realism. (Game 4 was only meant to introduce the M-graph.)

In fact, the full model in their paper has a few more variables and looks like the diagram in Figure 4.7. Note that Game 5 is embedded in this model in the sense that the variables A, B, C, X, and Y have exactly the same relationships. So we can transfer our conclusions over and conclude that we have to control for A and B or for C; but C is an unobservable and therefore uncontrollable variable. In addition we have four new confounding variables: $D =$ parental asthma, $E =$ chronic bronchitis, $F =$ sex, and $G =$ socio-economic status. The reader might enjoy figuring out that we must control for E, F, and G, but there is no need to control for D. So a sufficient set of variables for deconfounding is A, B, E, F, and G.

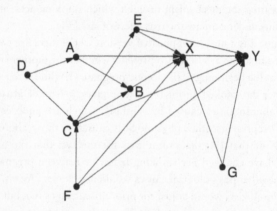

FIGURE 4.7. Andrew Forbes and Elizabeth Williamson's model of smoking (X) and asthma (Y).

In the end, Forbes and Williamson's found that smoking had a small and statistically insignificant association with adult asthma in the raw data, and the effect became even smaller and more insignificant after adjusting for the confounders. The null result should not detract, however, from the fact that their paper is a model for the "skillful interrogation of Nature."

One final comment about these "games": when you start identifying the variables as smoking, miscarriage, and so forth, they are quite obviously not games but serious business. I have referred to them as games because the joy of being able to solve them swiftly and meaningfully is akin to the pleasure a child feels on figuring out that he can crack puzzles that stumped him before.

Few moments in a scientific career are as satisfying as taking a problem that has puzzled and confused generations of predecessors and reducing it to a straightforward game or algorithm. I consider the complete solution of the confounding problem one of the main highlights of the Causal Revolution because it ended an era of confusion that has probably resulted in many wrong decisions in the past. It has been a quiet revolution, raging primarily in research laboratories and scientific meetings. Yet, armed with these new tools and insights, the scientific community is now tackling harder problems, both theoretical and practical, as subsequent chapters will show.

"Abe and Yak" (left and right, respectively) took opposite positions on the hazards of cigarette smoking. As was typical of the era, both were smokers (though Abe used a pipe). The smoking-cancer debate was unusually personal for many of the scientists who participated in it. (*Source:* Drawing by Dakota Harr.)

5

THE SMOKE-FILLED DEBATE: CLEARING THE AIR

At last the sailors said to each other, Come and let us cast lots to find out who is to blame for this ordeal.

—Jonah 1:7

IN the late 1950s and early 1960s, statisticians and doctors clashed over one of the highest-profile medical questions of the century: Does smoking cause lung cancer? Half a century after this debate, we take the answer for granted. But at that time, the issue was by no means clear. Scientists, and even families, were divided.

Jacob Yerushalmy's was one such divided family. A biostatistician at the University of California, Berkeley, Yerushalmy (1904–1973) was one of the last of the pro-tobacco holdouts in academia. "Yerushalmy opposed the notion that cigarette smoking caused cancer until his dying day," wrote his nephew David Lilienfeld many years later. On the other hand, Lilienfeld's father, Abe Lilienfeld, was an epidemiologist at Johns Hopkins University and one of the most outspoken proponents of the theory that smoking did cause cancer. The younger Lilienfeld recalled how Uncle "Yak" (short for Jacob) and his father would sit around and debate the effects of smoking, wreathed all the while in a "haze of smoke from Yak's cigarette and Abe's pipe" (see chapter frontispiece).

If only Abe and Yak had been able to summon the Causal Revolution to clear the air! As this chapter shows, one of the most important scientific arguments against the smoking-cancer hypothesis was the possible existence of unmeasured factors that cause both craving for nicotine and lung cancer. We have just discussed such confounding patterns and noted that today's causal diagrams have driven the menace of confounding out of existence. But we are now in the 1950s and 1960s, two decades before Sander Greenland and Jamie Robins and three decades before anyone had heard of the *do*-operator. It is interesting to examine, therefore, how scientists of that era dealt with the issue and showed that the confounding argument is all smoke and mirrors.

No doubt the subject of many of Abe and Yak's smoke-filled debates was neither tobacco nor cancer. It was that innocuous word "caused." It wasn't the first time that physicians confronted perplexing causal questions: some of the greatest milestones in medical history dealt with identifying causative agents. In the mid-1700s, James Lind had discovered that citrus fruits could prevent scurvy, and in the mid-1800s, John Snow had figured out that water contaminated with fecal matter caused cholera. (Later research identified a more specific causative agent in each case: vitamin C deficiency for scurvy, the cholera bacillus for cholera.) These brilliant pieces of detective work had in common a fortunate one-to-one relation between cause and effect. The cholera bacillus is the only cause of cholera; or as we would say today, it is both necessary and sufficient. If you aren't exposed to it, you won't get the disease. Likewise, a vitamin C deficiency is necessary to produce scurvy, and given enough time, it is also sufficient.

The smoking-cancer debate challenged this monolithic concept of causation. Many people smoke their whole lives and never get lung cancer. Conversely, some people get lung cancer without ever lighting up a cigarette. Some people may get it because of a hereditary disposition, others because of exposure to carcinogens, and some for both reasons.

Of course, statisticians already knew of one excellent way to establish causation in a more general sense: the randomized controlled trial (RCT). But such a study would be neither feasible nor ethical in the case of smoking. How could you assign people chosen at random to smoke for decades, possibly ruining their health, just to see if they would get lung cancer after thirty years? It's impossible to imagine anyone outside North Korea "volunteering" for such a study.

Without a randomized controlled trial, there was no way to convince skeptics like Yerushalmy and R. A. Fisher, who were committed to the idea that the observed association between smoking and lung cancer was spurious. To them, some lurking third factor could be producing the observed association. For example, there could be a smoking gene that caused people to crave cigarettes and also, at the same time, made them more likely to develop lung cancer (perhaps because of other lifestyle choices). The confounders they suggested were implausible at best. Still, the onus was on the antismoking contingent to prove there was no confounder—to prove a negative, which Fisher and Yerushalmy well knew is almost impossible.

The final breach of this stalemate is a tale at once of a great triumph and a great opportunity missed. It was a triumph for public health because the epidemiologists did get it right in the end. The US surgeon general's report, in 1964, stated in no uncertain terms, "Cigarette smoking is causally related to lung cancer in men." This blunt statement forever shut down the argument that smoking was "not proven" to cause cancer. The rate of smoking in the United States among men began to decrease the following year and is now less than half what it was in 1964. No doubt millions of lives have been saved and lifespans lengthened.

On the other hand, the triumph is incomplete. The period it took to reach the above conclusion, roughly from 1950 to 1964, might have been shorter if scientists had been able to call upon a more principled theory of causation. And most significantly from the point of view of this book, the scientists of the 1960s did not really put together such a theory. To justify the claim that smoking caused cancer, the surgeon general's committee relied on an infor-mal series of guidelines, called Hill's criteria, named for Univer-sity of London statistician Austin Bradford Hill. Every one of these criteria has demonstrable exceptions, although collectively they have a compelling commonsense value and even wisdom. From the overly methodological world of Fisher, the Hill guidelines take us to the opposite realm, to a methodology-free world where cau-sality is decided on the basis of qualitative patterns of statistical trends. The Causal Revolution builds a bridge between these two extremes, empowering our intuitive sense of causality with math-ematical rigor. But this job would be left to the next generation.

TOBACCO: A MANMADE EPIDEMIC

In 1902, cigarettes comprised only 2 percent of the US tobacco mar-ket; spittoons rather than ashtrays were the most ubiquitous symbol of tobacco consumption. But two powerful forces worked together to change Americans' habits: automation and advertising. Machine-made cigarettes easily outcompeted handcrafted cigars and pipes on the basis of availability and cost. Meanwhile, the tobacco industry invented and perfected many tricks of the trade of advertising (see Figure 5.1). People who watched TV in the 1960s can easily remem-ber any number of catchy cigarette jingles, from "You get a lot to like in a Marlboro" to "You've come a long way, baby."

By 1952, cigarettes' share of the tobacco market had rocketed from 2 to 81 percent, and the market itself had grown dramatically. This sea change in the habits of a country had unexpected ramifica-tions for public health. Even in the early years of the twentieth cen-tury, there had been suspicions that smoking was unhealthy, that it "irritated" the throat and caused coughing. Around mid-century, the evidence started to become a good deal more ominous. Before cigarettes, lung cancer had been so rare that a doctor might en-counter it only once in a lifetime of practice. But between 1900 and 1950, the formerly rare disease quadrupled in frequency, and by 1960 it would become the most common form of cancer among men. Such a huge change in the incidence of a lethal disease begged for an explanation.

FIGURE 5.1. Highly manipulative advertisements were intended to reassure the public that cigarettes were not injurious to their health—including this 1948 ad from the *Journal of the American Medical Association* targeting the doctors themselves. (*Source:* From the collection of Stanford Research into the Impact of Tobacco Advertising.)

With hindsight, it is easy to point the finger of blame at smoking. If we plot the rates of lung cancer and tobacco consumption on a graph (see Figure 5.2), the connection is impossible to miss. But time series data are poor evidence for causality. Many other things had changed between 1900 and 1950 and were equally plausible culprits: the paving of roads, the inhalation of leaded gasoline fumes, and air pollution in general. British epidemiologist Richard Doll said in 1991, "Motor cars . . . were a new factor and if I had had to put money on anything at the time, I should have put it on motor exhausts or possibly the tarring of roads."

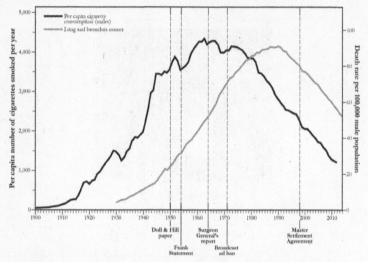

FIGURE 5.2. A graph of the per capita cigarette consumption rate in the United States (black) and the lung cancer death rate among men (gray) shows a stunning similarity: the cancer curve is almost a replica of the smoking curve, delayed by about thirty years. Nevertheless, this evidence is circumstantial, not proof of causation. Certain key dates are noted here, including the publication of Richard Doll and Austin Bradford Hill's paper in 1950, which first alerted many medical professionals to the association between smoking and lung cancer. (*Source:* Graph by Maayan Harel, using data from the American Cancer Society, Centers for Disease Control, and Office of the Surgeon General.)

The job of science is to put supposition aside and look at the facts. In 1948, Doll and Austin Bradford Hill teamed up to see if they could learn anything about the causes of the cancer epidemic. Hill had been the chief statistician on a highly successful randomized controlled trial, published earlier that year, which had proved that streptomycin—one of the first antibiotics—was effective against tuberculosis. The study, a landmark in medical history, not only introduced doctors to "wonder drugs" but also cemented the reputation of randomized controlled trials, which soon became the standard for clinical research in epidemiology.

Of course Hill knew that an RCT was impossible in this case, but he had learned the advantages of comparing a treatment group to a control group. So he proposed to compare patients who had already been diagnosed with cancer to a control group of healthy volunteers. Each group's members were interviewed on their past behaviors and medical histories. To avoid bias, the interviewers were not told who had cancer and who was a control.

The results of the study were shocking: out of 649 lung cancer patients interviewed, all but two had been smokers. This was a statistical improbability so extreme that Doll and Hill couldn't resist working out the exact odds against it: 1.5 million to 1. Also, the lung cancer patients had been heavier smokers on average than the controls, but (in an inconsistency that R. A. Fisher would later pounce on) a smaller percentage reported inhaling their smoke.

The type of study Doll and Hill conducted is now called a case-control study because it compares "cases" (people with a disease) to controls. It is clearly an improvement over time series data, because researchers can control for confounders like age, sex, and exposure to environmental pollutants. Nevertheless, the case-control design has some obvious drawbacks. It is retrospective; that means we study people known to have cancer and look backward to discover why. The probability logic is backward too. The data tell us the probability that a cancer patient is a smoker instead of the probability that a smoker will get cancer. It is the latter probability that really matters to a person who wants to know whether he should smoke or not.

In addition, case-control studies admit several possible sources of bias. One of them is called recall bias: although Doll and Hill ensured that the interviewers didn't know the diagnoses, the patients certainly knew whether they had cancer or not. This could have affected their recollections. Another problem is selection bias. Hospitalized cancer patients were in no way a representative sample of the population, or even of the smoking population.

In short, Doll and Hill's results were extremely suggestive but could not be taken as proof that smoking causes cancer. The two researchers were careful at first to call the correlation an "association." After dismissing several confounders, they ventured a stronger assertion that "smoking is a factor, and an important factor, in the production of carcinoma of the lung."

Over the next few years, nineteen case-control studies conducted in different countries all arrived at basically the same conclusion. But as R. A. Fisher was only too happy to point out, repeating a biased study nineteen times doesn't prove anything. It's still biased. Fisher wrote in 1957 that these studies "were mere repetitions of evidence of the same kind, and it is necessary to try to examine whether that kind is sufficient for any scientific conclusion."

Doll and Hill realized that if there were hidden biases in the case-control studies, mere replication would not overcome them. Thus, in 1951 they began a prospective study, for which they sent out questionnaires to 60,000 British physicians about their smoking habits and followed them forward in time. (The American Cancer Society launched a similar and larger study around the same time.) Even in just five years, some dramatic differences emerged. Heavy smokers had a death rate from lung cancer twenty-four times that of nonsmokers. In the American Cancer Society study, the results were even grimmer: smokers died from lung cancer twenty-nine times more often than nonsmokers, and heavy smokers died ninety times more often. On the other hand, people who had smoked and then stopped reduced their risk by a factor of two. The consistency of all these results—more smoking leads to a higher risk of cancer, stopping leads to a lower risk—was another strong piece of evidence for causality. Doctors call it a "dose-response effect": if substance A causes a biological effect B, then usually (though not always) a larger dose of A causes a stronger response B.

Nevertheless, skeptics like Fisher and Yerushalmy would not be convinced. The prospective studies still failed to compare smokers to otherwise identical nonsmokers. In fact, it is not clear that such a comparison can be made. Smokers are self-selecting. They may be genetically or "constitutionally" different from nonsmokers in a number of ways—more risk taking, likelier to drink heavily. Some of these behaviors might cause adverse health effects that might otherwise be attributed to smoking. This was an especially convenient argument for a skeptic to make because the constitutional hypothesis was almost untestable. Only after the sequencing of the human genome in 2000 did it become possible to look for genes linked to lung cancer. (Ironically, Fisher was proven right, albeit in a very limited way: such genes do exist.) However, in 1959 Jerome Cornfield, writing with Abe Lilienfeld, published a point-by-point rebuttal of Fisher's arguments that, in many physicians' eyes, settled the issue. Cornfield, who worked at the National Institutes of Health, was an unusual participant in the smoking-cancer debate. Neither a statistician nor a biologist by training, he had majored in history and learned statistics at the US Department of Agriculture. Though somewhat self-taught, he eventually became a highly sought consultant and president of the American Statistical Association. He also had been a 2.5-pack-a-day smoker but gave up his habit when he started seeing the data on lung cancer. (It is interesting to see how personal the smoking debate was for the scientists involved. Fisher never gave up his pipe, and Yerushalmy never gave up his cigarettes.)

Cornfield took direct aim at Fisher's constitutional hypothesis, and he did so on Fisher's own turf: mathematics. Suppose, he argued, that there is a confounding factor, such as a smoking gene, that completely accounts for the cancer risk of smokers. If smokers have nine times the risk of developing lung cancer, the confounding factor needs to be at least nine times more common in smokers to explain the difference in risk. Think of what this means. If 11 percent of nonsmokers have the "smoking gene," then 99 percent of the smokers would have to have it. And if even 12 percent of nonsmokers happen to have the cancer gene, then it becomes mathematically impossible for the cancer gene to account fully for the association between smoking and cancer. To biologists, this argument, called Cornfield's inequality, reduced Fisher's constitutional hypothesis to smoking ruins. It is inconceivable that a genetic variation could be so tightly linked to something as complex and unpredictable as a person's choice to smoke.

Cornfield's inequality was actually a causal argument in embryonic form: it gives us a criterion for adjudicating between Diagram 5.1 (in which the constitutional hypothesis cannot fully explain the association between smoking and lung cancer) and Diagram 5.2 (in which the smoking gene would fully account for the observed association).

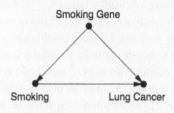

DIAGRAM 5.1.

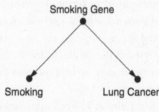

DIAGRAM 5.2.

As explained above, the association between smoking and lung cancer was much too strong to be explained by the constitutional hypothesis.

In fact, Cornfield's method planted the seeds of a very powerful technique called "sensitivity analysis," which today supplements the conclusions drawn from the inference engine described in the Introduction. Instead of drawing inferences by assuming the absence of certain causal relationships in the model, the analyst challenges such assumptions and evaluates how strong alternative relationships must be in order to explain the observed data. The quantitative result is then submitted to a judgment of plausibility, not unlike the crude judgments invoked in positing the absence of those causal relationships. Needless to say, if we want to extend Cornfield's approach to a model with more than three or four variables, we need algorithms and estimation techniques that are unthinkable without the advent of graphical tools.

Epidemiologists in the 1950s faced the criticism that their evidence was "only statistical." There was allegedly no "laboratory proof." But even a look at history shows that this argument was specious. If the standard of "laboratory proof" had been applied to scurvy, then sailors would have continued dying right up until the 1930s, because until the discovery of vitamin C, there was no "laboratory proof" that citrus fruits prevented scurvy. Furthermore, in the 1950s some types of laboratory proof of the effects of smoking did start to appear in medical journals. Rats painted with cigarette tar developed cancer. Cigarette smoke was proven to contain benzopyrenes, a previously known carcinogen. These experiments increased the biological plausibility of the hypothesis that smoking could cause cancer.

By the end of the decade, the accumulation of so many different kinds of evidence had convinced almost all experts in the field that smoking indeed caused cancer. Remarkably, even researchers at the tobacco companies were convinced—a fact that stayed deeply hidden until the 1990s, when litigation and whistle-blowers forced tobacco companies to release many thousands of previously secret documents. In 1953, for example, a chemist at R.J. Reynolds, Claude Teague, had written to the company's upper management that tobacco was "an important etiologic factor in the induction of primary cancer of the lung," nearly a word-for-word repetition of Hill and Doll's conclusion.

In public, the cigarette companies sang a different tune. In January 1954, the leading tobacco companies (including Reynolds) published a nationwide newspaper advertisement, "A Frank Statement to Cigarette Smokers," that said, "We believe the products we make are not injurious to health. We always have and always will cooperate closely with those whose task it is to safeguard the public health." In a speech given in March 1954, George Weissman, vice president of Philip Morris and Company, said, "If we had any thought or knowledge that in any way we were selling a product harmful to consumers, we would stop business tomorrow." Sixty years later, we are still waiting for Philip Morris to keep that promise.

This brings us to the saddest episode in the whole smoking-cancer controversy: the deliberate efforts of the tobacco companies to deceive the public about the health risks. If Nature is like a genie that answers a question truthfully but only exactly as it is asked, imagine how much more difficult it is for scientists to face an adversary that intends to deceive us. The cigarette wars were science's first confrontation with organized denialism, and no one was prepared. The tobacco companies magnified any shred of scientific controversy they could. They set up their own Tobacco Industry Research Committee, a front organization that gave money to scientists to study issues related to cancer or tobacco—but somehow never got around to the central question. When they could find legitimate skeptics of the smoking-cancer connection—such as R. A. Fisher and Jacob Yerushalmy—the tobacco companies paid them consulting fees.

The case of Fisher is particularly sad. Of course, skepticism has its place. Statisticians are paid to be skeptics; they are the conscience of science. But there is a difference between reasonable and unreasonable skepticism. Fisher crossed that line and then some. Always unable to admit his own mistakes, and surely influenced by his lifetime pipe-smoking habit, he could not acknowledge that the tide of evidence had turned against him. His arguments became desperate. He seized on one counterintuitive result in Doll and Hill's first paper—the finding (which barely reached the level of statistical significance) that lung cancer patients described themselves as inhalers less often than the controls—and would not let it go. None of the subsequent studies found any such effect. Although Fisher knew as well as anybody that "statistically significant" results sometimes fail to be replicated, he resorted to mockery. He argued that their study had showed that inhaling cigarette smoke might be beneficial and called for further research on this "extremely important point." Perhaps the only positive thing we can say about Fisher's role in the debate is that it is very unlikely that tobacco money corrupted him in any way. His own obstinacy was sufficient.

For all these reasons, the link between smoking and cancer remained controversial in the public mind long after it had ended among epidemiologists. Even doctors, who should have been more attuned to the science, remained unconvinced: a poll conducted by the American Cancer Society in 1960 showed that only a third of American doctors agreed with the statement that smoking was "a major cause of lung cancer," and 43 percent of doctors were themselves smokers.

While we may justly blame Fisher for his obduracy and the tobacco companies for their deliberate deception, we must also acknowledge that the scientific community was laboring in an ideological straightjacket. Fisher had been right to promote randomized controlled trials as a highly effective way to assess a causal effect. However, he and his followers failed to realize that there is much we can learn from observational studies. That is the benefit of a causal model: it leverages the experimenter's scientific knowledge. Fisher's methods assume that the experimenter begins with no prior knowledge of or opinions about the hypothesis to be tested. They impose ignorance on the scientist, a situation that the denialists eagerly took advantage of.

Because scientists had no straightforward definition of the word "cause" and no way to ascertain a causal effect without a randomized controlled trial, they were ill prepared for a debate over whether smoking caused cancer. They were forced to fumble their way toward a definition in a process that lasted throughout the 1950s and reached a dramatic conclusion in 1964.

THE SURGEON GENERAL'S COMMISSION AND HILL'S CRITERIA

The paper by Cornfield and Lilienfeld had paved the way for a definitive statement by health authorities about the effects of smoking. The Royal College of Physicians in the United Kingdom took the lead, issuing a report in 1962 concluding that cigarette smoking was a causative agent in lung cancer. Shortly thereafter, US Surgeon General Luther Terry (quite possibly on the urging of President John F. Kennedy) announced his intention to appoint a special advisory committee to study the matter (see Figure 5.3).

The committee was carefully balanced to include five smokers and five nonsmokers, two people suggested by the tobacco industry,

FIGURE 5.3. In 1963, a surgeon general's advisory committee wrestled with the problem of how to assess the causal effects of smoking. Depicted here are William Cochran (the committee's statistician), Surgeon General Luther Terry, and chemist Louis Fieser. (*Source:* Drawing by Dakota Harr.)

and nobody who had previously made public statements for or against smoking. For that reason, people like Lilienfeld and Cornfield were ineligible. The members of the committee were distinguished experts in medicine, chemistry, or biology. One of them, William Cochran of Harvard University, was a statistician. In fact, Cochran's credentials in statistics were the best possible: he was a student of a student of Karl Pearson.

The committee labored for more than a year on its report, and a major issue was the use of the word "cause." The committee members had to put aside nineteenth-century deterministic conceptions of causality, and they also had to put aside statistics. As they (probably Cochran) wrote in the report, "Statistical methods cannot establish proof of a causal relationship in an association. The causal significance of an association is a matter of judgment which goes beyond any statement of statistical probability. To judge or evaluate the causal significance of the association between the attribute or agent and the disease, or effect upon health, a number of criteria must be utilized, no one of which is an all-sufficient basis for judgment." The committee listed five such criteria: *consistency* (many studies, in different populations, show similar results); *strength of association* (including the dose-response effect: more smoking is associated with a higher risk); *specificity of the association* (a particular agent should have a particular effect and not a long litany of effects); *temporal relationship* (the effect should follow the cause); and *coherence* (biological plausibility and consistency with other types of evidence such as laboratory experiments and time series).

In 1965, Austin Bradford Hill, who was not on the committee, attempted to summarize the arguments in a way that could be applied to other public health problems and added four more criteria to the list; as a result, the whole list of nine criteria have become known as "Hill's criteria." Actually Hill called them "viewpoints," not requirements, and emphasized that any of them might be lacking in any particular case. "None of my nine viewpoints can bring indisputable evidence for or against the cause-and-effect hypothesis, and none can be required as a *sine qua non*," he wrote.

Indeed, it is quite easy to find arguments against each of the criteria on either Hill's list or the advisory committee's shorter list. Consistency by itself proves nothing; if thirty studies each ignore the same confounder, all can easily be biased. Strength of association is vulnerable for the same reason; as pointed out earlier, children's shoe sizes are strongly associated with but not causally related to their reading aptitude. Specificity has always been a particularly controversial criterion. It makes sense in the context of infectious disease, where one agent typically produces one illness, but less so in the context of environmental exposure. Smoking leads to an increased risk of a variety of other diseases, such as emphysema and cardiovascular disease. Does this really weaken the evidence that it causes cancer? Temporal relation has some exceptions, as mentioned before—for example, a rooster crow does not cause the sun to rise, even though it always precedes the sun.

Finally, coherence with established theory or facts is certainly desirable, but the history of science is filled with overturned theories and mistaken laboratory findings.

Hill's "viewpoints" are still useful as a description of how a discipline comes to accept a causal hypothesis, using a variety of evidence, but they came with no methodology to implement them. For example, biological plausibility and consistency with experiments are supposedly good things. But how, precisely, are we supposed to weigh these kinds of evidence? How do we bring preexisting knowledge into the picture? Apparently each scientist just has to decide for him- or herself. But gut decisions can be wrong, especially if there are political pressures or monetary considerations or if the scientist is addicted to the substance being studied.

None of these comments is intended to denigrate the work of the committee. Its members did the best they could in an environment that provided them with no mechanism for discussing causality. Their recognition that nonstatistical criteria were necessary was a great step forward. And the difficult personal decisions that the smokers on the committee made attest to the seriousness of their conclusions. Luther Terry, who had been a cigarette smoker, switched to a pipe. Leonard Schuman announced that he was quitting. William Cochran acknowledged that he could reduce his risk of cancer by quitting but felt that the "comfort of my cigarettes" was sufficient compensation for the risk. Most painfully, Louis Fieser, a four-pack-a-day smoker, was diagnosed with lung cancer less than a year after the report. He wrote to the committee, "You may recall that although fully convinced by the evidence, I continued heavy smoking throughout the deliberations of our committee and invoked all the usual excuses. . . . My case seems to me more convincing than any statistics." Minus one lung, he finally stopped smoking.

Viewed from the perspective of public health, the report of the advisory committee was a landmark. Within two years, Congress had required manufacturers to place health warnings on all cigarette packs. In 1971, cigarette advertisements were banned from radio and television. The percentage of US adults who smoke declined from its all-time maximum of 45 percent in 1965 to 19.3 percent in 2010. The antismoking campaign has been one of the largest and most successful, though painfully slow and incomplete, public health interventions in history. The committee's work also provided a valuable template for achieving scientific consensus and served as a model for future surgeon general's reports on smoking and many other topics in the years to come (including secondhand smoke, which became a major issue in the 1980s).

Viewed from the perspective of causality, the report was at best a modest success. It clearly established the gravity of causal questions and that data alone could not answer them. But as a roadmap for future discovery, its guidelines were uncertain and flimsy. Hill's criteria are best read as a historical document, summarizing the types of evidence that had emerged in the 1950s, and ultimately convinced the medical community. But as a guide to future research, they are inadequate. For all but the broadest causal questions, we need a more precise instrument. In retrospect, Cornfield's inequality, which planted the seeds of sensitivity analysis, was a step in that direction.

SMOKING FOR NEWBORNS

Even after the smoking and cancer debate died down, one major paradox lingered. In the mid-1960s, Jacob Yerushalmy pointed out that a mother's smoking during pregnancy seemed to benefit the health of her newborn baby, if the baby happened to be born underweight. This puzzle, called the birth-weight paradox, flew in the face of the emerging medical consensus about smoking and was not satisfactorily explained until 2006—more than forty years after Yerushalmy's original paper. I am absolutely convinced it took so long because the language of causality was not available from 1960 to 1990.

In 1959, Yerushalmy had launched a long-term public health study that collected pre- and postnatal data on more than 15,000 children in the San Francisco Bay Area. The data included information on mothers' smoking habits, as well as the birth weights and mortality rates of their babies in the first month of life.

Several studies had already shown that the babies of smoking mothers weighed less at birth on average than the babies of

nonsmokers, and it was natural to suppose that this would translate to poorer survival. Indeed, a nationwide study of low-birth-
weight infants (defined as those who weigh less than 5.5 pounds at
birth) had shown that their death rate was more than twenty times
higher than that of normal-birth-weight infants. Thus, epidemiologists posited a chain of causes and effects: Smoking → Low Birth
Weight → Mortality.

What Yerushalmy found in the data was unexpected even to
him. It was true that the babies of smokers were lighter on average
than the babies of nonsmokers (by seven ounces). However, the
low-birth-weight babies of smoking mothers had a better survival
rate than those of nonsmokers. It was as if the mother's smoking
actually had a protective effect.

If Fisher had discovered something like this, he probably
would have loudly proclaimed it as one of the benefits of smoking. Yerushalmy, to his credit, did not. He wrote, much more cau-
tiously, "These paradoxical findings raise doubts and argue against
the proposition that cigarette smoking acts as an exogenous factor
which interferes with intrauterine development of the fetus." In
short, there is no causal path from Smoking to Mortality.

Modern epidemiologists believe that Yerushalmy was wrong.
Most believe that smoking does increase neonatal mortality—for
example, because it interferes with oxygen transfer across the pla-
centa. But how can we reconcile this hypothesis with the data?

Statisticians and epidemiologists insisted on analyzing the paradox in probabilistic terms and seeing it as an anomaly peculiar to
birth weight. As it turns out, it has little to do with birth weight
and everything to do with colliders. When viewed in that light, it is
not paradoxical but instructive.

In fact, Yerushalmy's data were completely consistent with the
model Smoking → Low Birth Weight → Mortality once we add a
little bit more to it. Smoking may be harmful in that it contributes
to low birth weight, but certain other causes of low birth weight,
such as serious or life-threatening genetic abnormalities, are much
more harmful. There are two possible explanations for low birth
weight in one particular baby: it might have a smoking mother, or

it might be affected by one of those other causes. If we find out that the mother is a smoker, this explains away the low weight and consequently reduces the likelihood of a serious birth defect. But if the mother does not smoke, we have stronger evidence that the cause of the low birth weight is a birth defect, and the baby's prognosis becomes worse.

As before, a causal diagram makes everything clearer. When we incorporate the new assumptions, the causal diagram looks like Figure 5.4. We can see that the birth-weight paradox is a perfect example of collider bias. The collider is Birth Weight. By looking only at babies with low birth weight, we are conditioning on that collider. This opens up a back-door path between Smoking and Mortality that goes Smoking → Birth Weight ← Birth Defect → Mortality. This path is noncausal because one of the arrows goes the wrong way. Nevertheless, it induces a spurious correlation between Smoking and Mortality and biases our estimate of the actual (direct) causal effect, Smoking → Mortality. In fact, it biases the estimate to such a large extent that smoking actually appears beneficial.

The beauty of causal diagrams is that they make the source of bias obvious. Lacking such diagrams, epidemiologists argued about the paradox for forty years. In fact, they are still discussing it: the October 2014 issue of the *International Journal of Epidemiology* contains several articles on this topic. One of them, by Tyler VanderWeele of Harvard, nails the explanation perfectly and contains a diagram just like the one below.

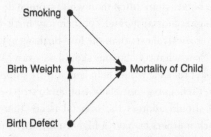

FIGURE 5.4. Causal diagram for the birth-weight paradox.

Of course, this diagram is likely too simple to capture the full story behind smoking, birth weight, and infant mortality. However, the principle of collider bias is robust. In this case the bias was detected because the apparent phenomenon was too implausible, but just imagine how many cases of collider bias go undetected because the bias does not conflict with theory.

PASSIONATE DEBATES: SCIENCE VS. CULTURE

After I began work on this chapter, I had occasion to contact Allen Wilcox, the epidemiologist probably most identified with this paradox. He has asked a very inconvenient question about the diagram in Figure 5.4: How do we know that low birth weight is actually a direct cause of mortality? In fact, he believes that doctors have misinterpreted low birth weight all along. Because it is strongly associated with infant mortality, doctors have interpreted it as a cause. In fact, that association could be due entirely to confounders (represented by "Birth Defect" in Figure 5.4, though Wilcox is not so specific).

Two points are worth making about Wilcox's argument. First, even if we delete the arrow Birth Weight → Mortality, the collider remains. Thus the causal diagram continues to account for the birth-weight paradox successfully. Second, the causal variable that Wilcox has studied the most is not smoking but race. And race still incites passionate debate in our society.

In fact, the same birth-weight paradox is observed in children of black mothers as in children of smokers. Black women give birth to underweight babies more often than white women do, and their babies have a higher mortality rate. Yet their low-birth-weight babies have a better survival rate than the low-birth-weight babies of white women. Now what conclusions should we draw? We can tell a pregnant smoker that she would help her baby by stopping smoking. But we can't tell a pregnant black woman to stop being black.

Instead, we should address the societal issues that cause the children of black mothers to have a higher mortality rate. This is surely not a controversial statement. But what causes should we address, and how should we measure our progress? For better or for worse, many advocates for racial justice have assumed birth weight as an intermediate step in the chain Race → Birth Weight → Mortality. Not only that, they have taken birth weight as a proxy for infant mortality, assuming that improvements in the one will automatically lead to improvements in the other. It's easy to understand why they did this. Measurements of average birth weights are easier to come by than measurements of infant mortality.

Now imagine what happens when someone like Wilcox comes along and asserts that low birth weight by itself is not a medical condition and has no causal relation to infant mortality. It upsets the entire applecart. Wilcox was accused of racism when he first suggested this idea back in the 1970s, and he didn't dare to publish it until 2001. Even then, two commentaries accompanied his article, and one of them brought up the race issue: "In the context of a society whose dominant elements justify their positions by arguing the genetic inferiority of those they dominate, it is hard to be neutral," wrote Richard David of Cook County Hospital in Chicago. "In the pursuit of 'pure science' a well-meaning investigator may be perceived as—and may be—aiding and abetting a social order he abhors."

This harsh accusation, conceived out of the noblest of motivations, is surely not the first instance in which a scientist has been reprimanded for elucidating truths that might have adverse social consequences. The Vatican's objections to Galileo's ideas surely arose out of genuine concerns for the social order of the time. The same can be said about Charles Darwin's evolution and Francis Galton's eugenics. However, the cultural shocks that emanate from new scientific findings are eventually settled by cultural realignments that accommodate those findings—not by concealment. A prerequisite for this realignment is that we sort out the science from the culture before opinions become inflamed. Fortunately, the language of causal diagrams now gives us a way to be dispassionate about causes and effects not only when it is easy but also when it is hard.

The "Monty Hall paradox," an enduring and for many people infuriating puzzle, highlights how our brains can be fooled by probabilistic reasoning when causal reasoning should apply. (*Source:* Drawing by Maayan Harel.)

6

PARADOXES GALORE!

He who confronts the paradoxical exposes himself to reality.

—Friedrich Dürrenmatt (1962)

THE birth-weight paradox, with which we ended Chapter 5, is representative of a surprisingly large class of paradoxes that reflect the tensions between causation and association. The tension starts because they stand on two different rungs of the Ladder of Causation and is aggravated by the fact that human intuition operates under the logic of causation, while data conform to the logic of probabilities and proportions. Paradoxes arise when we misapply the rules we have learned in one realm to the other.

We are going to devote a chapter to some of the most baffling and well-known paradoxes in probability and statistics because, first of all, they're fun. If you haven't seen the Monty Hall and Simpson's paradoxes before, I can promise that they will give your brain a workout. And even if you think you know all about them, I think that you will enjoy viewing them through the lens of causality, which makes everything look quite a bit different.

However, we study paradoxes not just because they are fun and games. Like optical illusions, they also reveal the way the brain works, the shortcuts it takes, and the things it finds conflicting.

Causal paradoxes shine a spotlight onto patterns of intuitive causal reasoning that clash with the logic of probability and statistics. To the extent that statisticians have struggled with them—and we'll see that they whiffed rather badly—it's a warning sign that something might be amiss with viewing the world without a causal lens.

THE PERPLEXING MONTY HALL PROBLEM

In the late 1980s, a writer named Marilyn vos Savant started a regular column in *Parade* magazine, a weekly supplement to the Sunday newspaper in many US cities. Her column, "Ask Marilyn," continues to this day and features her answers to various puzzles, brainteasers, and scientific questions submitted by readers. The magazine billed her as "the world's smartest woman," which undoubtedly motivated readers to come up with a question that would stump her.

Of all the questions she ever answered, none created a greater furor than this one, which appeared in a column in September 1990: "Suppose you're on a game show, and you're given the choice of three doors. Behind one door is a car, behind the others, goats. You pick a door, say #1, and the host, who knows what's behind the doors, opens another door, say #3, which has a goat. He says to you, 'Do you want to pick door #2?' Is it to your advantage to switch your choice of doors?"

For American readers, the question was obviously based on a popular televised game show called *Let's Make a Deal*, whose host, Monty Hall, used to play precisely this sort of mind game with the contestants. In her answer, vos Savant argued that contestants should switch doors. By not switching, they would have only a one-in-three probability of winning; by switching, they would double their chances to two in three.

Even the smartest woman in the world could never have anticipated what happened next. Over the next few months, she received more than 10,000 letters from readers, most of them disagreeing with her, and many of them from people who claimed to have PhDs in mathematics or statistics. A small sample of the comments from academics includes "You blew it, and you blew it big!" (Scott

Smith, PhD); "May I suggest that you obtain and refer to a standard textbook on probability before you try to answer a question of this type again?" (Charles Reid, PhD); "You blew it!" (Robert Sachs, PhD); and "You are utterly incorrect" (Ray Bobo, PhD). In general, the critics argued that it shouldn't matter whether you switch doors or not—there are only two doors left in the game, and you have chosen your door completely at random, so the probability that the car is behind your door must be one-half either way.

Who was right? Who was wrong? And why does the problem incite such passion? All three questions deserve closer examination.

Let's take a look first at how vos Savant solved the puzzle. Her solution is actually astounding in its simplicity and more compelling than any I have seen in many textbooks. She made a list (Table 6.1) of the three possible arrangements of doors and goats, along with the corresponding outcomes under the "Switch" strategy and the "Stay" strategy. All three cases assume that you picked Door 1. Because all three possibilities listed are (initially) equally likely, the probability of winning if you switch doors is two-thirds, and the probability of winning if you stay with Door 1 is only one-third. Notice that vos Savant's table does not explicitly state which door was opened by the host. That information is implicitly embedded in columns 4 and 5. For example, in the second row, we kept in mind that the host must open Door 3; therefore switching will land you on Door 2, a win. Similarly, in the first row, the door opened could be either Door 2 or Door 3, but column 4 states correctly that you lose in either case if you switch.

Even today, many people seeing the puzzle for the first time find the result hard to believe. Why? What intuitive nerve is jangled?

TABLE 6.1. The three possible arrangements of doors and goats in *Let's Make a Deal*, showing that switching doors is twice as attractive as not.

Door 1	Door 2	Door 3	Outcome If You Switch	Outcome If You Stay
Auto	Goat	Goat	Lose	Win
Goat	Auto	Goat	Win	Lose
Goat	Goat	Auto	Win	Lose

There are probably 10,000 different reasons, one for each reader, but I think the most compelling argument is this: vos Savant's solution seems to force us to believe in mental telepathy. If I should switch no matter what door I originally chose, then it means that the producers somehow read my mind. How else could they position the car so that it is more likely to be behind the door I did not choose?

The key element in resolving this paradox is that we need to take into account not only the data (i.e., the fact that the host opened a particular door) but also the data-generating process—in other words, the rules of the game. They tell us something about the data that could have been but has not been observed. No wonder statisticians in particular found this puzzle hard to comprehend. They are accustomed to, as R. A. Fisher (1922) put it, "the reduction of data" and ignoring the data-generating process.

For starters, let's try changing the rules of the game a bit and see how that affects our conclusion. Imagine an alternative game show, called *Let's Fake a Deal*, where Monty Hall opens one of the two doors you didn't choose, but his choice is completely random. In particular, he might open the door that has a car behind it. Tough luck!

As before, we will assume that you chose Door 1 to begin the game, and the host, again, opens Door 3, revealing a goat, and offers you an option to switch. Should you? We will show that, under the new rules, although the scenario is identical, you will not gain by switching.

To do that, we make a table like the previous one, taking into account that there are two random and independent events—the location of the car (three possibilities) and Monty Hall's choice of a door to open (two possibilities). Thus the table needs to have six rows, each of which is equally likely because the events are independent.

Now what happens if Monty Hall opens Door 3 and reveals a goat? This gives us some significant information: we must be in row 2 or 4 of the table. Focusing just on lines 2 and 4, we can see that the strategy of switching no longer offers us any advantage; we have a one-in-two probability of winning either way. So in the

TABLE 6.2. *Let's Fake a Deal* possibilities.

Door You Chose	Door with Auto	Door Opened by Host	Outcome If You Switch	Outcome If You Stay
1	1	2 (goat)	Lose	Win
1	1	3 (goat)	Lose	Win
1	2	2 (auto)	Lose	Lose
1	2	3 (goat)	Win	Lose
1	3	2 (goat)	Win	Lose
1	3	3 (auto)	Lose	Lose

game *Let's Fake a Deal*, all of Marilyn vos Savant's critics would be right! Yet the data are the same in both games. The lesson is quite simple: the way that we obtain information is no less important than the information itself.

Let's use our favorite trick and draw a causal diagram, which should illustrate immediately how the two games differ. First, Figure 6.1 shows a diagram for the actual *Let's Make a Deal* game, in which Monty Hall must open a door that does not have a car behind it. The absence of an arrow between Your Door and Location of Car means that your choice of a door and the producers' choice of where to put the car are independent. This means we are explicitly ruling out the possibility that the producers can read your mind (or that you can read theirs!). Even more important are the two arrows that are present in the diagram. They show that Door

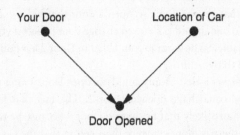

FIGURE 6.1. Causal diagram for *Let's Make a Deal*.

Opened is affected by both your choice and the producers' choice. That is because Monty Hall must pick a door that is different both from both Your Door and Location of Car; he has to take both factors into account.

As you can see from Figure 6.1, Door Opened is a collider. Once we obtain information on this variable, all our probabilities become conditional on this information. But when we condition on a collider, we create a spurious dependence between its parents. The dependence is borne out in the probabilities: if you chose Door 1, the car location is twice as likely to be behind Door 2 as Door 1; if you chose Door 2, the car location is twice as likely to be behind Door 1.

It is a bizarre dependence for sure, one of a type that most of us are unaccustomed to. It is a dependence that has no cause. It does not involve physical communication between the producers and us. It does not involve mental telepathy. It is purely an artifact of Bayesian conditioning: a magical transfer of information without causality. Our minds rebel at this possibility because from earliest infancy, we have learned to associate correlation with causation. If a car behind us takes all the same turns that we do, we first think it is following us (causation!). We next think that we are going to the same place (i.e., there is a common cause behind each of our turns). But causeless correlation violates our common sense. Thus, the Monty Hall paradox is just like an optical illusion or a magic trick: it uses our own cognitive machinery to deceive us.

Why do I say that Monty Hall's opening of Door 3 was a "transfer of information"? It didn't, after all, provide any evidence about whether your initial choice of Door 1 was correct. You knew in advance that he was going to open a door that hid a goat, and so he did. No one should ask you to change your beliefs if you witness the inevitable. So how come your belief in Door 2 has gone up from one-third to two-thirds?

The answer is that Monty could not open Door 1 after you chose it—but he could have opened Door 2. The fact that he did not makes it more likely that he opened Door 3 because he was forced to. Thus there is more evidence than before that the car is behind Door 2. This is a general theme of Bayesian analysis: any hypothesis that has survived some test that threatens its validity becomes more likely. The greater the threat, the more likely it becomes after surviving. Door 2 was vulnerable to refutation (i.e., Monty could have opened it), but Door 1 was not. Therefore, Door 2 becomes a more likely location, while Door 1 does not. The probability that the car is behind Door 1 remains one in three.

Now, for comparison, Figure 6.2 shows the causal diagram for *Let's Fake a Deal*, the game in which Monty Hall chooses a door that is different from yours but otherwise chosen at random. This diagram still has an arrow pointing from Your Door to Door Opened because he has to make sure that his door is different from yours. However, the arrow from Location of Car to Door Opened would be deleted because he no longer cares where the car is. In this diagram, conditioning on Door Opened has absolutely no effect: Your Door and Location of Car were independent to start with, and they remain independent after we see the contents of Monty's door. So in *Let's Fake a Deal*, the car is just as likely to be behind your door as the other door, as observed in Table 6.2.

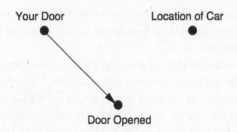

FIGURE 6.2. Causal diagram for *Let's Fake a Deal*.

From the Bayesian point of view, the difference between the two games is that in *Let's Fake a Deal*, Door 1 is vulnerable to refutation. Monty Hall could have opened Door 3 and revealed the car, which would have proven that your door choice was wrong. Because your door and Door 2 were equally vulnerable to refutation, they still have equal probability.

Although purely qualitative, this analysis could be made quantitative by using Bayes's rule or by thinking of the diagrams as a simple Bayesian network. Doing so places this problem in a unifying framework that we use for thinking about other problems. We don't have to invent a method for solving the puzzle; the belief propagation scheme described in Chapter 3 will deliver the correct answer: that is, $P(Door\ 2) = 2/3$ for *Let's Make a Deal*, and $P(Door\ 2) = 1/2$ for *Let's Fake a Deal*.

Notice that I have really given two explanations of the Monty Hall paradox. The first one uses causal reasoning to explain why we observe a spurious dependence between Your Door and Location of Car; the second uses Bayesian reasoning to explain why the probability of Door 2 goes up in *Let's Make a Deal*. Both explanations are valuable. The Bayesian one accounts for the phenomenon but does not really explain why we perceive it as so paradoxical. In my opinion, a true resolution of a paradox should explain why we see it as a paradox in the first place. Why did the people who read her column believe so strongly that vos Savant was wrong? It wasn't just the know-it-alls. Paul Erdos, one of the most brilliant mathematicians of modern times, likewise could not believe the solution until a computer simulation showed him that switching is advantageous. What deep flaw in our intuitive view of the world does this reveal?

"Our brains are just not wired to do probability problems very well, so I'm not surprised there were mistakes," said Persi Diaconis, a statistician at Stanford University, in a 1991 interview with the *New York Times*. True, but there's more to it. Our brains are not wired to do probability problems, but they are wired to do causal problems. And this causal wiring produces systematic probabilistic mistakes, like optical illusions. Because there is no causal connection between My Door and Location of Car, either directly or through a common cause, we find it utterly incomprehensible that there is a probabilistic association. Our brains are not prepared to accept causeless correlations, and we need special training—through examples like the Monty Hall paradox or the ones discussed in Chapter 3—to identify situations where they can arise. Once we have "rewired our brains" to recognize colliders, the paradox ceases to be confusing.

MORE COLLIDER BIAS: BERKSON'S PARADOX

In 1946, Joseph Berkson, a biostatistician at the Mayo Clinic, pointed out a peculiarity of observational studies conducted in a hospital setting: even if two diseases have no relation to each other in the general population, they can appear to be associated among patients in a hospital.

To understand Berkson's observation, let's start with a causal diagram (Figure 6.3). It's also helpful to think of a very extreme possibility: neither Disease 1 nor Disease 2 is ordinarily severe enough to cause hospitalization, but the combination is. In this case, we would expect Disease 1 to be highly correlated with Disease 2 in the hospitalized population.

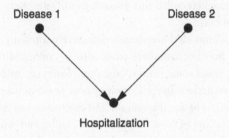

FIGURE 6.3. Causal diagram for Berkson's paradox.

By performing a study on patients who are hospitalized, we are controlling for Hospitalization. As we know, conditioning on a collider creates a spurious association between Disease 1 and Disease 2. In many of our previous examples the association was negative because of the explain-away effect, but here it is positive because both diseases have to be present for hospitalization (not just one).

However, for a long time epidemiologists refused to believe in this possibility. They still didn't believe it in 1979, when David

TABLE 6.3. Sackett's data illustrating Berkson's paradox.

Respiratory disease? ↓	General Population			Hospitalized in Last Six Months		
	Bone disease? ↓			Bone disease? ↓		
	Yes	No	% Yes	Yes	No	% Yes
Yes	17	207	7.6	5	15	25.0
No (control)	184	2,376	7.2	18	219	7.6

Sackett of McMaster University, an expert on all sorts of statistical bias, provided strong evidence that Berkson's paradox is real. In one example (see Table 6.3), he studied two groups of diseases: respiratory and bone. About 7.5 percent of people in the general population have a bone disease, and this percentage is independent of whether they have respiratory disease. But for hospitalized people with respiratory disease, the frequency of bone disease jumps to 25 percent! Sackett called this phenomenon "admission rate bias" or "Berkson bias."

Sackett admits that we cannot definitively attribute this effect to Berkson bias because there could also be confounding factors. The debate is, to some extent, ongoing. However, unlike in 1946 and 1979, researchers in epidemiology now understand causal diagrams and what biases they entail. The discussion has now moved on to finer points of how large the bias can be and whether it is large enough to observe in causal diagrams with more variables. This is progress!

Collider-induced correlations are not new. They have been found in work dating back to a 1911 study by the English economist Arthur Cecil Pigou, who compared children of alcoholic and nonalcoholic parents. They are also found, though not by that name, in the work of Barbara Burks (1926), Herbert Simon (1954), and of course Berkson. They are also not as esoteric as they may seem from my examples. Try this experiment: Flip two coins simultaneously one hundred times and write down the results only when at least one of them comes up heads. Looking at your table, which will probably contain roughly seventy-five entries, you will see that the outcomes of the two simultaneous coin flips are not independent. Every time Coin 1 landed tails, Coin 2 landed heads. How is this possible? Did the coins somehow communicate with each other at light speed? Of course not. In reality you conditioned on a collider by censoring all the tails-tails outcomes.

In *The Direction of Time*, published posthumously in 1956, philosopher Hans Reichenbach made a daring conjecture called the "common cause principle." Rebutting the adage "Correlation does not imply causation," Reichenbach posited a much stronger idea: "No correlation without causation." He meant that a correlation between two variables, X and Y, cannot come about by accident. Either one of the variables causes the other, or a third variable, say Z, precedes and causes them both.

Our simple coin-flip experiment proves that Reichenbach's dictum was too strong, because it neglects to account for the process by which observations are selected. There was no common cause of the outcome of the two coins, and neither coin communicated its result to the other. Nevertheless, the outcomes on our list were correlated. Reichenbach's error was his failure to consider collider structures—the structure behind the data selection. The mistake was particularly illuminating because it pinpoints the exact flaw in the wiring of our brains. We live our lives as if the common cause principle were true. Whenever we see patterns, we look for a causal explanation. In fact, we hunger for an explanation, in terms of stable mechanisms that lie outside the data. The most satisfying kind of explanation is direct causation: X causes Y. When that fails, finding a common cause of X and Y will usually satisfy us. By comparison, colliders are too ethereal to satisfy our causal appetites. We still want to know the mechanism through which the two coins coordinate their behavior. The answer is a crushing disappointment. They do not communicate at all. The correlation we observe is, in the purest and most literal sense, an illusion. Or perhaps even a delusion: that is, an illusion we brought upon ourselves by choosing which events to include in our data set and which to ignore. It is important to realize that we are not always conscious of making this choice, and this is one reason that collider bias can so easily trap the unwary. In the two-coin experiment, the choice was conscious: I told you not to record the trials with two tails. But on plenty of occasions we aren't aware of making the choice, or the choice is made for us. In the Monty Hall paradox, the host opens the door for us. In Berkson's paradox, an unwary researcher might choose to study hospitalized patients for reasons of convenience, without realizing that he is biasing his study.

The distorting prism of colliders is just as prevalent in everyday life. As Jordan Ellenberg asks in *How Not to Be Wrong*, have you ever noticed that, among the people you date, the attractive ones tend to be jerks? Instead of constructing elaborate psychosocial theories, consider a simpler explanation. Your choice of people to date depends on two factors: attractiveness and personality. You'll take a chance on dating a mean attractive person or a nice unattractive person, and certainly a nice attractive person, but not a mean unattractive person. It's the same as the two-coin example, when you censored tails-tails outcomes. This creates a spurious negative correlation between attractiveness and personality. The sad truth is that unattractive people are just as mean as attractive people—but you'll never realize it, because you'll never date somebody who is both mean and unattractive.

SIMPSON'S PARADOX

Now that we have shown that TV producers don't really have telepathic abilities and coins cannot communicate with one another, what other myths can we explode? Let's start with the myth of the bad/bad/good, or BBG, drug.

Imagine a doctor—Dr. Simpson, we'll call him—reading in his office about a promising new drug (Drug *D*) that seems to reduce the risk of a heart attack. Excitedly, he looks up the researchers' data online. His excitement cools a little when he looks at the data on male patients and notices that their risk of a heart attack is actually higher if they take Drug *D*. "Oh well, he says, "Drug *D* must be very effective for women."

But then he turns to the next table, and his disappointment turns to bafflement. "What is this?" Dr. Simpson exclaims. "It says here that women who took Drug *D* were also at higher risk of a heart attack. I must be losing my marbles! This drug seems to be bad for women, bad for men, but good for people."

Are you perplexed too? If so, you are in good company. This paradox, first discovered by a real-life statistician named Edward Simpson in 1951, has been bothering statisticians for more than sixty years—and it remains vexing to this very day. Even in 2016, as I was writing this book, four new articles (including a PhD dissertation) came out, attempting to explain Simpson's paradox from four different points of view.

In 1983 Melvin Novick wrote, "The apparent answer is that when we know that the gender of the patient is male or when we know that it is female we do not use the treatment, but if the gender is unknown we should use the treatment! Obviously that conclusion is ridiculous." I completely agree. It is ridiculous for a drug to be bad for men and bad for women but good for people. So one of these three assertions must be wrong. But which one? And why? And how is this confusion even possible?

To answer these questions, we of course need to look at the (fictional) data that puzzled our good Dr. Simpson so much. The study was observational, not randomized, with sixty men and sixty women. This means that the patients themselves decided whether to take or not to take the drug. Table 6.4 shows how many of each gender received Drug D and how many were subsequently diagnosed with heart attack.

Let me emphasize where the paradox is. As you can see, 5 percent (one in twenty) of the women in the control group later had a heart attack, compared to 7.5 percent of the women who took the

TABLE 6.4. Fictitious data illustrating Simpson's paradox.

	Control Group (No Drug)		Treatment Group (Took Drug)	
	Heart attack	No heart attack	Heart attack	No heart attack
Female	1	19	3	37
Male	12	28	8	12
Total	13	47	11	49

drug. So the drug is associated with a higher risk of heart attack for women. Among the men, 30 percent in the control group had a heart attack, compared to 40 percent in the treatment group. So the drug is associated with a higher risk of heart attack among men. Dr. Simpson was right.

But now look at the third line of the table. Among the control group, 22 percent had a heart attack, but in the treatment group only 18 percent did. So, if we judge on the basis of the bottom line, Drug D seems to decrease the risk of heart attack in the population as a whole. Welcome to the bizarre world of Simpson's paradox!

For almost twenty years, I have been trying to convince the scientific community that the confusion over Simpson's paradox is a result of incorrect application of causal principles to statistical proportions. If we use causal notation and diagrams, we can clearly and unambiguously decide whether Drug D prevents or causes heart attacks. Fundamentally, Simpson's paradox is a puzzle about confounding and can thus be resolved by the same methods we used to resolve that mystery. Curiously, three of the four 2016 papers that I mentioned continue to resist this solution.

Any claim to resolve a paradox (especially one that is decades old) should meet some basic criteria. First, as I said above in connection with the Monty Hall paradox, it should explain why people find the paradox surprising or unbelievable. Second, it should identify the class of scenarios in which the paradox can occur. Third, it should inform us of scenarios, if any, in which the paradox cannot occur. Finally, when the paradox does occur, and we have to make a choice between two plausible yet contradictory statements, it should tell us which statement is correct.

Let's start with the question of why Simpson's paradox is surprising. To explain this, we should distinguish between two things: Simpson's reversal and Simpson's paradox.

Simpson's reversal is a purely numerical fact: as seen in Table 6.4, it is a reversal in relative frequency of a particular event in two or more different samples upon merging the samples. In our example, we saw that $3/40 > 1/20$ (these were the frequencies of heart attack among women with and without Drug D) and $8/20 > 12/40$ (the frequencies among men). Yet when we combined women and men, the inequality reversed direction: $(3 + 8)/(40 + 20) < (1 + 12)/(20 + 40)$. If you thought such a reversal mathematically impossible, then you were probably basing your reaction on misapplied or misremembered properties of fractions. Many people seem to believe that if $A/B > a/b$ and $C/D > c/d$, then it follows that $(A + C)/(B + D) > (a + c)/(b + d)$. But this folk wisdom is simply wrong. The example we have just given refutes it.

Simpson's reversal can be found in real-world data sets. For baseball fans, here is a lovely example concerning two star baseball players, David Justice and Derek Jeter. In 1995, Justice had a higher batting average, .253 to .250. In 1996, Justice had a higher batting average again, .321 to .314. And in 1997, he had a higher batting average than Jeter for the third season in a row, .329 to .291. Yet over all three seasons combined, Jeter had the higher average! Table 6.5 shows the calculations for readers who would like to check them.

How can one player be a worse hitter than the other in 1995, 1996, and 1997 but better over the three-year period? This reversal seems just like the BBG drug. In fact it isn't possible; the problem is that we have used an overly simple word ("better") to describe a complex averaging process over uneven seasons. Notice that the at bats (the denominators) are not distributed evenly year to year. Jeter had very few at bats in 1995, so his rather low batting average that year had little effect on his overall average. On the other hand, Justice had many more at bats in his least productive year, 1995, and that brought his overall batting average down. Once you realize that "better hitter" is defined not by an actual head-to-head competition but by a weighted average that takes into account how often each player played, I think the surprise starts to wane.

TABLE 6.5. Data (not fictitious) illustrating Simpson's reversal.

| | Hits/At Bats | | | |
	1995	1996	1997	All Three Years
David Justice	104/411 = .253	45/140 = .321	163/495 = .329	312/1,046 = .298
Derek Jeter	12/48 = .250	183/582 = .314	190/654 = .291	385/1,284 = .300

There is no question that Simpson's reversal is surprising to some people, even baseball fans. Every year I have some students who cannot believe it at first. But then they go home, work out some examples like the two I have shown here, and come to terms with it. It simply becomes part of their new and slightly deeper understanding of how numbers (and especially aggregates of populations) work. I do not call Simpson's reversal a paradox because it is, at most, simply a matter of correcting a mistaken belief about the behavior of averages. A paradox is more than that: it should entail a conflict between two deeply held convictions.

For professional statisticians who work with numbers every day of their lives, there is even less reason to consider Simpson's reversal a paradox. A simple arithmetic inequality could not possibly puzzle and fascinate them to such an extent that they would still be writing articles about it sixty years later.

Now let's go back to our main example, the paradox of the BBG drug. I've explained why the three statements "bad for men," "bad for women," and "good for people," when interpreted as an increase or decrease in proportions, are not mathematically contradictory. Yet it may still seem to you that they are physically impossible. A drug can't simultaneously cause me and you to have a heart attack and at the same time prevent us both from having heart attacks. This intuition is universal; we develop it as two-year-olds, long before we start learning about numbers and fractions. So I think you will be relieved to find out that you do not have to abandon your intuition. A BBG drug indeed does not exist and will never be invented, and we can prove it mathematically.

The first person to bring attention to this intuitively obvious principle was the statistician Leonard Savage, who in 1954 called it the "sure-thing principle." He wrote,

A businessman contemplates buying a certain piece of property. He considers the outcome of the next presidential election relevant. So, to clarify the matter to himself, he asks whether he would buy if he knew that the Democratic candidate were going to win, and decides that he would. Similarly, he considers whether he would buy if he knew that the Republican candidate were going

to win, and again finds that he would. Seeing that he would buy
in either event, he decides that he should buy, even though he does
not know which event obtains, or will obtain, as we would ordinarily say. It is all too seldom that a decision can be arrived at on
the basis of this principle, but . . . I know of no other extra-logical
principle governing decisions that finds such ready acceptance.

Savage's last statement is particularly perceptive: he realizes that
the "sure-thing principle" is extralogical. In fact, when properly
interpreted it is based on causal, not classical, logic. Also, he says
he "know[s] of no other . . . principle . . . that finds such ready acceptance." Obviously he has talked about it with many people and
they found the line of reasoning very compelling.

To connect Savage's sure-thing principle to our previous discussion, suppose that the choice is actually between two properties,
A and B. If the Democrat wins, the businessman has a 5 percent
chance of making $1 on Property A and an 8 percent chance of
making $1 on Property B. So B is preferred to A. If the Republican
wins, he has a 30 percent chance of making $1 on Property A and
a 40 percent chance of making $1 on Property B. Again, B is pre-
ferred to A. According to the sure-thing principle, he should definitely buy Property B. But sharp-eyed readers may notice that the
numerical quantities are the same as in the Simpson story, and this
may alert us that buying Property B may be too hasty a decision.

In fact, the argument above has a glaring flaw. If the businessman's decision to buy can change the election's outcome (for example, if the media watched his actions), then buying Property A
may be in his best interest. The harm of electing the wrong president may outweigh whatever financial gain he might extract from
the deal, once a president is elected.

To make the sure-thing principle valid, we must insist that the
businessman's decision will not affect the outcome of the election.
As long as the businessman is sure his decision won't affect the
likelihood of a Democratic or Republican victory, he can go ahead
and buy Property B. Otherwise, all bets are off.

Note that the missing ingredient (which Savage neglected to
state explicitly) is a causal assumption. A correct version of his

principle would read as follows: an action that increases the probability of a certain outcome assuming either that Event C occurred or that Event C did not occur will also increase its probability if we don't know whether C occurred . . . provided that the action does not change the probability of C. In particular, there is no such thing as a BBG drug. This corrected version of Savage's sure-thing principle does not follow from classical logic: to prove it, you need a causal calculus invoking the *do*-operator. Our strong intuitive belief that a BBG drug is impossible suggests that humans (as well as machines programmed to emulate human thought) use something like the *do*-calculus to guide their intuition.

According to the corrected sure-thing principle, one of the following three statements must be false: Drug D increases the probability of heart attack in men and women; Drug D decreases the probability of heart attack in the population as a whole; and the drug does not change the number of men and women. Since it's very implausible that a drug would change a patient's sex, one of the first two statements must be false.

Which is it? In vain will you seek guidance from Table 6.4. To answer the question, we must look beyond the data to the data-generating process. As always, it is practically impossible to discuss that process without a causal diagram.

The diagram in Figure 6.4 encodes the crucial information that gender is unaffected by the drug and, in addition, gender affects the risk of heart attack (men being at greater risk) and whether the

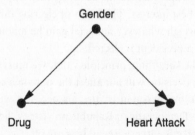

FIGURE 6.4. Causal diagram for the Simpson's paradox example.

patient chooses to take Drug *D*. In the study, women clearly had a preference for taking Drug *D* and men preferred not to. Thus Gender is a confounder of Drug and Heart Attack. For an unbiased estimate of the effect of Drug on Heart Attack, we must adjust for the confounder. We can do that by looking at the data for men and women separately, then taking the average:

- For women, the rate of heart attacks was 5 percent without Drug *D* and 7.5 percent with Drug *D*.
- For men, the rate of heart attacks was 30 percent without Drug *D* and 40 percent with.
- Taking the average (because men and women are equally frequent in the general population), the rate of heart attacks without Drug *D* is 17.5 percent (the average of 5 and 30), and the rate with Drug *D* is 23.75 percent (the average of 7.5 and 40).

This is the clear and unambiguous answer we were looking for. Drug *D* isn't BBG, it's BBB: bad for women, bad for men, and bad for people.

I don't want you to get the impression from this example that aggregating the data is always wrong or that partitioning the data is always right. It depends on the process that generated the data. In the Monty Hall paradox, we saw that changing the rules of the game also changed the conclusion. The same principle works here. I'll use a different story to demonstrate when pooling the data would be appropriate. Even though the data will be precisely the same, the role of the "lurking third variable" will differ and so will the conclusion.

Let's begin with the assumption that blood pressure is known to be a possible cause of heart attack, and Drug *B* is supposed to reduce blood pressure. Naturally, the Drug *B* researchers wanted to see if it might also reduce heart attack risk, so they measured their patients' blood pressure after treatment, as well as whether they had a heart attack.

Table 6.6 shows the data from the study of Drug *B*. It should look amazingly familiar: the numbers are the same as in Table

TABLE 6.6. Fictitious data for blood pressure example.

	Control Group (No Drug)		Treatment Group (Took Drug)	
	Heart attack	*No heart attack*	*Heart attack*	*No heart attack*
Low blood pressure	1	19	3	37
High blood pressure	12	28	8	12
Total	13	47	11	49

6.4! Nevertheless, the conclusion is exactly the opposite. As you can see, taking Drug *B* succeeded in lowering the patients' blood pressure: among the people who took it, twice as many had low blood pressure afterward (forty out of sixty, compared to twenty out of sixty in the control group). In other words, it did exactly what an anti–heart attack drug should do. It moved people from the higher-risk category into the lower-risk category. This factor outweighs everything else, and we can justifiably conclude that the aggregated part of Table 6.6 gives us the correct result.

As usual, a causal diagram will make everything clear and allow us to derive the result mechanically, without even thinking about the data or whether the drug lowers or increases blood pressure. In this case our "lurking third variable" is Blood Pressure, and the diagram looks like Figure 6.5. Here, Blood Pressure is a mediator rather than a confounder. A single glance at the diagram reveals that there is no confounder of the Drug → Heart Attack relationship (i.e., no back-door path), so stratifying the data is unnecessary. In fact, conditioning on Blood Pressure would disable one of the causal paths (maybe the main causal path) by which the drug works. For both these reasons, our conclusion is the exact opposite of what it was for Drug *D*: Drug *B* works, and the aggregate data reveal this fact.

From a historical point of view, it is noteworthy that Simpson, in his 1951 paper that started all the ruckus, did exactly the same thing that I have just done. He presented two stories with exactly

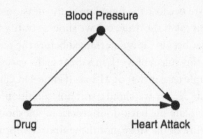

FIGURE 6.5. Causal diagram for the Simpson's
paradox example (second version).

the same data. In one example, it was intuitively clear that aggregating the data was, in his words, "the sensible interpretation"; in the other example, partitioning the data was more sensible. So Simpson understood that there was a paradox, not just a reversal. However, he suggested no resolution to the paradox other than common sense. Most importantly, he did not suggest that, if the story contains extra information that makes the difference between "sensible" and "not sensible," perhaps statisticians should embrace that extra information in their analysis.

Dennis Lindley and Melvin Novick considered this suggestion in 1981, but they could not reconcile themselves to the idea that the correct decision depends on the causal story, not on the data. They confessed, "One possibility would be to use the language of causation. . . . We have not chosen to do this; nor to discuss causation, because the concept, although widely used, does not seem to be well-defined." With these words, they summarized the frustration of five generations of statisticians, recognizing that causal information is badly needed but the language for expressing it is hopelessly lacking. In 2009, four years before his death at age ninety, Lindley confided in me that he would not have written those words if my book had been available in 1981.

Some readers of my books and articles have suggested that the rule governing data aggregation and separation rests simply on the temporal precedence of the treatment and the "lurking third variable." They argue that we should aggregate the data in the case of blood pressure because the blood pressure measurement comes after the patient takes the drug, but we should stratify the data in the case of gender because it is determined before the patient takes the drug. While this rule will work in a great many cases, it is not foolproof. A simple case is that of M-bias (Game 4 in Chapter 4). Here B can precede X; yet we should still not condition on B, because that would violate the back-door criterion. We should consult the causal structure of the story, not the temporal information.

Finally, you might wonder if Simpson's paradox occurs in the real world. The answer is yes. It is certainly not common enough for statisticians to encounter on a daily basis, but nor is it completely unknown, and it probably happens more often than journal articles report. Here are two documented cases:

- In an observational study published in 1996, open surgery to remove kidney stones had a better success rate than endoscopic surgery for small kidney stones. It also had a better success rate for large kidney stones. However, it had a lower success rate overall. Just as in our first example, this was a case where the choice of treatment was related to the severity of the patients' case: larger stones were more likely to lead to open surgery and also had a worse prognosis.
- In a study of thyroid disease published in 1995, smokers had a higher survival rate (76 percent) over twenty years than nonsmokers (69 percent). However, the nonsmokers had a better survival rate in six out of seven age groups, and the difference was minimal in the seventh. Age was clearly a confounder of Smoking and Survival: the average smoker was younger than the average nonsmoker (perhaps because the older smokers had already died). Stratifying the data by age, we conclude that smoking has a negative impact on survival.

Because Simpson's paradox has been so poorly understood, some statisticians take precautions to avoid it. All too often, these methods avoid the symptom, Simpson's reversal, without doing anything about the disease, confounding. Instead of suppressing the symptoms, we should pay attention to them. Simpson's paradox alerts us to cases where at least one of the statistical trends (either in the aggregated data, the partitioned data, or both) cannot represent the causal effects. There are, of course, other warning signs of confounding. The aggregated estimate of the causal effect could, for example, be larger than each of the estimates in each of the strata; this likewise should not happen if we have controlled properly for confounders. Compared to such signs, however, Simpson's reversal is harder to ignore precisely because it is a reversal, a qualitative change in the sign of the effect. The idea of a BBG drug would evoke disbelief even from a three-year-old child—and rightly so.

SIMPSON'S PARADOX IN PICTURES

So far most of our examples of Simpson's reversal and paradox have involved binary variables: a patient either got Drug D or didn't and either had a heart attack or didn't. However, the reversal can also occur with continuous variables and is perhaps easier to understand in that case because we can draw a picture.

Consider a study that measures weekly exercise and cholesterol levels in various age groups. When we plot hours of exercise on the x-axis and cholesterol on the y-axis, as in Figure 6.6(a), we see in each age group a downward trend, indicating perhaps that exercise reduces cholesterol. On the other hand, if we use the same scatter plot but don't segregate the data by age, as in Figure 6.6(b), then we see a pronounced upward trend, indicating that the more people exercise, the higher their cholesterol becomes. Once again we seem to have a BBG drug situation, where Exercise is the drug: it seems to have a beneficial effect in each age group but a harmful effect on the population as a whole.

To decide whether Exercise is beneficial or harmful, as always, we need to consult the story behind the data. The data show that older people in our population exercise more. Because it seems more likely that Age causes Exercise rather than vice versa, and since Age may have a causal effect on Cholesterol, we conclude that Age may

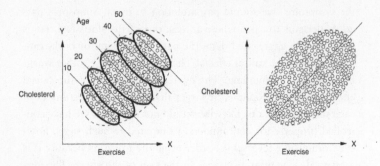

FIGURE 6.6. Simpson's paradox: exercise appears to be beneficial (downward slope) in each age group but harmful (upward slope) in the population as a whole.

be a confounder of Exercise and Cholesterol. So we should control for Age. In other words, we should look at the age-segregated data and conclude that exercise is beneficial, regardless of age.

A cousin of Simpson's paradox has also been lurking in the statistical literature for decades and lends itself nicely to a visual interpretation. Frederic Lord originally stated this paradox in 1967. It's again fictitious, but fictitious examples (like Einstein's thought experiments) always provide a good way to probe the limits of our understanding.

Lord posits a school that wants to study the effects of the diet it is providing in its dining halls and in particular whether it has different effects on girls and boys. To this end, the students' weight is measured in September and again the following June. Figure 6.7 plots the results, with the ellipses once again representing a scatter plot of data. The university retains two statisticians, who look at the data and come to opposite conclusions.

The first statistician looks at the weight distribution for girls as a whole and notes that the average weight of the girls is the same in June as in September. (This can be seen from the symmetry of the scatter plot around the line $W_F = W_I$, i.e., final weight = initial weight.) Individual girls may, of course, gain or lose weight, but the average weight gain is zero. The same observation is true for

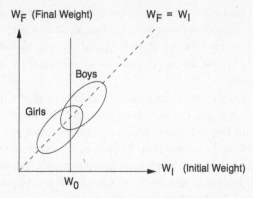

FIGURE 6.7. Lord's paradox. (Ellipses represent scatter plots of data.) As a whole, neither boys nor girls gain weight during the year, but in each stratum of the initial weight, boys tend to gain more than girls.

the boys. Therefore, the statistician concludes that the diet has no differential effect on the sexes.

The second statistician, on the other hand, argues that because the final weight of a student is strongly influenced by his or her initial weight, we should stratify the students by initial weight. If you make a vertical slice through both ellipses, which corresponds to looking only at the boys and girls with a particular value of the initial weight (say W_0 in Figure 6.7), you will notice that the vertical line intersects the Boys ellipse higher up than it does the Girls ellipse, although there is a certain amount of overlap. This means that boys who started with weight W_0 will have, on average, a higher final weight (W_F) than the girls who started with weight W_0. Accordingly, Lord writes, "the second statistician concludes, as is customary in such cases, that the boys showed significantly more gain in weight than the girls when proper allowance is made for differences in initial weight between the sexes."

What is the school's dietitian to do? Lord writes, "The conclusions of each statistician are visibly correct." That is, you don't have to crunch any numbers to see that two solid arguments are leading to two different conclusions. You need only look at the figure. In Figure 6.7, we can see that boys gain more weight than girls in every stratum (every vertical cross section). Yet it's equally obvious that both boys and girls gained nothing overall. How can that be? Is not the overall gain just an average of the stratum-specific gains?

Now that we are experienced pros at the fine points of Simpson's paradox and the sure-thing principle, we know what is wrong with that argument. The sure-thing principle works only in cases where the relative proportion of each subpopulation (each weight class) does not change from group to group. Yet, in Lord's case, the "treatment" (gender) very strongly affects the percentage of students in each weight class.

So we can't rely on the sure-thing principle, and that brings us back to square one. Who is right? Is there or isn't there a difference in the average weight gains between boys and girls when proper allowance is made for differences in the initial weight between the sexes? Lord's conclusion is very pessimistic: "The usual research study of this type is attempting to answer a question that simply cannot be answered in any rigorous way on the basis of available data." Lord's pessimism spread beyond statistics and has led to a rich and quite pessimistic literature in epidemiology and biostatistics on how to compare groups that differ in "baseline" statistics.

I will show now why Lord's pessimism is unjustified. The dietitian's question can be answered in a rigorous way, and as usual the starting point is to draw a causal diagram, as in Figure 6.8. In this diagram, we see that Sex (S) is a cause of initial weight (W_I) and final weight (W_F). Also, W_I affects W_F independently of gender, because students of either gender who weigh more at the beginning of the year tend to weigh more at the end of the year, as shown by the scatter plots in Figure 6.7. All these causal assumptions are commonsensical; I would not expect Lord to disagree with them.

The variable of interest to Lord is the weight gain, shown as Y in this diagram. Note that Y is related to W_I and W_F in a purely mathematical, deterministic way: $Y = W_F - W_I$. This means that the correlations between Y and W_I (or Y and W_F) are equal to -1 (or 1), and I have shown this information on the diagram with the coefficients -1 and $+1$.

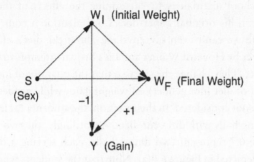

FIGURE 6.8. Causal diagram for Lord's paradox.

The first statistician simply compares the difference in weight gain between girls and boys. No back doors between S and Y need to be blocked, so the observed, aggregated data provide the answer: no effect, as the first statistician concluded.

By contrast, it is hard to even formulate the question that the second statistician is trying to answer (that is, the "correctly formulated query" described in the Introduction). He wants to ensure that "proper allowance is made for differences in initial weight between the two sexes," which is language you would usually use when controlling for a confounder. But W_I is not a confounder of S and Y. It is actually a mediating variable if we consider Sex to be the treatment. Thus, the query answered by controlling for W_I does not have the usual causal effect interpretation. Such control may at best provide an estimate of the "direct effect" of gender on weight, which we will discuss in Chapter 9. However, it seems unlikely that this is what the second statistician had in mind; more likely he was adjusting out of habit. And yet his argument is such an easy trap to fall into: "Is not the overall gain just an average of the stratum-specific gains?" Not if the strata themselves are shifting under treatment! Remember that Sex, not Diet, is the treatment, and Sex definitely changes the proportion of students in each stratum of W_I.

This last comment brings up one more curious point about Lord's paradox as originally phrased. Although the stated intention of the school dietitian is to "determine the effects of the diet," nowhere in his original paper does Lord mention a control diet. Therefore we can't even say anything about the diet's effects. A 2006 paper by Howard Wainer and Lisa Brown attempts to remedy this defect. They change the story so that the quantity of interest is the effect of diet (not gender) on weight gain, while gender differences are not considered. In their version, the students eat in one of two dining halls with different diets. Accordingly, the two ellipses of Figure 6.7 represent two dining halls, each serving a different diet, as depicted in Figure 6.9(a). Note that the students who weigh more in the beginning tend to eat in dining hall B, while the ones who weigh less eat in dining hall A.

Lord's paradox now surfaces with greater clarity, since the query is well defined as the effect of diet on gain. The first statistician claims, based on symmetry considerations, that switching from Diet A to B would have no effect on weight gain (the difference $W_F - W_I$ has the same distribution in both ellipses). The second statistician compares the final weights under Diet A to those of Diet B for a group of students starting with weight W_0 and concludes that the students on Diet B gain more weight.

As before, the data (Figure 6.9[a]) can't tell you whom to believe, and this is indeed what Wainer and Brown conclude. However, a causal diagram (Figure 6.9[b]) can settle the issue. There are two significant changes between Figure 6.8 and Figure 6.9(b). First, the causal variable becomes D (for "diet"), not S. Second, the arrow that originally pointed from S to W_I now reverses direction: the initial weight now affects the diet, so the arrow points from W_I to D.

In this diagram, W_I is a confounder of D and W_F, not a mediator. Therefore, the second statistician would be unambiguously correct here. Controlling for the initial weight is essential to deconfound D and W_F (as well as D and Y). The first statistician would be wrong, because he would only be measuring statistical associations, not causal effects.

To summarize, for us the main lesson of Lord's paradox is that it is no more of a paradox than Simpson's. In one paradox, the association reverses; in the other, it disappears. In either case, the causal

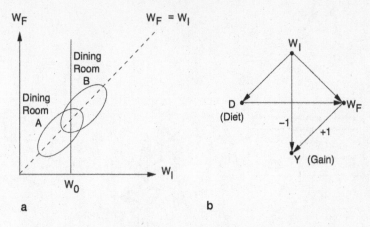

FIGURE 6.9. Wainer and Brown's revised version of Lord's paradox and the corresponding causal diagram.

diagram will tell us what procedure we need to use. However, for statisticians who are trained in "conventional" (i.e., model-blind) methodology and avoid using causal lenses, it is deeply paradoxical that the correct conclusion in one case would be incorrect in another, even though the data look exactly the same.

Now that we have a thorough grounding in colliders, confounders, and the perils that both pose, we are at last prepared to reap the fruits of our labor. In the next chapter we begin our ascent up the Ladder of Causation, beginning with rung two: intervention.

BACK-DOOR ADJUSTMENT =======
FRONT-DOOR ADJUSTMENT ————
GO YOUR OWN WAY – – – – –
(DO-CALCULUS)
INSTRUMENTAL VARIABLES

YOU ARE HERE

Scaling "Mount Intervention." The most familiar methods to estimate the effect of an intervention, in the presence of confounders, are the back-door adjustment and instrumental variables. The method of front-door adjustment was unknown before the introduction of causal diagrams. The *do*-calculus, which my students have fully automated, makes it possible to tailor the adjustment method to any particular causal diagram. (*Source:* Drawing by Dakota Harr.)

7

BEYOND ADJUSTMENT:
THE CONQUEST OF MOUNT INTERVENTION

*He whose actions exceed his theory, his theory shall
endure.*

—Rabbi Hanina ben Dosa (first century ad)

I N this chapter we finally make our bold ascent onto the second
level of the Ladder of Causation, the level of intervention—the
holy grail of causal thinking from antiquity to the present day.
This level is involved in the struggle to predict the effects of actions and policies that haven't been tried yet, ranging from medical
treatments to social programs, from economic policies to personal
choices. Confounding was the primary obstacle that caused us to
confuse *seeing* with *doing*. Having removed this obstacle with the
tools of "path blocking" and the back-door criterion, we can now
map the routes up Mount Intervention with systematic precision.
For the novice climber, the safest routes up the mountain are the
back-door adjustment and its various cousins, some going under
the rubric of "front-door adjustment" and some under "instrumental variables."

But these routes may not be available in all cases, so for the
experienced climber this chapter describes a "universal mapping

tool" called the *do*-calculus, which allows the researcher to explore and plot all possible routes up Mount Intervention, no matter how twisty. Once a route has been mapped, and the ropes and carabiners and pitons are in place, our assault on the mountain will assuredly result in a successful conquest!

THE SIMPLEST ROUTE: THE BACK-DOOR ADJUSTMENT FORMULA

For many researchers, the most (perhaps only) familiar method of predicting the effect of an intervention is to "control" for confounders using the adjustment formula. This is the method to use if you are confident that you have data on a sufficient set of variables (called deconfounders) to block all the back-door paths between the intervention and the outcome. To do this, we measure the average causal effect of an intervention by first estimating its effect at each "level," or stratum, of the deconfounder. We then compute a weighted average of those strata, where each stratum is weighted according to its prevalence in the population. If, for example, the deconfounder is gender, we first estimate the causal effect for males and females. Then we average the two, if the population is (as usual) half male and half female. If the proportions are different—say, two-thirds male and one-third female—then to estimate the average causal effect we would take a correspondingly weighted average.

The role that the back-door criterion plays in this procedure is to guarantee that the causal effect in each stratum of the deconfounder is none other than the observed trend in this stratum. So the causal effect can be estimated stratum by stratum from the data. Absent the back-door criterion, researchers have no guarantee that any adjustment is legitimate.

The fictitious drug example in Chapter 6 was the simplest situation possible: one treatment variable (Drug D), one outcome (Heart Attack), one confounder (Gender), and all three variables are binary. The example shows how we take a weighted average of the conditional probabilities $P(heart\ attack \mid drug)$ in each gender stratum. But the procedure described above can be adapted easily to handle more complicated situations, including multiple (de)confounders and multiple strata.

However, in many cases, the variables X, Y, or Z take numerical values—for example, income or height or birth weight. We saw this in our visual example of Simpson's paradox. Because the variable could take (at least, for all practical purposes) infinite possible values, we cannot make a table listing all the possibilities, as we did in Chapter 6.

An obvious remedy is to separate the numerical values into a finite and manageable number of categories. There is nothing in principle wrong with this option, but the choice of categories is a bit arbitrary. Worse, if we have more than a handful of adjusted variables, we get an exponential blowup in the number of categories. This will make the procedure computationally prohibitive; worse yet, many of the strata will end up devoid of samples and thus incapable of providing any probability estimates whatsoever.

Statisticians have devised ingenious methods for handling this "curse of dimensionality" problem. Most involve some sort of extrapolation, whereby a smooth function is fitted to the data and used to fill in the holes created by the empty strata.

The most widely used smoothing function is of course a linear approximation, which served as the workhorse of most quantitative work in the social and behavioral sciences in the twentieth century. We have seen how Sewall Wright embedded his path diagrams into the context of linear equations, and we noted there one computational advantage of this embedding: every causal effect can be represented by a single number (the path coefficient). A second and no less important advantage of linear approximations is the astonishing simplicity of computing the adjustment formula.

We have previously seen Francis Galton's invention of a regression line, which takes a cloud of data points and interpolates the best-fitting line through that cloud. In the case of one treatment variable (X) and one outcome variable (Y), the equation of the regression line will look like this: $Y = aX + b$. The parameter a (often denoted by r_{YX}, the regression coefficient of Y on X) tells us the average observed trend: a one-unit increase of X will, on average, produce an a-unit increase in Y. If there are no confounders of Y

and X, then we can use this as our estimate of an intervention to increase X by one unit.

But what if there is a confounder, Z? In this case, the correlation coefficient r_{YX} will not give us the average causal effect; it only gives us the average observed trend. That was the case in Wright's problem of the guinea pig birth weights, discussed in Chapter 2, where the apparent benefit (5.66 grams) of an extra day's gestation was biased because it was confounded with the effect of a smaller litter size. But there is still a way out: by plotting all three variables together, with each value of (X, Y, Z) describing one point in space. In this case, the data will form a cloud of points in XYZ-space. The analogue of a regression line is a regression plane, which has an equation that looks like $Y = aX + bZ + c$. We can easily compute a, b, c from the data. Here something wonderful happens, which Galton did not realize but Karl Pearson and George Udny Yule certainly did. The coefficient a gives us the regression coefficient of Y on X already adjusted for Z. (It is called a partial regression coefficient and written $r_{YX.Z}$.)

Thus we can skip the cumbersome procedure of regressing Y on X for each level of Z and computing the weighted average of the regression coefficients. Nature already does all the averaging for us! We need only compute the plane that best fits the data. A statistical package will do it in no time. The coefficient a in the equation of that plane, $Y = aX + bZ + c$, will automatically adjust the observed trend of Y on X to account for the confounder Z. If Z is the only confounder, then a is the average causal effect of X on Y. A truly miraculous simplification!

You can easily extend the procedure to deal with multiple variables as well. If the set of variables Z should happen to satisfy the back-door condition, then the coefficient of X in the regression equation, a, will be none other than the average causal effect of X on Y.

For this reason generations of researchers came to believe that adjusted (or partial) regression coefficients are somehow endowed with causal information that unadjusted regression coefficients lack. Nothing could be further from the truth. Regression coefficients, whether adjusted or not, are only statistical trends, conveying no causal information in themselves. $r_{YX.Z}$ represents the causal effect of X on Y, whereas r_{YX} does not, exclusively because we have a diagram showing Z as a confounder of X and Y.

In short, sometimes a regression coefficient represents a causal effect, and sometimes it does not—and you can't rely on the data alone to tell you the difference. Two additional ingredients are required to endow $r_{YX.Z}$ with causal legitimacy. First, the path diagram should represent a plausible picture of reality, and second, the adjusted variable(s) Z should satisfy the back-door criterion.

That is why it was so crucial that Sewall Wright distinguished path coefficients (which represent causal effects) from regression coefficients (which represent trends of data points). Path coefficients are fundamentally different from regression coefficients, although they can often be computed from the latter. Wright failed to realize, however, as did all path analysts and econometricians after him, that his computations were unnecessarily complicated. He could have gotten the path coefficients from partial correlation coefficients, if only he had known that the proper set of adjusting variables can be identified, by inspection, from the path diagram itself.

Keep in mind also that the regression-based adjustment works only for linear models, which involve a major modeling assumption. With linear models, we lose the ability to model nonlinear interactions, such as when the effect of X on Y depends on the level of Z. The back-door adjustment, on the other hand, still works fine even when we have no idea what functions are behind the arrows in the diagrams. But in this so-called nonparametric case, we need to employ other extrapolation methods to deal with the curse of dimensionality.

To sum up, the back-door adjustment formula and the back-door criterion are like the front and back of a coin. The back-door criterion tells us which sets of variables we can use to deconfound our data. The adjustment formula actually does the deconfounding. In the simplest case of linear regression, partial regression coefficients perform the back-door adjustment implicitly. In the nonparametric case, we must do the adjustment explicitly, either using the back-door adjustment formula directly on the data or on some extrapolated version of it.

You might think that our assault on Mount Intervention would end there with complete success. Unfortunately, though, adjustment does not work at all if there is a back-door path we cannot block because we don't have the requisite data. Yet we can still use certain tricks even in this situation. I will tell you about one of my favorite methods next, called the front-door adjustment. Even though it was published more than twenty years ago, only a handful of researchers have taken advantage of this shortcut up Mount Intervention, and I am convinced that its full potential remains untapped.

THE FRONT-DOOR CRITERION

The debate over the causal effect of smoking occurred at least two generations too early for causal diagrams to make any contribution. We have already seen how Cornfield's inequality helped persuade researchers that the smoking gene, or "constitutional hypothesis," was highly implausible. But a more radical approach, using causal diagrams, could have shed more light on the hypothetical gene and possibly eliminated it from further consideration.

Suppose that researchers had measured the tar deposits in smokers' lungs. Even in the 1950s, the formation of tar deposits was suspected as one of the possible intermediate stages in the development of lung cancer. Suppose also that, just like the Surgeon General's committee, we want to rule out R. A. Fisher's hypothesis that a smoking gene confounds smoking behavior and lung cancer. We might then arrive at the causal diagram in Figure 7.1.

Figure 7.1 incorporates two very important assumptions, which we'll suppose are valid for the purpose of our example. The first assumption is that the smoking gene has no effect on the formation of tar deposits, which are exclusively due to the physical action of cigarette smoke. (This assumption is indicated by the lack of an arrow between Smoking Gene and Tar; it does not rule out, however, random factors unrelated to Smoking Gene.) The second significant assumption is that Smoking leads to Cancer only through the accumulation of tar deposits. Thus we assume that

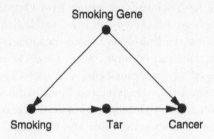

FIGURE 7.1. Hypothetical causal diagram for smoking and cancer, suitable for front-door adjustment.

no direct arrow points from Smoking to Cancer, and there are no other indirect pathways.

Suppose we are doing an observational study and have collected data on Smoking, Tar, and Cancer for each of the participants. Unfortunately, we cannot collect data on the Smoking Gene because we do not know whether such a gene exists. Lacking data on the confounding variable, we cannot block the back-door path Smoking ← Smoking Gene → Cancer. Thus we cannot use back-door adjustment to control for the effect of the confounder.

So we must look for another way. Instead of going in the back door, we can go in the front door! In this case, the front door is the direct causal path Smoking → Tar → Cancer, for which we do have data on all three variables. Intuitively, the reasoning is as follows. First, we can estimate the average causal effect of Smoking on Tar, because there is no unblocked back-door path from Smoking to Tar, as the Smoking ← Smoking Gene → Cancer ← Tar path is already blocked by the collider at Cancer. Because it is blocked already, we don't even need back-door adjustment. We can simply observe $P(tar \mid smoking)$ and $P(tar \mid no\ smoking)$, and the difference between them will be the average causal effect of Smoking on Tar.

Likewise, the diagram allows us to estimate the average causal effect of Tar on Cancer. To do this we can block the back-door path from Tar to Cancer, Tar ← Smoking ← Smoking Gene → Cancer, by adjusting for Smoking. Our lessons from Chapter 4 come in handy: we only need data on a sufficient set of deconfounders (i.e., Smoking). Then the back-door adjustment formula will give us $P(cancer \mid do(tar))$ and $P(cancer \mid do(no\ tar))$. The difference between these is the average causal effect of Tar on Cancer.

Now we know the average increase in the likelihood of tar deposits due to smoking and the average increase of cancer due to tar deposits. Can we combine these somehow to obtain the average increase in cancer due to smoking? Yes, we can. The reasoning goes as follows. Cancer can come about in two ways: in the presence of Tar or in the absence of Tar. If we force a person to smoke, then the probabilities of these two states are $P(tar \mid do(smoking))$ and $P(no\ tar \mid do(smoking))$, respectively. If a Tar state evolves, the likelihood of causing Cancer is $P(cancer \mid do(tar))$. If, on the other hand, a No-Tar state evolves, then it would result in a Cancer likelihood of $P(cancer \mid do(no\ tar))$. We can weight the two scenarios by their respective probabilities under $do(smoking)$ and in this way compute the total probability of cancer due to smoking. The same argument holds if we prevent a person from smoking, $do(no\ smoking)$. The difference between the two gives us the average causal effect on cancer of smoking versus not smoking.

As I have just explained, we can estimate each of the *do*-probabilities discussed from the data. That is, we can write them mathematically in terms of probabilities that do not involve the *do*-operator. In this way, mathematics does for us what ten years of debate and congressional testimony could not: quantify the causal effect of smoking on cancer—provided our assumptions hold, of course.

The process I have just described, expressing $P(cancer \mid do(smoking))$ in terms of *do*-free probabilities, is called the frontdoor adjustment. It differs from the back-door adjustment in that we adjust for two variables (Smoking and Tar) instead of one, and these variables lie on the front-door path from Smoking to Cancer rather than the back-door path. For those readers who "speak mathematics," I can't resist showing you the formula (Equation 7.1), which cannot be found in ordinary statistics textbooks. Here X stands for Smoking, Y stands for Cancer, Z stands for Tar, and

U (which is conspicuously absent from the formula) stands for the unobservable variable, the Smoking Gene.

$$P(Y \mid do(X)) = \Sigma_z \, P(Z = z, \mid X) \, \Sigma_x \, P(Y \mid X = x, Z = z) \, P(X = x)$$

(7.1)

Readers with an appetite for mathematics might find it interesting to compare this to the formula for the back-door adjustment, which looks like Equation 7.2.

$$P(Y \mid do(X)) = \Sigma_z \, P(Y \mid X, Z = z) \, P(Z = z)$$ (7.2)

Even for readers who do not speak mathematics, we can make several interesting points about Equation 7.1. First and most important, you don't see U (the Smoking Gene) anywhere. This was the whole point. We have successfully deconfounded U even without possessing any data on it. Any statistician of Fisher's generation would have seen this as an utter miracle. Second, way back in the Introduction I talked about an estimand as a recipe for computing the quantity of interest in a query. Equations 7.1 and 7.2 are the most complicated and interesting estimands that I will show you in this book. The left-hand side represents the query "What is the effect of X on Y?" The right-hand side is the estimand, a recipe for answering the query. Note that the estimand contains no *do*'s, only *see*'s, represented by the vertical bars, and this means it can be estimated from data.

At this point, I'm sure that some readers are wondering how close this fictional scenario is to reality. Could the smoking-cancer controversy have been resolved by one observational study and one causal diagram? If we assume that Figure 7.1 accurately reflects the causal mechanism for cancer, the answer is absolutely yes. However, we now need to discuss whether our assumptions are valid in the real world.

David Freedman, a longtime friend and a Berkeley statistician, took me to task over this issue. He argued that the model in Figure 7.1 is unrealistic in three ways. First, if there is a smoking gene, it might also affect how the body gets rid of foreign matter in the lungs, so that people with the gene are more vulnerable to the formation of tar deposits and people without it are more resistant. Therefore, he would draw an arrow from Smoking Gene to Tar, and in that case the front-door formula would be invalid.

Freedman also considered it unlikely that Smoking affects Cancer only through Tar. Certainly other mechanisms could be imagined; perhaps smoking produces chronic inflammation that leads to cancer. Finally, he said, tar deposits in a living person's lungs cannot be measured with sufficient accuracy anyway—so an observational study such as the one I have proposed cannot be conducted in the real world.

I have no quarrel with Freedman's criticism in this particular example. I am not a cancer specialist, and I would always have to defer to the expert opinion on whether such a diagram represents the real-world processes accurately. In fact, one of the major accomplishments of causal diagrams is to make the assumptions transparent so that they can be discussed and debated by experts and policy makers.

However, the point of my example was not to propose a new mechanism for the effect of smoking but to demonstrate how mathematics, given the right situation, can eliminate the effect of confounders even without data on the confounder. And the situation can be clearly recognized. Anytime the causal effect of X on Y is confounded by one set of variables (C) and mediated by another (M) (see Figure 7.2), and, furthermore, the mediating variables are shielded from the effects of C, then you can estimate X's effect from observational data. Once scientists are made aware of this fact, they should seek shielded mediators whenever they face incurable confounders. As Louis Pasteur said, "Fortune favors the prepared mind."

Fortunately, the virtues of front-door adjustment have not remained completely unappreciated. In 2014, Adam Glynn and Konstantin Kashin, both political scientists at Harvard (Glynn subsequently moved to Emory University), wrote a prize-winning paper that should be required reading for all quantitative social scientists. They applied the new method to a data set well scrutinized by social scientists, called the Job Training Partnership Act

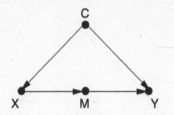

FIGURE 7.2. The basic setup for the front-door criterion.

(JTPA) Study, conducted from 1987 to 1989. As a result of the 1982 JTPA, the Department of Labor created a job-training program that, among other services, provided participants with occupational skills, job-search skills, and work experience. It collected data on people who applied for the program, people who actually used the services, and their earnings over the subsequent eighteen months. Notably, the study included both a randomized controlled trial (RCT), where people were randomly assigned to receive services or not, and an observational study, in which people could choose for themselves.

Glynn and Kashin did not draw a causal diagram, but from their description of the study, I would draw it as shown in Figure 7.3. The variable Signed Up records whether a person did or did not register for the program; the variable Showed Up records whether the enrollee did or did not actually use the services. Obviously the program can only affect earnings if the user actually shows up, so the absence of a direct arrow from Signed Up to Earnings is easy to justify.

Glynn and Kashin refrain from specifying the nature of the confounders, but I have summed them up as Motivation. Clearly, a person who is highly motivated to increase his or her earnings is more likely to sign up. That person is also more likely to earn more after eighteen months, regardless of whether he or she shows up. The goal of the study is, of course, to disentangle the effect of this

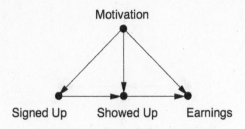

FIGURE 7.3. Causal diagram for the JTPA Study.

confounding factor and find out just how much the services themselves are helping.

Comparing Figure 7.2 to Figure 7.3, we can see that the frontdoor criterion would apply if there were no arrow from Motivation to Showed Up, the "shielding" I mentioned earlier. In many cases we could justify the absence of that arrow. For example, if the services were only offered by appointment and people only missed their appointments because of chance events unrelated to Motivation (a bus strike, a sprained ankle, etc.), then we could erase that arrow and use the front-door criterion.

Under the actual circumstances of the study, where the services were available all the time, such an argument is hard to make. However—and this is where things get really interesting—Glynn and Kashin tested out the front-door criterion anyway. We might think of this as a sensitivity test. If we suspect that the middle arrow is weak, then the bias introduced by treating it as absent may be very small. Judging from their results, that was the case.

By making certain reasonable assumptions, Glynn and Kashin derived inequalities saying whether the adjustment was likely to be too high or too low and by how much. Finally, they compared the front-door predictions and back-door predictions to the results from the randomized controlled experiment that was run at the same time. The results were impressive. The estimates from the backdoor criterion (controlling for known confounders like Age, Race, and Site) were wildly incorrect, differing from the experimental benchmarks by hundreds or thousands of dollars. This is exactly what you would expect to see if there is an unobserved confounder, such as Motivation. The back-door criterion cannot adjust for it.

On the other hand, the front-door estimates succeeded in removing almost all of the Motivation effect. For males, the front-door estimates were well within the experimental error of the randomized controlled trial, even with the small positive bias that Glynn and Kashin predicted. For females, the results were even better: The front-door estimates matched the experimental benchmark almost perfectly, with no apparent bias. Glynn and Kashin's work gives both empirical and methodological proof that as long as the effect of C on M (in Figure 7.2) is weak, front-door adjustment can give a reasonably good estimate of the effect of X on Y. It is much better than not controlling for C.

Glynn and Kashin's results show why the front-door adjustment is such a powerful tool: it allows us to control for confounders that we cannot observe (like Motivation), including those that we can't even name. RCTs are considered the "gold standard" of causal effect estimation for exactly the same reason. Because front-door estimates do the same thing, with the additional virtue of observing people's behavior in their own natural habitat instead of a laboratory, I would not be surprised if this method eventually becomes a useful alternative to randomized controlled trials.

THE *DO*-CALCULUS, OR MIND OVER MATTER

In both the front- and back-door adjustment formulas, the ultimate goal is to calculate the effect of an intervention, $P(Y \mid do(X))$, in terms of data such as $P(Y \mid X, A, B, Z, \dots)$ that do not involve a *do*-operator. If we are completely successful at eliminating the *do*'s, then we can use observational data to estimate the causal effect, allowing us to leap from rung one to rung two of the Ladder of Causation.

The fact that we were successful in these two cases (front- and back-door) immediately raises the question of whether there are other doors through which we can eliminate all the *do*'s. Thinking more generally, we can ask whether there is some way to decide in advance if a given causal model lends itself to such an elimination procedure. If so, we can apply the procedure and find ourselves in possession of the causal effect, without having to lift a finger to intervene. Otherwise, we would at least know that the assumptions imbedded in the model are not sufficient to uncover the causal effect from observational data, and no matter how clever we are, there is no escape from running an interventional experiment of some kind.

The prospect of making these determinations by purely mathematical means should dazzle anybody who understands the cost and difficulty of running randomized controlled trials, even when they are physically feasible and legally permissible. The idea dazzled me, too, in the early 1990s, not as an experimenter but as a computer scientist and part-time philosopher. Surely one of the most exhilarating experiences you can have as a scientist is to sit at your desk and realize that you can finally figure out what is possible or impossible in the real world—especially if the problem is important to society and has baffled those who have tried to solve it before you. I imagine this is how Hipparchus of Nicaea felt when he discovered he could figure out the height of a pyramid from its shadow on the ground, without actually climbing the pyramid. It was a clear victory of mind over matter.

Indeed, the approach I took was very much inspired by the ancient Greeks (including Hipparchus) and their invention of a formal logical system for geometry. At the center of the Greeks' logic, we find a set of axioms or self-evident truths, such as "Between any two points one can draw one and only one line." With the help of those axioms, the Greeks could construct complex statements, called theorems, whose truth is far from evident. Take, for instance, the statement that the sum of the angles in a triangle is 180 degrees (or two right angles), regardless of its size or shape. The truth of this statement is not self-evident by any means; yet the Pythagorean philosophers of the fifth century BC were able to prove its universal truth using those self-evident axioms as building blocks.

If you remember your high school geometry, even just the gist of it, you will recall that proofs of theorems invariably consist of auxiliary constructions: for example, drawing a line parallel to an edge of a triangle, marking certain angles as equal, drawing a circle with a given segment as its radius, and so on. These auxiliary constructions can be regarded as temporary mathematical sentences that make assertions (or claims) about properties of the figure drawn. Each new construction is licensed by the previous ones, as well as by the axioms of geometry and perhaps some already derived theorems. For example, drawing a line parallel to one edge of a triangle is licensed by Euclid's fifth axiom, that it is possible to draw one and only one parallel to a given line from a point outside that line. The act of drawing any of these auxiliary constructions is just a mechanical "symbol manipulation" operation; it takes the sentence previously written (or picture previously drawn) and rewrites it in a new format, whenever the rewriting is licensed by the axioms. Euclid's greatness was to identify a short list of five elementary axioms, from which all other true geometric statements can be derived.

Now let us return to our central question of when a model can replace an experiment, or when a "do" quantity can be reduced to a "see" quantity. Inspired by the ancient Greek geometers, we want to reduce the problem to symbol manipulation and in this way wrest causality from Mount Olympus and make it available to the average researcher.

First, let us rephrase the task of finding the effect of X on Y using the language of proofs, axioms, and auxiliary constructions, the language of Euclid and Pythagoras. We start with our target sentence, $P(Y \mid do(X))$. Our task will be complete if we can succeed in eliminating the do-operator from it, leaving only classical probability expressions, like $P(Y \mid X)$ or $P(Y \mid X, Z, W)$. We cannot, of course, manipulate our target expression at will; the operations must conform to what $do(X)$ means as a physical intervention. Thus, we must pass the expression through a sequence of legitimate manipulations, each licensed by the axioms and the assumptions of our model. The manipulations should preserve the meaning of the manipulated expression, only changing the format it is written in. An example of a "meaning preserving" transformation is the algebraic transformation that turns $y = ax + b$ into

$ax = y - b$. The relationship between x and y remains intact; only the format changes.

We are already familiar with some "legitimate" transformations on *do*-expressions. For example, Rule 1 says when we observe a variable W that is irrelevant to Y (possibly conditional on other variables Z), then the probability distribution of Y will not change. For example, in Chapter 3 we saw that the variable Fire is irrelevant to Alarm once we know the state of the mediator (Smoke). This assertion of irrelevance translates into a symbolic manipulation:

$$P(Y \mid do(X), Z, W) = P(Y \mid do(X), Z)$$

The stated equation holds provided that the variable set Z blocks all the paths from W to Y after we have deleted all the arrows leading into X. In the example of Fire $\rightarrow$ Smoke $\rightarrow$ Alarm, we have $W =$ Fire, $Z =$ Smoke, $Y =$ Alarm, and Z blocks all the paths from W to Y. (In this case we do not have a variable X.)

Another legitimate transformation is familiar to us from our back-door discussion. We know that if a set Z of variables blocks all back-door paths from X to Y, then conditional on Z, $do(X)$ is equivalent to *see*(X). We can, therefore, write

$$P(Y \mid do(X), Z) = P(Y \mid X, Z)$$

if Z satisfies the back-door criterion. We adopt this as Rule 2 of our axiomatic system. While this is perhaps less self-evident than Rule 1, in the simplest cases it is Hans Reichenbach's common-cause principle, amended so that we won't mistake colliders for confounders. In other words, we are saying that after we have controlled for a sufficient deconfounding set, any remaining correlation is a genuine causal effect.

Rule 3 is quite simple: it essentially says that we can remove $do(X)$ from $P(Y \mid do(X))$ in any case where there are no causal paths from X to Y. That is,

$$P(Y \mid do(X)) = P(Y)$$

if there is no path from X to Y with only forward-directed arrows. We can paraphrase this rule is follows: if we *do* something that does not affect Y, then the probability distribution of Y will not change. Aside from being just as self-evident as Euclid's axioms, Rules 1 to 3 can also be proven mathematically using our arrow-deleting definition of the *do*-operator and basic laws of probability.

Note that Rules 1 and 2 include conditional probabilities involving auxiliary variables Z other than X and Y. These variables can be thought of as a context in which the probability is being computed. Sometimes the presence of this context itself licenses the transformation. Rule 3 may also have auxiliary variables, but I omitted them for simplicity.

Note that each rule has a simple syntactic interpretation. Rule 1 permits the addition or deletion of observations. Rule 2 permits the replacement of an intervention with an observation, or vice versa. Rule 3 permits the deletion or addition of interventions. All of these permits are issued under appropriate conditions, which have to be verified in any particular case from the causal diagram.

We are ready now to demonstrate how Rules 1 to 3 allow us to transform one formula into another until, if we are smart, we obtain an expression to our liking. Although it's a bit elaborate, I think that nothing can substitute for actually showing you how the front-door formula is derived using a successive application of the rules of *do*-calculus (Figure 7.4). You do not need to follow all the steps, but I am showing you the derivation to give you the flavor of *do*-calculus. We begin the journey with a target expression $P(Y \mid do(X))$. We introduce auxiliary variables and transform the target expression into a *do*-free expression that coincides, of course, with the front-door adjustment formula. Each step of the argument gets its license from the causal diagram that relates X, Y, and the auxiliary variables or, in several cases, from subdiagrams that have had arrows erased to account for interventions. These licenses are displayed on the right-hand side.

I feel a special attachment to the *do*-calculus. With these three humble rules I was able to derive the front-door formula. This was the first causal effect estimated by means other than control

DO-CALCULUS AT WORK

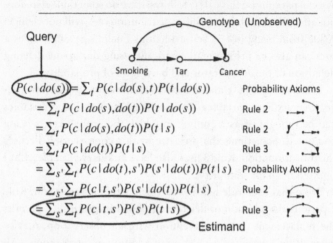

$$P(c \mid do(s)) = \sum_t P(c \mid do(s),t)P(t \mid do(s)) \quad \text{Probability Axioms}$$

$$= \sum_t P(c \mid do(s),do(t))P(t \mid do(s)) \quad \text{Rule 2}$$

$$= \sum_t P(c \mid do(s),do(t))P(t \mid s) \quad \text{Rule 2}$$

$$= \sum_t P(c \mid do(t))P(t \mid s) \quad \text{Rule 3}$$

$$= \sum_{s'} \sum_t P(c \mid do(t),s')P(s' \mid do(t))P(t \mid s) \quad \text{Probability Axioms}$$

$$= \sum_{s'} \sum_t P(c \mid t,s')P(s' \mid do(t))P(t \mid s) \quad \text{Rule 2}$$

$$= \sum_{s'} \sum_t P(c \mid t,s')P(s')P(t \mid s) \quad \text{Rule 3}$$

Estimand

FIGURE 7.4. Derivation of the front-door adjustment formula from the rules of *do*-calculus.

for confounders. I believed no one could do this without the *do*-calculus, so I presented it as a challenge in a statistics seminar at Berkeley in 1993 and even offered a $100 prize to anyone who could solve it. Paul Holland, who attended the seminar, wrote that he had assigned the problem as a class project and would send me the solution when ripe. (Colleagues tell me that he eventually presented a long solution at a conference in 1995, and I may owe him $100 if I could only find his proof.) Economists James Heckman and Rodrigo Pinto made the next attempt to prove the front-door formula using "standard tools" in 2015. They succeeded, albeit at the cost of eight pages of hard labor.

In a restaurant the evening before the talk, I had written the proof (very much like the one in Figure 7.4) on a napkin for David Freedman. He wrote me later to say that he had lost the napkin. He could not reconstruct the argument and asked if I had kept a copy. The next day, Jamie Robins wrote to me from Harvard, saying that he had heard about the "napkin problem" from Freedman, and he straightaway offered to fly to California to check the proof with me. I was thrilled to share with Robins the secrets of the *do*-calculus, and I believe that his trip to Los Angeles that year has been the key to his enthusiastic acceptance of causal diagrams. Through his and Sander Greenland's influence, diagrams have become a second language for epidemiologists. This explains why I am so fond of the "napkin problem."

The front-door adjustment formula was a delightful surprise and an indication that *do*-calculus had something important to offer. However, at this point I still wondered whether the three rules of *do*-calculus were enough. Was it possible that we had missed a fourth rule that would help us solve problems that are unsolvable with only three?

In 1994, when I first proposed the *do*-calculus, I selected these three rules because they were sufficient in any case that I knew of. I had no idea whether, like Ariadne's thread, they would always lead me out of the maze, or I would someday encounter a maze of such fiendish complexity that I could not escape. Of course, I hoped for the best. I conjectured that whenever a causal effect is estimable from data, a sequence of steps using these three rules would eliminate the *do*-operator. But I could not prove it.

This type of problem has many precedents in mathematics and logic. The property is usually called "completeness" in mathematical logic; an axiom system that is complete has the property that the axioms suffice to derive every true statement in that language. Some very good axiom systems are incomplete: for instance, Philip Dawid's axioms describing conditional independence in probability theory.

In this modern-day labyrinth tale, two groups of researchers played the role of Ariadne to my wandering Theseus: Yiming Huang and Marco Valtorta at the University of South Carolina and my own student, Ilya Shpitser, at the University of California, Los Angeles (UCLA). Both groups independently and simultaneously proved that Rules 1 to 3 suffice to get out of any *do*-labyrinth that has an exit. I am not sure whether the world was waiting breathlessly for their completeness result, because by then most researchers had become content with just using the front- and back-door criteria. Both teams were, however, recognized with best student paper awards at the Uncertainty in Artificial Intelligence conference in 2006.

I confess that I was the one waiting breathlessly for this result. It tells us that if we cannot find a way to estimate $P(Y \mid do(X))$ from Rules 1 to 3, then a solution does not exist. In that case, we know that there is no alternative to conducting a randomized controlled trial. It further tells us what additional assumptions or experiments might make the causal effect estimable.

Before declaring total victory, we should discuss one issue with the *do*-calculus. Like any other calculus, it enables the construction of a proof, but it does not help us find one. It is an excellent verifier of a solution but not such a good searcher for one. If you know the correct sequence of transformations, it is easy to demonstrate to others (who are familiar with Rules 1 to 3) that the *do*-operator can be eliminated. However, if you do not know the correct sequence, it is not easy to discover it, or even to determine whether one exists. Using the analogy with geometrical proofs, we need to decide which auxiliary construction to try next. A circle around point A? A line parallel to AB? The number of possibilities is limitless, and the axioms themselves provide no guidance about what to try next. My high school geometry teacher used to say that you need "mathematical eyeglasses."

In mathematical logic, this is known as the "decision problem." Many logical systems are plagued with intractable decision problems. For instance, given a pile of dominos of various sizes, we have no tractable way to decide if we can arrange them to fill a square of a given size. But once an arrangement is proposed, it takes no time at all to verify whether it constitutes a solution.

Luckily (again) for *do*-calculus, the decision problem turns out to be manageable. Ilya Shpitser, building on earlier work by one of my other students, Jin Tian, found an algorithm that decides if a solution exists in "polynomial time." This is a somewhat technical term, but continuing our analogy with solving a maze, it means that we have a much more efficient way out of the labyrinth than hunting at random through all possible paths.

Shpitser's algorithm for finding each and every causal effect does not eliminate the need for the *do*-calculus. In fact, we need it even more, and for several independent reasons. First, we need it in order to go beyond observational studies. Suppose that worst comes to worst, and our causal model does not permit estimation of the causal effect $P(Y \mid do(X))$ from observations alone. Perhaps we also cannot conduct a randomized experiment with random assignment of X. A clever researcher might ask whether we might estimate $P(Y \mid do(X))$ by randomizing some other variable, say Z, that is more accessible to control than X. For instance, if we want to assess the effect of cholesterol levels (X) on heart disease (Y), we might be able to manipulate the subjects' diet (Z) instead of exercising direct control over the cholesterol levels in their blood.

We then ask if we can find such a surrogate Z that will enable us to answer the causal question. In the world of *do*-calculus, the question is whether we can find a Z such that we can transform $P(Y \mid do(X))$ into an expression in which the variable Z, but not X, is subjected to a *do*-operator. This is a completely different problem not covered by Shpitser's algorithm. Luckily, it has a complete answer too, with a new algorithm discovered by Elias Bareinboim at my lab in 2012. Even more problems of this sort arise when we consider problems of transportability or external validity—assessing whether an experimental result will still be valid when transported to a different environment that may differ in several key ways from the one studied. This more ambitious set of questions touches on the heart of scientific methodology, for there is no science without generalization. Yet the question of generalization has been lingering for at least two centuries, without an iota of progress. The tools for producing a solution were simply not available. In 2015, Bareinboim and I presented a paper at the National Academy of Sciences that solves the problem, provided that you can express your assumptions about both environments with a causal diagram. In this case the rules of *do*-calculus provide a systematic method to determine whether causal effects found in the study environment can help us estimate effects in the intended target environment.

Yet another reason that the *do*-calculus remains important is transparency. As I wrote this chapter, Bareinboim (now a professor at Purdue) sent me a new puzzle: a diagram with just four observed variables, X, Y, Z, and W, and two unobservable variables,

U_1, U_2 (see Figure 7.5). He challenged me to figure out if the effect of X on Y was estimable. There was no way to block the back-door paths and no front-door condition. I tried all my favorite shortcuts and my otherwise trustworthy intuitive arguments, both pro and con, and I couldn't see how to do it. I could not find a way out of the maze. But as soon as Bareinboim whispered to me, "Try the *do*-calculus," the answer came shining through like a baby's smile. Every step was clear and meaningful. This is now the simplest model known to us in which the causal effect needs to be estimated by a method that goes beyond the front- and back-door adjustments.

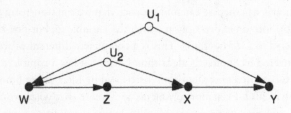

FIGURE 7.5. A new napkin problem?

In order not to leave the reader with the impression that the *do*-calculus is good only for theory and to serve as a recreational brainteaser, I will end this section with a practical problem recently brought up by two leading statisticians, Nanny Wermuth and David Cox. It demonstrates how a friendly whisper, "Try the *do*-calculus," can help expert statisticians solve difficult practical problems.

Around 2005, Wermuth and Cox became interested in a problem called "sequential decisions" or "time-varying treatments," which are common, for example, in the treatment of AIDS. Typically treatments are administered over a length of time, and in each time period physicians vary the strength and dosage of a follow-up treatment according to the patient's condition. The patient's condition, on the other hand, is influenced by the treatments taken in the

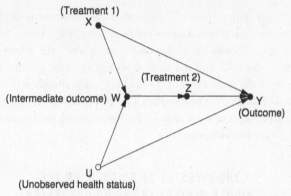

FIGURE 7.6. Wermuth and Cox's example of a sequential treatment.

past. We thus end up with a scenario like the one depicted in Figure 7.6, showing two time periods and two treatments. The first treatment is randomized (X), and the second (Z) is given in response to an observation (W) that depends on X. Given data collected under such a treatment regime, Cox and Wermuth's task was to predict the effect of X on the outcome Y, assuming that they were to keep Z constant through time, independent of the observation W.

Jamie Robins first brought the problem of time-varying treatments to my attention in 1994, and with the help of *do*-calculus, we were able to derive a general solution invoking a sequential version of the back-door adjustment formula. Wermuth and Cox, unaware of this method, called their problem "indirect confounding" and published three papers on its analysis (2008, 2014, and 2015). Unable to solve it in general, they resorted to a linear approximation, and even in the linear case they found it difficult to handle, because it is not solvable by standard regression methods.

Fortunately, when a muse whispered in my ear, "Try the *do*-calculus," I noticed that their problem can be solved in three lines of calculation. The logic goes as follows. Our target quantity is $P(Y \mid do(X), do(Z))$, while the data we have available to us are of the form $P(Y \mid do(X), Z, W)$ and $P(W \mid do(X))$. These reflect the fact that, in the study from which we have data, Z is not controlled externally but follows W through some (unknown) protocol. Thus, our task is to transform the target expression to another expression, reflecting the study conditions in which the *do*-operator applies only to X and not to Z. It so happens that a single application of the three rules of *do*-calculus can accomplish this. The moral of the story is nothing but a deep appreciation of the power of mathematics to solve difficult problems, which occasionally entail practical consequences.

THE TAPESTRY OF SCIENCE, OR THE HIDDEN PLAYERS IN THE *DO*-ORCHESTRA

I've already mentioned the role of some of my students in weaving this beautiful *do*-calculus tapestry. Like any tapestry, it gives a sense of completeness that may conceal how painstaking making it was and how many hands contributed to the process. In this case, it took more than twenty years and contributions from several students and colleagues.

The first was Thomas Verma, whom I met when he was a sixteen-year-old boy. His father brought him to my office one day and said, essentially, "Give him something to do." He was too talented for any of his high school math teachers to keep him interested. What he eventually accomplished was truly amazing. Verma finally proved what became known as the *d*-separation property (i.e., the fact that you can use the rules of path blocking to determine which independencies should hold in the data). Astonishingly, he told me that he proved the *d*-separation property thinking it was a homework problem, not an unsolved conjecture! Sometimes it pays to be young and naive. You can still see his legacy in Rule 1 of the *do*-calculus and in any imprint that path blocking leaves on rung one of the Ladder of Causation.

The power of Verma's proof would have remained only partially appreciated without a complementary result to show that it cannot be improved. That is, no other independencies are implied by a causal diagram except those revealed through path blocking. This step was completed by another student, Dan Geiger. He had switched to my research lab from another group at UCLA, after I promised to give him an "instant PhD" if he could prove two theorems. He did, and I did! He is now Dean of computer science at the Technion in Israel, my alma mater.

But Dan was not the only student I raided from another department. One day in 1997, as I was getting dressed in the locker room of the UCLA pool, I struck up a conversation with a Chinese fellow next to me. He was a PhD student in physics, and, as was my usual habit at the time, I tried to convince him to switch over to artificial intelligence, where the action was. He was not completely convinced, but the very next day I received an email from a friend of his, Jin Tian, saying that he would like to switch from physics to computer science and did I have a challenging summer project for him? Two days later, he was working in my lab.

Four years later, in April 2001, he stunned the world with a simple graphical criterion that generalizes the front door, the back door, and all doors we could think of at the time. I recall presenting Tian's criterion at a Santa Fe conference. One by one, leaders in the research community stared at my poster and shook their heads in disbelief. How could such a simple criterion work for all diagrams?

Tian (now a professor at Iowa State University) came to our lab with a style of thinking that was foreign to us then, in the 1990s. Our conversations were always loaded with wild metaphors and half-baked conjectures. But Tian would never utter a word unless it was rigorous, proven, and baked five times over. The mixture of the two styles proved its merit. Tian's method, called c-decomposition, enabled Ilya Shpitser to develop his complete algorithm for the do-calculus. The moral: never underestimate the power of a locker-room conversation!

Ilya Shpitser came in at the end of the ten-year battle to understand interventions. He arrived during a very difficult period, when I had to take time off to set up a foundation in honor of my son, Daniel, a victim of anti-Western terrorism. I have always expected my students to be self-reliant, but for my students at that time, this expectation was pushed to the extreme. They gave me the best of all possible gifts by putting the final but crucial touches on the tapestry of do-calculus, which I could not have done myself. In fact, I tried to discourage Ilya from trying to prove the completeness of *do*-calculus. Completeness proofs are notoriously difficult and are best avoided by any student who aims to finish his PhD on time. Luckily, Ilya did it behind my back.

Colleagues, too, exert a profound effect on your thinking at crucial moments. Peter Spirtes, a professor of philosophy at Carnegie-Mellon, preceded me in the network approach to causality, and his influence was pivotal. At a lecture of his in Uppsala, Sweden, I first learned that performing interventions could be thought of as deleting arrows from a causal diagram. Until then I had been laboring under the same burden as generations of statisticians, trying to think of causality in terms of only one diagram representing one static probability distribution.

The idea of arrow deletion was not entirely Spirtes's, either. In 1960, two Swedish economists, Robert Strotz and Herman Wold, proposed essentially the same idea. In the world of economics at the time, diagrams were never used; instead, economists relied on structural equation models, which are Sewall Wright's equations without the diagrams. Arrow deletion in a path diagram corresponds to deleting an equation from a structural equation model. So, in a rough sense, Strotz and Wold had the idea first, unless we want to go even further back in history: they were preceded by Trygve Haavelmo (a Norwegian economist and Nobel laureate), who in 1943 advocated equation modification to represent interventions.

Nevertheless, Spirtes's translation of equation deletion into the world of causal diagrams unleashed an avalanche of new insights and new results. The back-door criterion was one of the first beneficiaries of the translation, while the *do*-calculus came second. The avalanche, however, is not yet over. Advances in such areas as counterfactuals, generalizability, missing data, and machine learning are still coming up.

If I were less modest, I would close here with Isaac Newton's famous saying about "standing on the shoulders of giants." But given who I am, I am tempted to quote from the Mishnah instead: "Harbe lamadeti mirabotai um'haverai yoter mehem, umitalmidai yoter mikulam"—that is, "I have learned much from my teachers, and more so from my colleagues, and most of all from my students" (Taanit 7a). The *do*-operator and *do*-calculus would not exist as they do today without the contributions of Verma, Geiger, Tian, and Shpitser, among others.

THE CURIOUS CASE(S) OF DR. SNOW

In 1853 and 1854, England was in the grips of a cholera epidemic. In that era, cholera was as terrifying as Ebola is today; a healthy person who drinks cholera-tainted water can die within twenty-four hours. We know today that cholera is caused by a bacterium that attacks the intestines. It spreads through the "rice water" diarrhea of its victims, who excrete this diarrhea in copious amounts before dying.

But in 1853, disease-causing germs had never yet been seen under a microscope for any illness, let alone cholera. The prevailing wisdom held that a "miasma" of unhealthy air caused cholera, a theory seemingly supported by the fact that the epidemic hit harder in the poorer sections of London, where sanitation was worse.

Dr. John Snow, a physician who had taken care of cholera victims for more than twenty years, was always skeptical of the miasma theory. He argued, sensibly, that since the symptoms manifested themselves in the intestinal tract, the body must first come into contact with the pathogen there. But because he couldn't see the culprit, he had no way to prove this—until the epidemic of 1854.

The John Snow story has two chapters, one much more famous than the other. In what we could call the "Hollywood" version, he painstakingly goes from house to house, recording where victims of cholera died, and notices a cluster of dozens of victims near a pump in Broad Street. Talking with people who live in the area, he discovers that almost all the victims had drawn their water from that particular pump. He even learns of a fatal case that occurred far away, in Hampstead, to a woman who liked the taste of the water from the Broad Street pump. She and her niece drank the water from Broad Street and died, while no one else in her area even got sick. Putting all this evidence together, Snow asks the local authorities to remove the pump handle, and on September 8 they agree. As

Snow's biographer wrote, "The pump-handle was removed, and the plague was stayed."

All of this makes a wonderful story. Nowadays a John Snow Society even reenacts the removal of the famous pump handle every year. Yet, in truth, the removal of the pump handle hardly made a dent in the citywide cholera epidemic, which went on to claim nearly 3,000 lives.

In the non-Hollywood chapter of the story, we again see Dr. Snow walking the streets of London, but this time his real object is to find out where Londoners get their water. There were two main water companies at the time: the Southwark and Vauxhall Company and the Lambeth Company. The key difference between the two, as Snow knew, was that the former drew its water from the area of the London Bridge, which was downstream from London's sewers. The latter had moved its water intake several years earlier so that it would be upstream of the sewers. Thus, Southwark customers were getting water tainted by the excrement of cholera victims. Lambeth customers, on the other hand, were getting uncontaminated water. (None of this has anything to do with the contaminated Broad Street water, which came from a well.)

The death statistics bore out Snow's grim hypothesis. Districts supplied by the Southwark and Vauxhall Company were especially hard-hit by cholera and had a death rate eight times higher. Even so, the evidence was merely circumstantial. A proponent of the miasma theory could argue that the miasma was strongest in those districts, and there would be no way to disprove it. In terms of a causal diagram, we have the situation diagrammed in Figure 7.7. We have no way to observe the confounder Miasma (or other confounders like Poverty), so we can't control for it using back-door adjustment.

Here Snow had his most brilliant idea. He noticed that in those districts served by both companies, the death rate was still much higher in the households that received Southwark water. Yet these households did not differ in terms of miasma or poverty. "The mixing of the supply is of the most intimate kind," Snow wrote. "The pipes of each Company go down all the streets, and into nearly all the courts and alleys. . . . Each company supplies both rich and

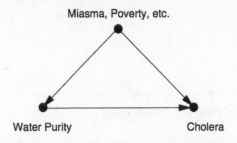

FIGURE 7.7. Causal diagram for cholera (before discovery of the cholera bacillus).

poor, both large houses and small; there is no difference either in the condition or occupation of the persons receiving the water of the different Companies." Even though the notion of an RCT was still in the future, it was very much as if the water companies had conducted a randomized experiment on Londoners. In fact, Snow even notes this: "No experiment could have been devised which would more thoroughly test the effect of water supply on the progress of cholera than this, which circumstances placed ready made before the observer. The experiment, too, was on the grandest scale. No fewer than three hundred thousand people of both sexes, of every age and occupation, and of every rank and station, from gentlefolks down to the very poor, were divided into two groups without their choice, and in most cases, without their knowledge." One group had received pure water; the other had received water tainted with sewage.

Snow's observations introduced a new variable into the causal diagram, which now looks like Figure 7.8. Snow's painstaking detective work had showed two important things: (1) there is no arrow between Miasma and Water Company (the two are independent), and (2) there is an arrow between Water Company and Water Purity. Left unstated by Snow, but equally important, is a third assumption: (3) the absence of a direct arrow from Water Company to Cholera, which is fairly obvious to us today because

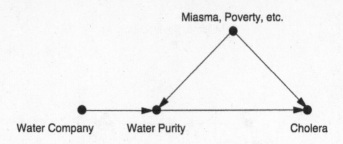

FIGURE 7.8. Diagram for cholera after introduction of an instrumental variable.

we know the water companies were not delivering cholera to their customers by some alternate route.

A variable that satisfies these three properties is today called an instrumental variable. Clearly Snow thought of this variable as similar to a coin flip, which simulates a variable with no incoming arrows. Because there are no confounders of the relation between Water Company and Cholera, any observed association must be causal. Likewise, since the effect of Water Company on Cholera must go through Water Purity, we conclude (as did Snow) that the observed association between Water Purity and Cholera must also be causal. Snow stated his conclusion in no uncertain terms: if the Southwark and Vauxhall Company had moved its intake point upstream, more than 1,000 lives would have been saved.

Few people took note of Snow's conclusion at the time. He printed a pamphlet of the results at his own expense, and it sold a grand total of fifty-six copies. Nowadays, epidemiologists view his pamphlet as the seminal document of their discipline. It showed that through "shoe-leather research" (a phrase I have borrowed from David Freedman) and causal reasoning, you can track down a killer.

Although the miasma theory has by now been discredited, poverty was undoubtedly a confounder, as was location. But even without measuring these (because Snow's door-to-door detective work only went so far), we can still use instrumental variables to determine how many lives would have been saved by purifying the water supply.

Here's how the trick works. For simplicity we'll go back to the names Z, X, Y, and U for our variables and redraw Figure 7.8 as seen in Figure 7.9. I have included path coefficients (a, b, c, d) to represent the strength of the causal effects. This means we are assuming that the variables are numerical and the functions relating them are linear. Remember that the path coefficient a means that an intervention to increase Z by one standard unit will cause X to increase by a standard units. (I will omit the technical details of what the "standard units" are.)

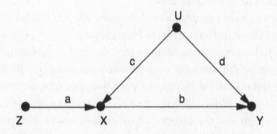

FIGURE 7.9. General setup for instrumental variables.

Because Z and X are unconfounded, the causal effect of Z on X (that is, a) can be estimated from the slope r_{XZ} of the regression line of X on Z. Likewise, the variables Z and Y are unconfounded, because the path $Z \to X \leftarrow U \to Y$ is blocked by the collider at X. So the slope of the regression line of Z on Y (r_{ZY}) will equal the causal effect on the direct path $Z \to X \to Y$, which is the product of the path coefficients: ab. Thus we have two equations: $ab = r_{ZY}$ and $a = r_{ZX}$. If we divide the first equation by the second, we get the causal effect of X on Y: $b = r_{ZY}/r_{ZX}$.

In this way, instrumental variables allow us to perform the same kind of magic trick that we did with front-door adjustment: we have found the effect of X on Y even without being able to control for, or collect data on, the confounder, U. We can therefore provide decision makers with a conclusive argument that they should move their water supply—even if those decision makers still believe in the miasma theory. Also notice that we have gotten information on the second rung of the Ladder of Causation (b) from information about the first rung (the correlations, r_{ZY} and r_{ZX}). We were able to do this because the assumptions embodied in the path diagram are causal in nature, especially the crucial assumption that there is no arrow between U and Z. If the causal diagram were different—for example, if Z were a confounder of X and Y—the formula $b = r_{ZY}/r_{ZX}$ would not correctly estimate the causal effect of X on Y. In fact, these two models cannot be told apart by any statistical method, regardless of how big the data.

Instrumental variables were known before the Causal Revolution, but causal diagrams have brought new clarity to how they work. Indeed, Snow was using an instrumental variable implicitly, although he did not have a quantitative formula. Sewall Wright certainly understood this use of path diagrams; the formula $b = r_{ZY}/r_{ZX}$ can be derived directly from his method of path coefficients. And it seems that the first person other than Sewall Wright to use instrumental variables in a deliberate way was . . . Sewall Wright's father, Philip!

Recall that Philip Wright was an economist who worked at what later became the Brookings Institution. He was interested in predicting how the output of a commodity would change if a tariff were imposed, which would raise the price and therefore, in theory, encourage production. In economic terms, he wanted to know the elasticity of supply.

In 1928 Wright wrote a long monograph dedicated to computing the elasticity of supply for flaxseed oil. In a remarkable appendix, he analyzed the problem using a path diagram. This was a brave thing to do: remember that no economist had ever seen or heard of such a thing before. (In fact, he hedged his bets and verified his calculations using more traditional methods.)

Figure 7.10 shows a somewhat simplified version of Wright's diagram. Unlike most diagrams in this book, this one has "two-way"

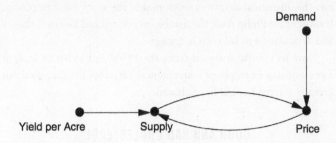

FIGURE 7.10. Simplified version of Wright's supply-price causal diagram.

arrows, but I would ask the reader not to lose too much sleep over it. With some mathematical trickery we could equally well replace the Demand → Price → Supply chain with a single arrow Demand → Supply, and the figure would then look like Figure 7.9 (though it would be less acceptable to economists). The important point to note is that Philip Wright deliberately introduced the variable Yield per Acre (of flaxseed) as an instrument that directly affects supply but has no correlation to demand. He then used an analysis like the one I just gave to deduce both the effect of supply on price and the effect of price on supply.

Historians quarrel about who invented instrumental variables, a method that became extremely popular in modern econometrics. There is no question in my mind that Philip Wright borrowed the idea of path coefficients from his son. No economist had ever before insisted on the distinction between causal coefficients and regression coefficients; they were all in the Karl Pearson–Henry Niles camp that causation is nothing more than a limiting case of correlation. Also, no one before Sewall Wright had ever given a recipe for computing regression coefficients in terms of path coefficients, then reversing the process to get the causal coefficients from the regression. This was Sewall's exclusive invention.

Naturally, some economic historians have suggested that Sewall wrote the whole mathematical appendix himself. However, stylometric analysis has shown that Philip was indeed the author. To me, this historical detective work makes the story more beautiful. It shows that Philip took the trouble to understand his son's theory and articulate it in his own language.

Now let's move forward from the 1850s and 1920s to look at a present-day example of instrumental variables in action, one of literally dozens I could have chosen.

GOOD AND BAD CHOLESTEROL

Do you remember when your family doctor first started talking to you about "good" and "bad" cholesterol? It may have happened in the 1990s, when drugs that lowered blood levels of "bad" cholesterol, low-density lipoprotein (LDL), first came on the market. These drugs, called statins, have turned into multibillion-dollar revenue generators for pharmaceutical companies.

An early cholesterol-modifying drug subjected to a randomized controlled trial was cholestyramine. The Coronary Primary Prevention Trial, begun in 1973 and concluded in 1984, showed a 12.6 percent reduction in cholesterol among men given the drug cholestyramine and a 19 percent reduction in the risk of heart attack.

Because this was a randomized controlled trial, you might think we wouldn't need any of the methods in this chapter, because they are specifically designed to replace RCTs in situations where you only have observational data. But that is not true. This trial, like many RCTs, faced the problem of noncompliance, when subjects randomized to receive a drug don't actually take it. This will reduce the apparent effectiveness of the drug, so we may want to adjust the results to account for the noncompliers. But as always, confounding rears its ugly head. If the noncompliers are different from the compliers in some relevant way (maybe they are sicker to start with?), we cannot predict how they would have responded had they adhered to instructions.

In this situation, we have a causal diagram that looks like Figure 7.11. The variable Assigned (Z) will take the value 1 if the patient is randomly assigned to receive the drug and 0 if he is randomly assigned a placebo. The variable Received will be 1 if the patient actually took the drug and 0 otherwise. For convenience, we'll also

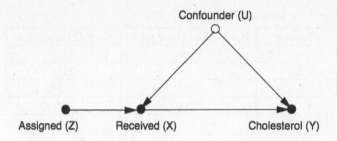

FIGURE 7.11. Causal diagram for an RCT with noncompliance.

use a binary definition for Cholesterol, recording an outcome of 1 if the cholesterol levels were reduced by a certain fixed amount.

Notice that in this case our variables are binary, not numerical. This means right away that we cannot use a linear model, and therefore we cannot apply the instrumental variables formula that we derived earlier. However, in such cases we can often replace the linearity assumption with a weaker condition called monotonicity, which I'll explain below.

But before we do that, let's make sure our other necessary assumptions for instrumental variables are valid. First, is the instrumental variable Z independent of the confounder? The randomization of Z ensures that the answer is yes. (As we saw in Chapter 4, randomization is a great way to make sure that a variable isn't affected by any confounders.) Is there any direct path from Z to Y? Common sense says that there is no way that receiving a particular random number (Z) would affect cholesterol (Y), so the answer is no. Finally, is there a strong association between Z and X? This time the data themselves should be consulted, and the answer again is yes. We must always ask the above three questions before we apply instrumental variables. Here the answers are obvious, but we should not be blind to the fact that we are using causal intuition to answer them, intuition that is captured, preserved, and elucidated in the diagram.

Table 7.1 shows the observed frequencies of outcomes X and Y. For example, 91.9 percent of the people who were not assigned the drug had the outcome $X = 0$ (didn't take drug) and $Y = 0$ (no reduction in cholesterol). This makes sense. The other 8.1 percent

TABLE 7.1. Data from cholestyramine trial.

Outcome	Not Assigned Drug ($Z = 0$)	Assigned Drug ($Z = 1$)
$X = 0, Y = 0$	0.919	0.315
$X = 1, Y = 0$	0.000	0.139
$X = 0, Y = 1$	0.081	0.073
$X = 1, Y = 1$	0.000	0.473

had the outcome $X = 0$ (didn't take drug) and $Y = 1$ (did have a reduction in cholesterol). Evidently they improved for other reasons than taking the drug. Notice also that there are two zeros in the table: there was nobody who was not assigned the drug ($Z = 0$) but nevertheless procured some ($X = 1$). In a well-run randomized study, especially in the medical field where the physicians have exclusive access to the experimental drug, this will typically be true. The assumption that there are no individuals with $Z = 0$ and $X = 1$ is called monotonicity.

Now let's see how we can estimate the effect of the treatment. First let's take the worst-case scenario: none of the noncompliers would have improved if they had complied with treatment. In that case, the only people who would have taken the drug and improved would be the 47.3 percent who actually did comply and improve. But we need to correct this estimate for the placebo effect, which is in the third row of the table. Out of the people who were assigned the placebo and took the placebo, 8.1 percent improved. So the net improvement above and beyond the placebo effect is 47.3 percent minus 8.1 percent, or 39.2 percent.

What about the best-case scenario, in which all the noncompliers would have improved if they had complied? In this case we add the noncompliers' 31.5 percent plus 7.3 percent to the 39.2 percent baseline we just computed, for a total of 78.0 percent.

Thus, even in the worst-case scenario, where the confounding goes completely against the drug, we can still say that the drug improves cholesterol for 39 percent of the population. In the best-case scenario, where the confounding works completely in favor of the drug, 78 percent of the population would see an improvement.

Even though the bounds are quite far apart, due to the large number of noncompliers, the researcher can categorically state that the drug is effective for its intended purpose.

This strategy of taking the worst case and then the best case will usually give us a range of estimates. Obviously it would be nice to have a point estimate, as we did in the linear case. There are ways to narrow the range if necessary, and in some cases it is even possible to get point estimates. For example, if you are interested only in the complying subpopulation (those people who will take X if and only if assigned), you can derive a point estimate known as the Local Average Treatment Effect (LATE). In any event, I hope this example shows that our hands are not tied when we leave the world of linear models.

Instrumental variable methods have continued to develop since 1984, and one particular version has become extremely popular: Mendelian randomization. Here's an example. Although the effect of LDL, or "bad," cholesterol is now settled, there is still considerable uncertainty about high-density lipoprotein (HDL), or "good," cholesterol. Early observational studies, such as the Framingham Heart Study in the late 1970s, suggested that HDL had a protective effect against heart attacks. But high HDL often goes hand in hand with low LDL, so how can we tell which lipid is the true causal factor?

To answer this question, suppose we knew of a gene that caused people to have higher HDL levels, with no effect on LDL. Then we could set up the causal diagram in Figure 7.12, where I have used Lifestyle as a possible confounder. Remember that it is always advantageous, as in Snow's example, to use an instrumental variable that is randomized. If it's randomized, no causal arrows point toward it. For this reason, a gene is a perfect instrumental variable. Our genes are randomized at the time of conception, so it's just as if Gregor Mendel himself had reached down from heaven and assigned some people a high-risk gene and others a low-risk gene. That's the reason for the term "Mendelian randomization."

Could there be an arrow going the other way, from HDL Gene to Lifestyle? Here we again need to do "shoe-leather work" and think causally. The HDL gene could only affect people's lifestyle

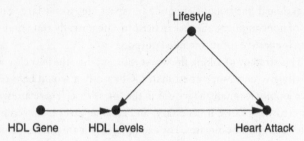

Figure 7.12. Causal diagram for Mendelian randomization example.

if they knew which version they had, the high-HDL version or the low-HDL one. But until 2008 no such genes were known, and even today, people do not routinely have access to this information. So it's highly likely that no such arrow exists.

At least two studies have taken this Mendelian randomization approach to the cholesterol question. In 2012, a giant collaborative study led by Sekar Kathiresan of Massachusetts General Hospital showed that there was no observable benefit from higher HDL levels. On the other hand, the researchers found that LDL has a very large effect on heart attack risk. According to their figures, decreasing your LDL count by 34 mg/dl would reduce your chances of a heart attack by about 50 percent. So lowering your "bad" cholesterol levels, whether by diet or exercise or statins, seems to be a smart idea. On the other hand, increasing your "good" cholesterol levels, despite what some fish-oil salesmen might tell you, does not seem likely to change your heart attack risk at all.

As always, there is a caveat. The second study, published in the same year, pointed out that people with the lower-risk variant of the LDL gene have had lower cholesterol levels for their entire lives. Mendelian randomization tells us that decreasing your LDL by thirty-four units over your entire lifetime will decrease your heart attack risk by 50 percent. But statins can't lower your LDL cholesterol over your entire lifetime—only from the day you start taking the drug. If you're sixty years old, your arteries have already sustained sixty years of damage. For that reason it's very likely that Mendelian randomization overestimates the true benefits of statins.

On the other hand, starting to reduce your cholesterol when you're young—whether through diet or exercise or even statins—will have big effects later.

From the point of view of causal analysis, this teaches us a good lesson: in any study of interventions, we need to ask whether the variable we're actually manipulating (lifetime LDL levels) is the same as the variable we think we are manipulating (current LDL levels). This is part of the "skillful interrogation of nature."

To sum up, instrumental variables are an important tool in that they help us uncover causal information that goes beyond the *do*-calculus. The latter insists on point estimates rather than inequalities and would give up on cases like Figure 7.12, in which all we can get are inequalities. On the other hand, it's also important to realize that the *do*-calculus is vastly more flexible than instrumental variables. In *do*-calculus we make no assumptions whatsoever regarding the nature of the functions in the causal model. But if we can justify an assumption like monotonicity or linearity on scientific grounds, then a more special-purpose tool like instrumental variables is worth considering.

Instrumental variable methods can be extended beyond simple four-variable models like Figure 7.9 (or 7.11 or 7.12), but it is not possible to go very far without guidance from causal diagrams. For example, in some cases an imperfect instrument (e.g., one that is not independent of the confounder) can be used after conditioning on a cleverly chosen set of auxiliary variables, which block the paths between the instrument and the confounder. My former student Carlos Brito, now a professor at the Federal University of Ceara, Brazil, fully developed this idea of turning noninstrumental variables into instrumental variables.

In addition, Brito studied many cases where a set of variables can be used successfully as an instrument. Although the identification of instrumental sets goes beyond *do*-calculus, it still uses the tools of causal diagrams. For researchers who understand this language, the possible research designs are rich and varied; they need not feel constrained to use only the four-variable model shown in Figures 7.9, 7.11, and 7.12. The possibilities are limited only by our imaginations.

"And sorry I could not travel both
And be one traveler, long I stood..."

Robert Frost's famous lines show a poet's acute insight into counterfactuals. We cannot travel both roads, and yet our brains are
equipped to judge what would have happened if we had taken the
other path. Armed with this judgment, Frost ends the poem pleased
with his choice, realizing that it "made all the difference." (*Source:*
Drawing by Maayan Harel.)

8

COUNTERFACTUALS:
MINING WORLDS THAT COULD HAVE BEEN

*Had Cleopatra's nose been shorter, the whole face
of the world would have changed.*

—BLAISE PASCAL (1669)

A S we prepare to move up to the top rung of the Ladder of
Causation, let's recapitulate what we have learned from the
second rung. We have seen several ways to ascertain the effect of
an intervention in various settings and under a variety of conditions. In Chapter 4, we discussed randomized controlled trials, the
widely cited "gold standard" for medical trials. We have also seen
methods that are suitable for observational studies, in which the
treatment and control groups are not assigned at random. If we can
measure variables that block all the back-door paths, we can use
the back-door adjustment formula to obtain the needed effect. If
we can find a front-door path that is "shielded" from confounders,
we can use front-door adjustment. If we are willing to live with the
assumption of linearity or monotonicity, we can use instrumental
variables (assuming that an appropriate variable can be found in
the diagram or created by an experiment). And truly adventurous

researchers can plot other routes to the top of Mount Intervention using the *do*-calculus or its algorithmic version.

In all these endeavors, we have dealt with effects on a population or a typical individual selected from a study population (the average causal effect). But so far we are missing the ability to talk about personalized causation at the level of particular events or individuals. It's one thing to say, "Smoking causes cancer," but another to say that my uncle Joe, who smoked a pack a day for thirty years, would have been alive had he not smoked. The difference is both obvious and profound: none of the people who, like Uncle Joe, smoked for thirty years and died can ever be observed in the alternate world where they did not smoke for thirty years.

Responsibility and blame, regret and credit: these concepts are the currency of a causal mind. To make any sense of them, we must be able to compare what did happen with what would have happened under some alternative hypothesis. As argued in Chapter 1, our ability to conceive of alternative, nonexistent worlds separated us from our protohuman ancestors and indeed from any other creature on the planet. Every other creature can see what is. Our gift, which may sometimes be a curse, is that we can see what might have been.

This chapter shows how to use observational and experimental data to extract information about counterfactual scenarios. It explains how to represent individual-level causes in the context of a causal diagram, a task that will force us to explain some nuts and bolts of causal diagrams that we have not talked about yet. I also discuss a highly related concept called "potential outcomes," or the Neyman-Rubin causal model, initially proposed in the 1920s by Jerzy Neyman, a Polish statistician who later became a professor at Berkeley. But only after Donald Rubin began writing about potential outcomes in the mid-1970s did this approach to causal analysis really begin to flourish.

I will show how counterfactuals emerge naturally in the framework developed over the last several chapters—Sewall Wright's path diagrams and their extension to structural causal models (SCMs). We got a good taste of this in Chapter 1, in the example of the firing squad, which showed how to answer counterfactual questions such as "Would the prisoner be alive if rifleman *A* had not shot?" I will compare how counterfactuals are defined in the Neyman-Rubin paradigm and in SCMs, where they enjoy the benefit of causal diagrams. Rubin has steadfastly maintained over the years that diagrams serve no useful purpose. So we will examine how students of the Rubin causal model must navigate causal problems blindfolded, lacking a facility to represent causal knowledge or to derive its testable implications.

Finally, we will look at two applications where counterfactual reasoning is essential. For decades or even centuries, lawyers have used a relatively straightforward test of a defendant's culpability called "but-for causation": the injury would not have occurred *but for* the defendant's action. We will see how the language of counterfactuals can capture this elusive notion and how to estimate the probability that a defendant is culpable.

Next, I will discuss the application of counterfactuals to climate change. Until recently, climate scientists have found it very difficult and awkward to answer questions like "Did global warming cause this storm [or this heat wave, or this drought]?" The conventional answer has been that individual weather events cannot be attributed to global climate change. Yet this answer seems rather evasive and may even contribute to public indifference about climate change.

Counterfactual analysis allows climate scientists to make much more precise and definite statements than before. It requires, however, a slight addition to our everyday vocabulary. It will be helpful to distinguish three different kinds of causation: *necessary* causation, *sufficient* causation, and *necessary-and-sufficient* causation. (Necessary causation is the same as but-for causation.) Using these words, a climate scientist can say, "There is a 90 percent probability that man-made climate change was a necessary cause of this heat wave," or "There is an 80 percent probability that climate change will be sufficient to produce a heat wave this strong at least once every 50 years." The first sentence has to do with attribution: Who was responsible for the unusual heat? The second has to do with policy. It says that we had better prepare for such heat waves because they are likely to occur sooner or later. Either of these statements is more informative than shrugging our shoulders and saying nothing about the causes of individual weather events.

FROM THUCYDIDES AND ABRAHAM TO HUME AND LEWIS

Given that counterfactual reasoning is part of the mental apparatus that makes us human, it is not surprising that we can find counterfactual statements as far back as we want to go in human history. For example, in Thucydides's *History of the Peloponnesian War*, the ancient Greek historian, often described as the pioneer of a "scientific" approach to history, describes a tsunami that occurred in 426 BC:

> About the same time that these earthquakes were so common, the sea at Orobiae, in Euboea, retiring from the then line of coast, returned in a huge wave and invaded a great part of the town, and retreated leaving some of it still under water; so that what was once land is now sea; such of the inhabitants perishing as could not run up to the higher ground in time. . . . The cause, in my opinion, of this phenomenon must be sought in the earthquake. At the point where its shock has been the most violent the sea is driven back, and suddenly recoiling with redoubled force, causes the inundation. Without an earthquake I do not see how such an accident could happen.

This is a truly remarkable passage when you consider the era in which it was written. First, the precision of Thucydides's observations would do credit to any modern scientist, and all the more so because he was working in an era when there were no satellites, no video cameras, no 24/7 news organizations broadcasting images of the disaster as it unfolded. Second, he was writing at a time in human history when natural disasters were ordinarily ascribed to the will of the gods. His predecessor Homer or his contemporary Herodotus would undoubtedly have attributed this event to the wrath of Poseidon or some other deity. Yet Thucydides proposes a causal model without any supernatural processes: the earthquake drives back the sea, which recoils and inundates the land. The last sentence of the quote is especially interesting because it expresses the notion of necessary or but-for causation: but for the earthquake, the tsunami could not have occurred. This counterfactual judgment promotes the earthquake from a mere antecedent of the tsunami to an actual cause.

Another fascinating and revealing instance of counterfactual reasoning occurs in the book of Genesis in the Bible. Abraham is talking with God about the latter's intention to destroy the cities of Sodom and Gomorrah as retribution for their evil ways.

> And Abraham drew near, and said, Wilt thou really destroy the righteous with the wicked?
>
> Suppose there be fifty righteous within the city: wilt thou also destroy and not spare the place for the sake of the fifty righteous that are therein? ...
>
> And the Lord said, If I find in Sodom fifty righteous within the city, then I will spare all the place for their sakes.

But the story does not end there. Abraham is not satisfied and asks the Lord, what if there are only forty-five righteous men? Or forty? Or thirty? Or twenty? Or even ten? Each time he receives an affirmative answer, and God ultimately assures him that he will spare Sodom even for the sake of ten righteous men, if he can find that many.

What is Abraham trying to accomplish with this haggling and bargaining? Surely he does not doubt God's ability to count. And of course, Abraham knows that God knows how many righteous men live in Sodom. He is, after all, omniscient.

Knowing Abraham's obedience and devotion, it is hard to believe that the questions are meant to convince the Lord to change his mind. Instead, they are meant for Abraham's own comprehension. He is reasoning just as a modern scientist would, trying to understand the laws that govern collective punishment. What level of wickedness is sufficient to warrant destruction? Would thirty righteous men be enough to save a city? Twenty? We do not have a complete causal model without such information. A modern scientist might call it a dose-response curve or a threshold effect.

While Thucydides and Abraham probed counterfactuals through individual cases, the Greek philosopher Aristotle investigated more generic aspects of causation. In his typically systematic style, Aristotle set up a whole taxonomy of causation, including "material causes," "formal causes," "efficient causes," and "final causes." For example, the material cause of the shape of a statue is the bronze from which it is cast and its properties; we could not make the same statue out of Silly Putty. However, Aristotle nowhere makes a statement about causation as a counterfactual, so his ingenious classification lacks the simple clarity of Thucydides's account of the cause of the tsunami.

To find a philosopher who placed counterfactuals at the heart of causality, we have to move ahead to David Hume, the Scottish philosopher and contemporary of Thomas Bayes. Hume rejected Aristotle's classification scheme and insisted on a single definition of causation. But he found this definition quite elusive and was in fact torn between two different definitions. Later these would turn into two incompatible ideologies, which ironically could both cite Hume as their source!

In his *Treatise of Human Nature* (Figure 8.1), Hume denies that any two objects have innate qualities or "powers" that make one a cause and the other an effect. In his view, the cause-effect relationship is entirely a product of our own memory and experience. "Thus we remember to have seen that species of object we call *flame*, and to have felt that species of sensation we call *heat*," he writes. "We likewise call to mind their constant conjunction in all past instances. Without any further ceremony, we call the one *cause* and the other *effect*, and infer the existence of the one from the other." This is now known as the "regularity" definition of causation.

The passage is breathtaking in its chutzpah. Hume is cutting off the second and third rungs of the Ladder of Causation and saying that the first rung, observation, is all that we need. Once we observe flame and heat together a sufficient number of times (and note that flame has temporal precedence), we agree to call flame the cause of heat. Like most twentieth-century statisticians, Hume in 1739 seems happy to consider causation as merely a species of correlation.

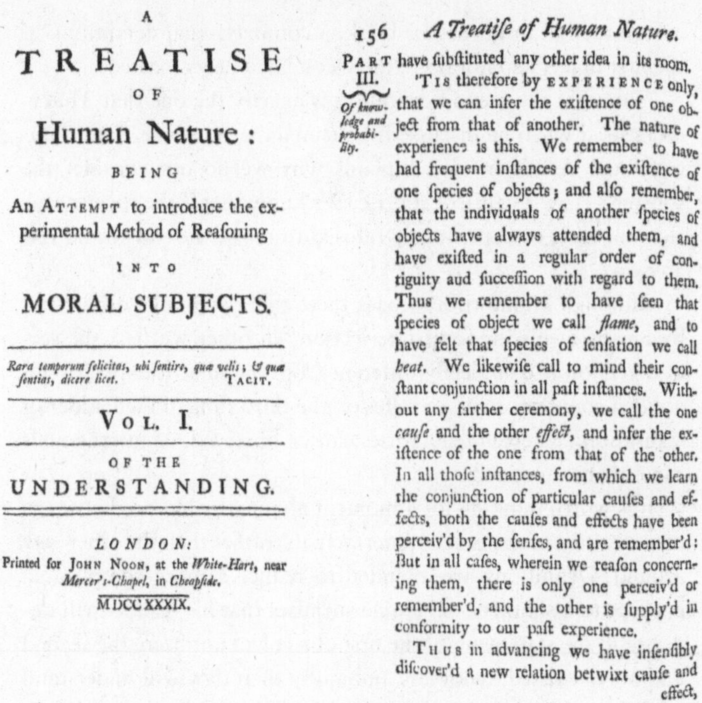

A

TREATISE

OF

Human Nature :

BEING

An ATTEMPT to introduce the experimental Method of Reasoning

INTO

MORAL SUBJECTS.

Rara temporum felicitas, ubi sentire, quæ velis ; & quæ
sentias, dicere licet.   TACIT.

VOL. I.

OF THE

UNDERSTANDING.

LONDON:
Printed for JOHN NOON, at the *White-Hart*, near
Mercer's-Chapel, in *Cheapside.*

MDCCXXXIX.

156   *A Treatise of Human Nature.*

PART have substituted any other idea in its room.
III.   'TIS therefore by EXPERIENCE only,
Of know- that we can infer the existence of one ob-
ledge and ject from that of another. The nature of
probabi-
lity.   experience is this. We remember to have
had frequent instances of the existence of
one species of objects ; and also remember,
that the individuals of another species of
objects have always attended them, and
have existed in a regular order of con-
tiguity and succession with regard to them.
Thus we remember to have seen that
species of object we call *flame,* and to
have felt that species of sensation we call
heat. We likewise call to mind their constant conjunction in all past instances. Without any farther ceremony, we call the one
cause and the other *effect,* and infer the existence of the one from that of the other.
In all those instances, from which we learn
the conjunction of particular causes and effects, both the causes and effects have been
perceiv'd by the senses, and are remember'd :
But in all cases, wherein we reason concerning them, there is only one perceiv'd or
remember'd, and the other is supply'd in
conformity to our past experience.
     THUS in advancing we have insensibly
discover'd a new relation betwixt cause and
                                           effect,

FIGURE 8.1. Hume's "regularity" definition of cause and effect, proposed
in 1739.

And yet Hume, to his credit, did not remain satisfied with this
definition. Nine years later, in *An Enquiry Concerning Human Un-*
derstanding, he wrote something quite different: "We may define
a cause to be *an object followed by another, and where all the ob-*
jects, similar to the first, are followed by objects similar to the sec-
ond. Or, in other words, where, if the first object had not been, the
second never had existed" (emphasis in the original). The first sentence, the version where *A* is consistently observed together with
B, simply repeats the regularity definition. But by 1748, he seems
to have some misgivings and finds it in need of some repair. As au-
thorized Whiggish historians, we can understand why. According
to his earlier definition, the rooster's crow would cause sunrise. To
patch over this difficulty, he adds a second definition that he never

even hinted at in his earlier book, a counterfactual definition: "if the first object had not been, the second had never existed."

Note that the second definition is exactly the one that Thucydides used when he discussed the tsunami at Orobiae. The counterfactual definition also explains why we do not consider the rooster's crow a cause of sunrise. We know that if the rooster was sick one day, or capriciously refused to crow, the sun would rise anyway.

Although Hume tries to pass these two definitions off as one, by means of his innocent interjection "in other words," the second version is completely different from the first. It explicitly invokes a counterfactual, so it lies on the third rung of the Ladder of Causation. Whereas regularities can be observed, counterfactuals can only be imagined.

It is worth thinking for a moment about why Hume chooses to define causes in terms of counterfactuals, rather than the other way around. Definitions are intended to reduce a more complicated concept to a simpler one. Hume surmises that his readers will understand the statement "if the first object had not been, the second had never existed" with less ambiguity than they will understand "the first object caused the second." He is absolutely right. The latter statement invites all sorts of fruitless metaphysical speculation about what quality or power inherent in the first object brings about the second one. The former statement merely asks us to perform a simple mental test: imagine a world without the earthquake and ask whether it also contains a tsunami. We have been making judgments like this since we were children, and the human species has been making them since Thucydides (and probably long before).

Nevertheless, philosophers ignored Hume's second definition for most of the nineteenth and twentieth centuries. Counterfactual statements, the "would haves," have always appeared too squishy and uncertain to satisfy academics. Instead, philosophers tried to rescue Hume's first definition through the theory of probabilistic causation, as discussed in Chapter 1.

One philosopher who defied convention, David Lewis, called in his 1973 book *Counterfactuals* for abandoning the regularity account altogether and for interpreting "*A* has caused *B*" as "*B* would not have occurred if not for *A*." Lewis asked, "Why not take counterfactuals at face value: as statements about possible alternatives to the actual situation?"

Like Hume, Lewis was evidently impressed by the fact that humans make counterfactual judgments without much ado, swiftly, comfortably, and consistently. We can assign them truth values and probabilities with no less confidence than we do for factual statements. In his view, we do this by envisioning "possible worlds" in which the counterfactual statements are true.

When we say, "Joe's headache would have gone away if he had taken aspirin," we are saying (according to Lewis) that there are other possible worlds in which Joe did take an aspirin and his headache went away. Lewis argued that we evaluate counterfactuals by comparing our world, where he did not take aspirin, to the most similar world in which he did take an aspirin. Upon finding no headache in that world, we declare the counterfactual statement to be true. "Most similar" is key. There may be some "possible worlds" in which his headache did not go away—for example, a world in which he took the aspirin and then bumped his head on the bathroom door. But that world contains an extra, adventitious circumstance. Among all possible worlds in which Joe took aspirin, the one most similar to ours would be one not where he bumped his head but where his headache is gone.

Many of Lewis's critics pounced on the extravagance of his claims for the literal existence of many other possible worlds. "Mr. Lewis was once dubbed a 'mad-dog modal realist' for his idea that any logically possible world you can think of actually exists," said his *New York Times* obituary in 2001. "He believed, for instance, that there was a world with talking donkeys."

But I think that his critics (and perhaps Lewis himself) missed the most important point. We do not need to argue about whether such worlds exist as physical or even metaphysical entities. If we aim to explain what people mean by saying "*A* causes *B*," we need only postulate that people are capable of generating alternative worlds in their heads, judging which world is "closer" to ours and, most importantly, doing it coherently so as to form a consensus.

Surely we could not communicate about counterfactuals if one person's "closer" was another person's "farther." In this view, Lewis's appeal "Why not take counterfactuals at face value?" called not for metaphysics but for attention to the amazing uniformity of the architecture of the human mind.

As a licensed Whiggish philosopher, I can explain this consistency quite well: it stems from the fact that we experience the same world and share the same mental model of its causal structure. We talked about this all the way back in Chapter 1. Our shared mental models bind us together into communities. We can therefore judge closeness not by some metaphysical notion of "similarity" but by how much we must take apart and perturb our shared model before it satisfies a given hypothetical condition that is contrary to fact (Joe not taking aspirin).

In structural models we do a very similar thing, albeit embellished with more mathematical detail. We evaluate expressions like "had X been x" in the same way that we handled interventions $do(X = x)$, by deleting arrows in a causal diagram or equations in a structural model. We can describe this as making the minimal alteration to a causal diagram needed to ensure that X equals x. In this respect, structural counterfactuals are compatible with Lewis's idea of the most similar possible world.

Structural models also offer a resolution of a puzzle Lewis kept silent about: How do humans represent "possible worlds" in their minds and compute the closest one, when the number of possibilities is far beyond the capacity of the human brain? Computer scientists call this the "representation problem." We must have some extremely economical code to manage that many worlds. Could structural models, in some shape or form, be the actual shortcut that we use? I think it is very likely, for two reasons. First, structural causal models are a shortcut that works, and there aren't any competitors around with that miraculous property. Second, they were modeled on Bayesian networks, which in turn were modeled on David Rumelhart's description of message passing in the brain. It is not too much of a stretch to think that 40,000 years ago, humans co-opted the machinery in their brain that already existed for pattern recognition and started to use it for causal reasoning.

Philosophers tend to leave it to psychologists to make statements about how the mind does things, which explains why the questions above were not addressed until quite recently. However, artificial intelligence (AI) researchers could not wait. They aimed to build robots that could communicate with humans about alternate scenarios, credit and blame, responsibility and regret. These are all counterfactual notions that AI researchers had to mechanize before they had the slightest chance of achieving what they call "strong AI"—humanlike intelligence.

With these motivations I entered counterfactual analysis in 1994 (with my student Alex Balke). Not surprisingly, the algorithmization of counterfactuals made a bigger splash in artificial intelligence and cognitive science than in philosophy. Philosophers tended to view structural models as merely one of many possible implementations of Lewis's possible-worlds logic. I dare to suggest that they are much more than that. Logic void of representation is metaphysics. Causal diagrams, with their simple rules of following and erasing arrows, must be close to the way that our brains represent counterfactuals.

This assertion must remain unproven for the time being, but the upshot of the long story is that counterfactuals have ceased to be mystical. We understand how humans manage them, and we are ready to equip robots with similar capabilities to the ones our ancestors acquired 40,000 years ago.

POTENTIAL OUTCOMES, STRUCTURAL EQUATIONS, AND THE ALGORITHMIZATION OF COUNTERFACTUALS

Just a year after the release of Lewis's book, and independently of it, Donald Rubin (Figure 8.2) began writing a series of papers that introduced potential outcomes as a language for asking causal questions. Rubin, at that time a statistician for the Educational Testing Service, single-handedly broke the silence about causality that had persisted in statistics for seventy-five years and legitimized the concept of counterfactuals in the eyes of many health scientists. It is impossible to overstate the importance of this development. It provided researchers with a flexible language to express almost

FIGURE 8.2. Donald Rubin (right) with the author in 2014. (*Source:* Photo courtesy of Grace Hyun Kim.)

every causal question they might wish to ask, at both the population and individual levels.

In the Rubin causal model, a potential outcome of a variable Y is simply "the value that Y would have taken for individual u, had X been assigned the value x." That's a lot of words, so it's often convenient to write this quantity more compactly as $Y_{X=x}(u)$. Often we abbreviate this further as $Y_x(u)$ if it is apparent from the context what variable is being set to the value x.

To appreciate how audacious this notation is, you have to step back from the symbols and think about the assumptions they embody. By writing down the symbol Y_x, Rubin asserted that Y definitely would have taken some value if X had been x, and this has just as much objective reality as the value Y actually did take. If you don't buy this assumption (and I'm pretty sure Heisenberg wouldn't), you can't use potential outcomes. Also, note that the potential outcome, or counterfactual, is defined at the level of an individual, not a population.

The very first scientific appearance of a potential outcome came in the master's thesis of Jerzy Neyman, written in 1923. Neyman, a descendant of Polish nobility, had grown up in exile in Russia and did not set foot in his native land until 1921, when he was twenty-seven years old. He had received a very strong mathematical education in Russia and would have liked to continue research in pure mathematics, but it was easier for him to find employment as a statistician. Much like R. A. Fisher in England, he did his first statistical research at an agricultural institute, a job for which he was hugely overqualified. Not only was he the only statistician in the institute, but he was really the only person in the country thinking about statistics as a discipline.

Neyman's first mention of potential outcomes came in the context of an agricultural experiment, where the subscript notation represents the "unknown potential yield of the i-th variety [of a given seed] on the respective plot." The thesis remained unknown and untranslated into English until 1990. However, Neyman himself did not remain unknown. He arranged to spend a year at Karl Pearson's statistical laboratory at University College London, where he made friends with Pearson's son Egon. The two kept in touch for the next seven years, and their collaboration paid great dividends: the Neyman-Pearson approach to statistical hypothesis testing was a milestone that every beginning statistics student learns about.

In 1933, Karl Pearson's long autocratic leadership finally came to an end with his retirement, and Egon was his logical successor—or would have been, if not for the singular problem of R. A. Fisher, by then the most famous statistician in England. The university came up with a unique and disastrous solution, dividing Pearson's position into a chair of statistics (Egon Pearson) and a chair of eugenics (Fisher). Egon wasted no time hiring his Polish friend. Neyman arrived in 1934 and almost immediately locked horns with Fisher.

Fisher was already spoiling for a fight. He knew he was the world's leading statistician and had practically invented large parts of the subject, yet was forbidden from teaching in the statistics department. Relations were extraordinarily tense. "The Common Room was carefully shared," writes Constance Reid in her biography of Neyman. "Pearson's group had tea at 4; and at 4:30, when they were safely out of the way, Fisher and his group trooped in."

In 1935, Neyman gave a lecture at the Royal Statistical Society titled "Statistical Problems in Agricultural Experimentation," in which he called into question some of Fisher's own methods and also, incidentally, discussed the idea of potential outcomes. After Neyman was done, Fisher stood up and told the society that "he had hoped that Dr. Neyman's paper would be on a subject with which the author was fully acquainted."

"[Neyman had] asserted that Fisher was wrong," wrote Oscar Kempthorne years later about the incident. "This was an unforgivable offense—Fisher was never wrong and indeed the suggestion that he might be was treated by him as a deadly assault. Anyone who did not accept Fisher's writing as the God-given truth was at best stupid and at worst evil." Neyman and Pearson saw the extent of Fisher's fury a few days later, when they went to the department in the evening and found Neyman's wooden models, with which he had illustrated his lecture, strewn all over the floor. They concluded that only Fisher could have been responsible for the wreckage.

While Fisher's fit of rage may seem amusing now, his attitude did have serious consequences. Of course he could not swallow his pride and use Neyman's potential outcome notation, even though it would have helped him later with problems of mediation. The lack of potential outcome vocabulary led him and many other people into the so-called Mediation Fallacy, which we will discuss in Chapter 9.

At this point some readers might still find the concept of counterfactuals somewhat mystical, so I'd like to show how some of Rubin's followers would infer potential outcomes and contrast this model-free approach with the structural causal model approach.

Suppose that we are looking at a certain firm to see whether education or years of experience is a more important factor in determining an employee's salary. We have collected some data on the existing salaries at this firm, reproduced in Table 8.1. We're letting EX represent years of experience, ED represent education, and S represent salary. We're also assuming, for simplicity, just three levels of education: 0 = high school degree, 1 = college degree, 2 = graduate degree. Thus $S_{ED=0}(u)$, or $S_0(u)$, represents the salary of individual u if u were a high school graduate but not a college graduate, and $S_1(u)$ represents u's salary if u were a college graduate. A typical counterfactual question we might want to ask is, "What would Alice's salary be if she had a college degree?" In other words, what is $S_1(\text{Alice})$?

The first thing to notice about Table 8.1 is all the missing data, indicated by question marks. We can never observe more than one potential outcome in the same individual. Although obvious, nevertheless this statement is important. Statistician Paul Holland once called it the "fundamental problem of causal inference," a name that has stuck. If we could only fill in the question marks, we could answer all our causal questions.

I have never agreed with Holland's characterization of the missing values in Table 8.1 as a "fundamental problem," perhaps because I have rarely described causal problems in terms of a table. But more fundamentally, viewing causal inference as a missing-data problem can be terribly misleading, as we will soon see. Observe that, aside from the decorative headings of the last three columns, Table 8.1 is totally devoid of causal information about ED, EX, and S—for example, whether education affects salary or the other way around. Worse yet, it does not allow us to represent such information when available. But for statisticians who perceive the "fundamental problem" to be missing data, such a table appears to present endless opportunities. Indeed, if we look at S_0, S_1, and S_2 not as potential outcomes but as ordinary variables, we have dozens of interpolation techniques to fill in the blanks or, as statisticians would say, "impute the missing data," in some optimal way.

TABLE 8.1. Fictitious data for potential outcomes example.

Employee (u)	EX(u)	ED(u)	$S_0(u)$	$S_1(u)$	$S_2(u)$
Alice	6	0	$81,000	?	?
Bert	9	1	?	$92,500	?
Caroline	9	2	?	?	$97,000
David	8	1	?	$91,000	?
Ernest	12	1	?	$100,000	?
Frances	13	0	$97,000	?	?
etc.					

One common approach is matching. We look for pairs of individuals who are well matched in all variables except the one of interest and then fill in their rows to match each other. The clearest case here is that of Bert and Caroline, who match perfectly on experience. So we assume that Bert's salary, if he had a graduate degree, would be the same as Caroline's ($97,000), and Caroline's salary, if she had only an undergraduate degree, would be the same as Bert's ($92,500). Note that matching invokes the same idea as conditioning (or stratifying): we select for comparison groups that share an observed characteristic and use the comparison to infer characteristics that they do not seem to share.

It is hard to estimate Alice's salary this way because there is no good match for her in the data I have given. Nevertheless, statisticians have developed techniques of considerable subtlety to impute missing data from approximate matches, and Rubin has been a pioneer of this approach. Unfortunately, even the most gifted matchmaker in the world cannot turn data into potential outcomes, not even approximately. I will show below that the correct answer depends critically on whether education affects experience or the other way around, information nowhere to be found in the table.

A second possible method is linear regression (not to be conflated with structural equations). In this approach we pretend that the data came from some unknown random source and use standard statistical methods to find the line (or, in this case, plane) that best fits the data. The output of such an approach might be an equation that looks like this:

$$S = \$65,000 + 2,500 \times EX + 5,000 \times ED \qquad (8.1)$$

Equation 8.1 tells us that (on average) the base salary of an employee with no experience and only a high school diploma is $65,000. For each year of experience, the salary increases by $2,500, and for each additional educational degree (up to two), the salary increases by $5,000. Accordingly, a regression analyst would claim, our estimate of Alice's salary, if she had a college degree, is $65,000 + $2,500 × 6 + $5,000 × 1 = $85,000.

The ease and familiarity of such imputation techniques explain why Rubin's conception of causal inference as a missing-data problem has enjoyed broad popularity. Alas, as innocuous as these interpolation methods appear, they are fundamentally flawed. They are data driven, not model driven. All the missing data are filled in by examining other values in the table. As we have learned from the Ladder of Causation, any such method is doomed to start with; no methods based only on data (rung one) can answer counterfactual questions (rung three).

Before contrasting these methods with the structural causal model approach, let us examine intuitively what goes wrong with model-blind imputation. In particular, let us explain why Bert and Caroline, who match perfectly in experience, may in fact be quite incomparable when it comes to comparing their potential outcomes. More surprising, a reasonable causal story (fitting Table 8.1) would show that the best match for Caroline for Salary would be someone who does not match her on Experience.

The first key point to realize is that Experience is likely to depend on Education. After all, those employees who got an extra educational degree took four years of their lives to do so. Thus, if Caroline had only one degree of education (like Bert), she would have been able to use that extra time to gain more experience compared to what she now has. This would have given her equal education to and greater experience than Bert. We can thus conclude that $S_1(\text{Caroline}) > S_1(\text{Bert})$, contrary to what naive matching would predict. We see that, once we have a causal story in which Education affects Experience, it is inevitable that "matching" on Experience will create a mismatch on potential Salary.

Ironically, equal Experience, which started out as an invitation for matching, has now turned into a loud warning against it. Table 8.1 will, of course, continue its silence about such dangers. For this reason I cannot share Holland's enthusiasm for casting causal inference as a missing-data problem. Quite the contrary. Recent work of Karthika Mohan, a former student of mine, reveals that even standard problems of missing data require causal modeling for their solution.

Now let's see how a structural causal model would treat the same data. First, before we even look at the data, we draw a causal diagram (Figure 8.3). The diagram encodes the causal story behind the data, according to which Experience listens to Education and Salary listens to both. In fact, we can already tell something very important just by looking at the diagram. If our model were wrong and *EX* were a cause of *ED*, rather than vice versa, then Experience would be a confounder, and matching employees with similar experience would be completely appropriate. With *ED* as the cause of *EX*, Experience is a mediator. As you surely know by now, mistaking a mediator for a confounder is one of the deadliest sins in causal inference and may lead to the most outrageous errors. The latter invites adjustment; the former forbids it.

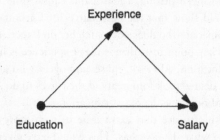

FIGURE 8.3. Causal diagram for the effect of education (*ED*) and experience (*EX*) on salary (*S*).

So far in this book, I have used a very informal word—"listening"—to express what I mean by the arrows in a causal diagram. But now it's time to put a little bit of mathematical meat on this concept, and this is in fact where structural causal models differ from Bayesian networks or regression models. When I say that Salary listens to Education and Experience, I mean that it is a mathematical function of those variables: $S = f_s(EX, ED)$. But we need to allow for individual variations, so we extend this function to read $S = f_s(EX, ED, U_s)$, where U_s stands for "unobserved variables that affect salary." We know these variables exist (e.g., Alice is a friend of the company's president), but they are too diverse and too numerous to incorporate explicitly into our model.

Let's see how this would play out in our education/experience/ salary example, assuming linear functions throughout. We can use the same statistical methods as before to find the best-fitting linear equation. The result would look just like Equation 8.1 with one small difference:

$$S = \$65,000 + 2,500 \times EX + 5,000 \times ED + U_S \qquad (8.2)$$

However, the formal similarity between Equations 8.1 and 8.2 is profoundly deceptive; their interpretations differ like night and day. The fact that we chose to regress S on ED and EX in Equation 8.1 in no way implies that S listens to ED and EX in the real world. That choice was purely ours, and nothing in the data would prevent us from regressing EX on ED and S or following any other order. (Remember Francis Galton's discovery in Chapter 2 that regressions are cause blind.) We lose this freedom once we proclaim an equation to be "structural." In other words, the author of Equation 8.2 must commit to writing equations that mirror his belief about who listens to whom in the world. In our case, he believes that S truly listens to EX and ED. More importantly, the absence of an equation $ED = f_{ED}(EX, S, U_{ED})$ from the model means that ED is believed to be oblivious to changes in EX or S. This difference in commitment gives structural equations the power to support counterfactuals, a power denied to regression equations.

In compliance with Figure 8.3, we must also have a structural equation for EX, but now we will force the coefficient of S to zero, to reflect the absence of an arrow from S to EX. Once we estimate the coefficients from the data, the equation might look something like this:

$$EX = 10 - 4 \times ED + U_{EX} \qquad (8.3)$$

This equation says that the average experience for people with no advanced degrees is ten years, and each degree of education (up to two) decreases EX by four years on average. Again, note the key difference between structural and regression equations: variable S does not enter into Equation 8.3, despite the fact that S and EX are likely to be highly correlated. This reflects the analyst's belief that the experience EX acquired by any individual is totally unaffected by his current salary.

Now let's demonstrate how to derive counterfactuals from a structural model. To estimate Alice's salary if she had a college education, we perform three steps:

1. (Abduction) Use the data about Alice and about the other employees to estimate Alice's idiosyncratic factors, U_S(Alice) and U_{EX}(Alice).
2. (Action) Use the *do*-operator to change the model to reflect the counterfactual assumption being made, in this case that she has a college degree: ED(Alice) = 1.
3. (Prediction) Calculate Alice's new salary using the modified model and the updated information about the exogenous variables U_S(Alice), U_{EX}(Alice), and ED(Alice). This newly calculated salary is equal to $S_{ED=1}$(Alice).

For step 1, we observe from the data that EX(Alice) = 6 and ED(Alice) = 0. We substitute these values into Equations 8.2 and 8.3. The equations then tell us Alice's idiosyncratic factors: U_S(Alice) = \$1,000 and U_{EX}(Alice) = –4. This represents everything that is unique, special, and wonderful about Alice. Whatever that is, it adds \$1,000 to her predicted salary.

Step 2 tells us to use the *do*-operator to erase the arrows pointing to the variable that is being set to a counterfactual value (Education) and set Alice's Education to a college degree (Education = 1). In this example, Step 2 is trivial, because there are no arrows pointing to Education and hence no arrows to erase. In more complicated models, though, this step of erasing the arrows cannot be left out, because it affects the computation in Step 3. Variables that might have affected the outcome through the intervened variable will no longer be allowed to do so.

Finally, Step 3 says to update the model to reflect the new information that U_S = \$1,000, U_{EX} = –4, and ED = 1. First we use

Equation 8.3 to recompute what Alice's Experience would be if she had gone to college: $EX_{ED=1}(\text{Alice}) = 10 - 4 - 4 = 2$ years. Then we use Equation 8.2 to recompute her potential Salary:

$$S_{ED=1}(\text{Alice}) = \$65{,}000 + 2{,}500 \times 2 + 5{,}000 \times 1 + 1{,}000 = \$76{,}000.$$

Our result, $S_1(\text{Alice}) = \$76{,}000$, is a valid estimate of Alice's would-be salary; that is, the two will coincide if the model assumptions are valid. Because this example entails a very simple causal model and very simple (linear) functions, the differences between it and the data-driven regression method may seem rather minor. But the minor differences on the surface reflect vast differences underneath. Whatever counterfactual (potential) outcome we obtain from the structural method follows logically from the assumptions displayed in the model, while the answer obtained by the data-driven method is as whimsical as spurious correlations because it leaves important modeling assumptions unaccounted for.

This example has forced us to go further into the "nuts and bolts" of causal models than we have previously done in this book. But let me step back a little to celebrate and appreciate the miracle that came into being through Alice's example. Using a combination of data and model, we were able to predict the behavior of an individual (Alice) under totally hypothetical conditions. Of course, there is no such thing as a free lunch: we got these strong results because we made strong assumptions. In addition to asserting the causal relationships between the observed variables, we also assumed that the functional relationships were linear. But the linearity matters less here than knowing what those specific functions are. That enabled us to compute Alice's idiosyncrasies from her observed characteristics and update the model as required in the three-step procedure.

At the risk of adding a sober note to our celebration, I have to tell you that this functional information will not always be available to us in practice. In general, we call a model "completely specified" if the functions behind the arrows are known and "partially specified" otherwise. For instance, as in Bayesian networks, we may only know probabilistic relationships between parents and children in the graph. If the model is partially specified, we may not be able to estimate Alice's salary exactly; instead we may have to make a probability-interval statement, such as "There is a 10 to 20 percent chance that her salary would be $76,000." But even such probabilistic answers are good enough for many applications. Moreover, it is truly remarkable how much information we can extract from the causal diagram even when we have no information on the specific functions lying behind the arrows or only very general information, such as the "monotonicity" assumption we encountered in the last chapter.

Steps 1 to 3 above can be summed up in what I call the "first law of causal inference": $Y_x(u) = Y_{M_x}(u)$. This is the same rule that we used in the firing squad example in Chapter 1, except that the functions are different. The first law says that the potential outcome $Y_x(u)$ can be imputed by going to the model M_x (with arrows into X deleted) and computing the outcome $Y(u)$ there. All estimable quantities on rungs two and three of the Ladder of Causation follow from there. In short, the reduction of counterfactuals to an algorithm allows us to conquer as much territory from rung three as mathematics will permit—but, of course, not a bit more.

THE VIRTUE OF SEEING YOUR ASSUMPTIONS

The SCM method I have shown for computing counterfactuals is not the same method that Rubin would use. A major point of difference between us is the use of causal diagrams. They allow researchers to represent causal assumptions in terms that they can understand and then treat all counterfactuals as derived properties of their world model. The Rubin causal model treats counterfactuals as abstract mathematical objects that are managed by algebraic machinery but not derived from a model.

Deprived of a graphical facility, the user of the Rubin causal model is usually asked to accept three assumptions. The first one, called the "stable unit treatment value assumption," or SUTVA, is reasonably transparent. It says that each individual (or "unit," the preferred term of causal modelers) will have the same effect of treatment regardless of what treatment the other individuals (or "units") receive. In many cases, barring epidemics and other collective interactions, this makes perfectly good sense. For example, assuming headache is not contagious, my response to aspirin will not depend on whether Joe receives aspirin.

The second assumption in Rubin's model, also benign, is called "consistency." It says that a person who took aspirin and recovered would also recover if given aspirin by experimental design. This reasonable assumption, which is a theorem in the SCM framework, says in effect that the experiment is free of placebo effects and other imperfections.

But the major assumption that potential outcome practitioners are invariably required to make is called "ignorability." It is more technical, but it's the crucial part of the transaction, for it is in essence the same thing as Jamie Robins and Sander Greenland's condition of exchangeability discussed in Chapter 4. Ignorability expresses this same requirement in terms of the potential outcome variable Y_x. It requires that Y_x be independent of the treatment actually received, namely X, given the values of a certain set of (de) confounding variables Z. Before exploring its interpretation, we should acknowledge that any assumption expressed as conditional independence inherits a large body of familiar mathematical machinery developed by statisticians for ordinary (noncounterfactual) variables. For example, statisticians routinely use rules for deciding when one conditional independence follows from another. To Rubin's credit, he recognized the advantages of translating the causal notion of "nonconfoundedness" into the syntax of probability theory, albeit on counterfactual variables. The ignorability assumption makes the Rubin causal model actually a model; Table 8.1 in itself is not a model because it contains no assumptions about the world.

Unfortunately, I have yet to find a single person who can explain what ignorability means in a language spoken by those who need to make this assumption or assess its plausibility in a given problem. Here is my best try. The assignment of patients to either treatment or control is ignorable if, within any stratum of the confounder Z, patients who would have one potential outcome, $Y_x = y$, are just as likely to be in the treatment or control group as the patients who would have a different potential outcome, $Y_x = y'$.

This definition is perfectly legitimate for someone in possession of a probability function over counterfactuals. But how is a biologist or economist with only scientific knowledge for guidance supposed to assess whether this is true or not? More concretely, how is a scientist to assess whether ignorability holds in any of the examples discussed in this book?

To understand the difficulty, let us attempt to apply this explanation to our example. To determine if ED is ignorable (conditional on EX), we are supposed to judge whether employees who would have one potential salary, say $S_1 = s$, are just as likely to have one level of education as the employees who would have a different potential salary, say $S_1 = s'$. If you think that this sounds circular, I can only agree with you! We want to determine Alice's potential salary, and even before we start—even before we get a hint about the answer—we are supposed to speculate on whether the result is dependent or independent of ED, in every stratum of EX. It is quite a cognitive nightmare.

As it turns out, ED in our example is not ignorable with respect to S, conditional on EX, and this is why the matching approach (setting Bert and Caroline equal) would yield the wrong answer for their potential salaries. In fact, their estimates should differ by an amount $S_1(\text{Bert}) - S_1(\text{Caroline}) = -\$9,500$. (The reader should be able to show this from the numbers in Table 8.1 and the three-step procedure.) I will now show that with the help of a causal diagram, a student could see immediately that ED is not ignorable and would not attempt matching here. Lacking a diagram, a student would be tempted to assume that ignorability holds by default and would fall into this trap. (This is not a speculation. I borrowed the idea for this example from an article in *Harvard Law Review* where the story was essentially the same as in Figure 8.3 and the author did use matching.)

Here is how we can use a causal diagram to test for (conditional) ignorability. To determine if X is ignorable relative to outcome Y, conditional on a set Z of matching variables, we need only test to see if Z blocks all the back-door paths between X and Y and no member of Z is a descendant of X. It is as simple as that! In our example, the proposed matching variable (Experience) blocks all the back-door paths (because there aren't any), but it fails the test because it is a descendant of Education. Therefore *ED* is not ignorable, and *EX* cannot be used for matching. No elaborate mental gymnastics are needed, just a look at a diagram. Never is a researcher required to mentally assess how likely a potential outcome is given one treatment or another.

Unfortunately, Rubin does not consider causal diagrams to "aid the drawing of causal inferences." Therefore, researchers who follow his advice will be deprived of this test for ignorability and will either have to perform formidable mental gymnastics to convince themselves that the assumption holds or else simply accept the assumption as a "black box." Indeed, a prominent potential outcome researcher, Marshall Joffe, wrote in 2010 that ignorability assumptions are usually made because they justify the use of available statistical methods, not because they are truly believed.

Closely related to transparency is the notion of testability, which has come up several times in this book. A model cast as a causal diagram can easily be tested for compatibility with the data, whereas a model cast in potential outcome language lacks this feature. The test goes like this: whenever all paths between X and Y in the diagram are blocked by a set of nodes Z, then in the data X and Y should be independent, conditional on Z. This is the d-separation property mentioned in Chapter 7, which allows us to reject a model whenever the independence fails to show up in the data. In contrast, if the same model is expressed in the language of potential outcomes (i.e., as a collection of ignorability statements), we lack the mathematical machinery to unveil the independencies that the model entails, and researchers are unable to subject the model to a test. It is hard to understand how potential outcome researchers managed to live with this deficiency without rebelling. My only explanation is that they were kept away from graphical tools for so long that they forgot that causal models can and should be testable.

Now I must apply the same standards of transparency to myself and say a little bit more about the assumptions embodied in a structural causal model.

Remember the story of Abraham that I related earlier? Abraham's first response to the news of Sodom's imminent destruction was to look for a dose-response relationship, or a response function, relating the wickedness of the city to its punishment. It was a sound scientific instinct, but I suspect few of us would have been calm enough to react that way.

The response function is the key ingredient that gives SCMs the power to handle counterfactuals. It is implicit in Rubin's potential outcome paradigm but a major point of difference between SCMs and Bayesian networks, including causal Bayesian networks. In a probabilistic Bayesian network, the arrows into Y mean that the probability of Y is governed by the conditional probability tables for Y, given observations of its parent variables. The same is true for causal Bayesian networks, except that the conditional probability tables specify the probability of Y given interventions on the parent variables. Both models specify probabilities for Y, not a specific value of Y. In a structural causal model, there are no conditional probability tables. The arrows simply mean Y is a function of its parents, as well as the exogenous variable U_Y:

$$Y = f_Y(X, A, B, C, \ldots, U_Y) \qquad (8.4)$$

Thus, Abraham's instinct was sound. To turn a noncausal Bayesian network into a causal model—or, more precisely, to make it capable of answering counterfactual queries—we need a dose-response relationship at each node.

This realization did not come to me easily. Even before delving into counterfactuals, I tried for a very long time to formulate causal models using conditional probability tables. One obstacle I faced was cyclic models, which were totally resistant to conditional probability formulations. Another obstacle was that of coming up with a notation to distinguish probabilistic Bayesian networks from causal ones. In 1991, it suddenly hit me that all the difficulties would vanish if we made Y a function of its parent variables and let the U_Y term handle all the uncertainties concerning Y. At the time, it seemed like a heresy against my own teaching. After devoting several years to the cause of probabilities in artificial intelligence, I was now proposing to take a step backward and use a nonprobabilistic, quasi-deterministic model. I can still remember my student at the time, Danny Geiger, asking incredulously, "Deterministic equations? Truly deterministic?" It was as if Steve Jobs had just told him to buy a PC instead of a Mac. (This was 1990!)

On the surface, there was nothing revolutionary about these equations. Economists and sociologists had been using such models since the 1950s and 1960s and calling them structural equation models (SEMs). But this name signaled controversy and confusion over the causal interpretation of the equations. Over time, economists lost sight of the fact that the pioneers of these models, Trygve Haavelmo in economics and Otis Dudley Duncan in sociology, had intended them to represent causal relationships. They began to confuse structural equations with regression lines, thus stripping the substance from the form. For example, in 1988, when David Freedman challenged eleven SEM researchers to explain how to apply interventions to a structural equation model, not one of them could. They could tell you how to estimate the coefficients from data, but they could not tell you why anyone should bother. If the response-function interpretation I presented between 1990 and 1994 did anything new, it was simply to restore and formalize Haavelmo's and Duncan's original intentions and lay before their disciples the bold conclusions that follow from those intentions if you take them seriously.

Some of these conclusions would be considered astounding, even by Haavelmo and Duncan. Take for example the idea that from every SEM, no matter how simple, we can compute all the counterfactuals that one can imagine among the variables in the model. Our ability to compute Alice's potential salary, had she had college education, followed from this idea. Even today modern-day economists have not internalized this idea.

One other important difference between SEMs and SCMs, besides the middle letter, is that the relationship between causes and effects in an SCM is not necessarily linear. The techniques that emerge from SCM analysis are valid for nonlinear as well as linear functions, discrete as well as continuous variables.

Linear structural equation models have many advantages and many disadvantages. From the viewpoint of methodology, they are seductively simple. They can be estimated from observational data by linear regression, and you can choose between dozens of statistical software packages to do this for you.

On the other hand, linear models cannot represent dose-response curves that are not straight lines. They cannot represent threshold effects, such as a drug that has increasing effects up to a certain dosage and then no further effect. They also cannot represent interactions between variables. For instance, a linear model cannot describe a situation in which one variable enhances or inhibits the effect of another variable. (For example, Education might enhance the effect of Experience by putting the individual in a faster-track job that gets bigger annual raises.)

While debates about the appropriate assumptions to make are inevitable, our main message is quite simple: Rejoice! With a fully specified structural causal model, entailing a causal diagram and all the functions behind it, we can answer any counterfactual query. Even with a partial SCM, in which some variables are hidden or the dose-response relationships are unknown, we can still in many cases answer our query. The next two sections give some examples.

COUNTERFACTUALS AND THE LAW

In principle, counterfactuals should find easy application in the courtroom. I say "in principle" because the legal profession is very conservative and takes a long time to accept new mathematical methods. But using counterfactuals as a mode of argument is actually very old and known in the legal profession as "but-for causation."

The Model Penal Code expresses the "but-for" test as follows: "Conduct is the cause of a result when: (a) it is an antecedent but for which the result in question would not have occurred." If the defendant fired a gun and the bullet struck and killed the victim, the firing of the gun is a but-for, or necessary, cause of the death, since the victim would be alive if not for the firing. But-for causes can also be indirect. If Joe blocks a building's fire exit with furniture, and Judy dies in a fire after she could not reach the exit, then

Joe is legally responsible for her death even though he did not light the fire.

How can we express necessary or but-for causes in terms of potential outcomes? If we let the outcome Y be "Judy's death" (with $Y = 0$ if Judy lives and $Y = 1$ if Judy dies) and the treatment X be "Joe's blocking the fire escape" (with $X = 0$ if he does not block it and $X = 1$ if he does), then we are instructed to ask the following question:

> Given that we know the fire escape was blocked ($X = 1$) and Judy died ($Y = 1$), what is the probability that Judy would have lived ($Y = 0$) if X had been 0?

Symbolically, the probability we want to evaluate is $P(Y_{X=0} = 0 \mid X = 1, Y = 1)$. Because this expression is rather cumbersome, I will later abbreviate it as "*PN*," the *probability of necessity* (i.e., the probability that $X = 1$ is a necessary or but-for cause of $Y = 1$).

Note that the probability of necessity involves a contrast between two different worlds: the actual world where $X = 1$ and the counterfactual world where $X = 0$ (expressed by the subscript $X = 0$). In fact, hindsight (knowing what happened in the actual world) is a critical distinction between counterfactuals (rung three of the Ladder of Causation) and interventions (rung two). Without hindsight, there is no difference between $P(Y_{X=0} = 0)$ and $P(Y = 0 \mid do(X = 0))$. Both express the probability that, under normal conditions, Judy will be alive if we ensure that the exit is not blocked; they do not mention the fire, Judy's death, or the blocked exit. But hindsight may change our estimate of the probabilities. Suppose we observe that $X = 1$ and $Y = 1$ (hindsight). Then $P(Y_{X=0} = 0 \mid X = 1, Y = 1)$ is not the same as $P(Y_{X=0} = 0 \mid X = 1)$. Knowing that Judy died ($Y = 1$) gives us information on the circumstances that we would not get just by knowing that the door was blocked ($X = 1$). For one thing, it is evidence of the strength of the fire.

In fact, it can be shown that there is no way to capture $P(Y_{X=0} = 0 \mid X = 1, Y = 1)$ in a *do*-expression. While this may seem like a rather arcane point, it does give mathematical proof that counterfactuals (rung three) lie above interventions (rung two) on the Ladder of Causation.

In the last few paragraphs, we have almost surreptitiously introduced probabilities into our discussion. Lawyers have long understood that mathematical certainty is too high a standard of proof. For criminal cases in the United States, the Supreme Court in 1880 established that guilt has to be proven "to the exclusion of all reasonable doubt." The court said not "beyond all doubt" or "beyond a shadow of a doubt" but beyond reasonable doubt. The Supreme Court has never given a precise definition of that term, but one might conjecture that there is some threshold, perhaps 99 percent or 99.9 percent probability of guilt, above which doubt becomes unreasonable and it is in society's interest to lock the defendant up. In civil rather than criminal proceedings, the standard of proof is somewhat clearer. The law requires a "preponderance of evidence" that the defendant caused the injury, and it seems reasonable to interpret this to mean that the probability is greater than 50 percent.

Although but-for causation is generally accepted, lawyers have recognized that in some cases it might lead to a miscarriage of justice. One classic example is the "falling piano" scenario, where the defendant fires a shot at the victim and misses, and in the process of fleeing the scene, the victim happens to run under a falling piano and is killed. By the but-for test the defendant would be guilty of murder, because the victim would not have been anywhere near the falling piano if he hadn't been running away. But our intuition says that the defendant is not guilty of murder (though he may be guilty of attempted murder), because there was no way that he could have anticipated the falling piano. A lawyer would say that the piano, not the gunshot, is the *proximate cause* of death.

The doctrine of proximate cause is much more obscure than but-for cause. The Model Penal Code says that the outcome should not be "too remote or accidental in its occurrence to have a [just] bearing on the actor's liability or the gravity of his offense." At present this determination is left to the intuition of the judge. I would suggest that it is a form of *sufficient cause*. Was the defendant's action sufficient to bring about, with high enough probability, the event that actually caused the death?

While the meaning of proximate cause is very vague, the meaning of sufficient cause is quite precise. Using counterfactual notation, we can define the probability of sufficiency, or PS, to be $P(Y_{X=1} = 1 \mid X = 0, Y = 0)$. This tells us to imagine a situation where $X = 0$ and $Y = 0$: the shooter did not fire at the victim, and the victim did not run under a piano. Then we ask how likely it is that in such a situation, firing the shot ($X = 1$) would result in outcome $Y = 1$ (running under a piano)? This calls for counterfactual judgment, but I think that most of us would agree that the likelihood of such an outcome would be extremely small. Both intuition and the Model Penal Code suggest that if PS is too small, we should not convict the defendant of causing $Y = 1$.

Because the distinction between necessary and sufficient causes is so important, I think it may help to anchor these two concepts in simple examples. Sufficient cause is the more common of the two, and we have already encountered this concept in the firing squad example of Chapter 1. There, the firing of either Soldier A or Soldier B is sufficient to cause the prisoner's death, and neither (in itself) is necessary. So $PS = 1$ and $PN = 0$.

Things get a bit more interesting when uncertainty strikes—for example, if each soldier has some probability of disobeying orders or missing the target. For example, if Soldier A has a probability p_A of missing the target, then his PS would be $1 - p_A$, since this is his probability of hitting the target and causing death. His PN, however, would depend on how likely Soldier B is to refrain from shooting or to miss the target. Only under such circumstances would the shooting of Soldier A be necessary; that is, the prisoner would be alive had Soldier A not shot.

A classic example demonstrating necessary causation tells the story of a fire that broke out after someone struck a match, and the question is "What caused the fire, striking the match or the presence of oxygen in the room?" Note that both factors are equally necessary, since the fire would not have occurred absent one of them. So, from a purely logical point of view, the two factors are equally responsible for the fire. Why, then, do we consider lighting the match a more reasonable explanation of the fire than the presence of oxygen?

To answer this, consider the two sentences:

1. The house would still be standing if only the match had not been struck.
2. The house would still be standing if only the oxygen had not been present.

Both sentences are true. Yet the overwhelming majority of readers, I'm sure, would come up with the first scenario if asked to explain what caused the house to burn down, the match or the oxygen. So, what accounts for the difference?

The answer clearly has something to do with normality: having oxygen in the house is quite normal, but we can hardly say that about striking a match. The difference does not show up in the logic, but it does show up in the two measures we discussed above, PS and PN.

If we take into account that the probability of striking a match is much lower than that of having oxygen, we find quantitatively that for Match, both PN and PS are high, while for Oxygen, PN is high but PS is low. Is this why, intuitively, we blame the match and not the oxygen? Quite possibly, but it may be only part of the answer.

In 1982, psychologists Daniel Kahneman and Amos Tversky investigated how people choose an "if only" culprit to "undo" an undesired outcome and found consistent patterns in their choices. One was that people are more likely to imagine undoing a rare event than a common one. For example, if we are undoing a missed appointment, we are more likely to say, "If only the train had left on schedule," than "If only the train had left early." Another pattern was people's tendency to blame their own actions (e.g., striking a match) rather than events not under their control. Our ability to estimate PN and PS from our model of the world suggests a systematic way of accounting for these considerations and eventually teaching robots to produce meaningful explanations of peculiar events.

We have seen that PN captures the rationale behind the "but-for" criterion in a legal setting. But should PS enter legal considerations in criminal and tort law? I believe that it should, because attention to sufficiency implies attention to the consequences of one's action. The person who lit the match ought to have anticipated the presence of oxygen, whereas nobody is generally expected to pump all the oxygen out of the house in anticipation of a match-striking ceremony.

What weight, then, should the law assign to the necessary versus sufficient components of causation? Philosophers of law have not discussed the legal status of this question, perhaps because the notions of *PS* and *PN* were not formalized with such precision. However, from an AI perspective, clearly *PN* and *PS* should take part in generating explanations. A robot instructed to explain why a fire broke out has no choice but to consider both. Focusing on *PN* only would yield the untenable conclusion that striking a match and having oxygen are equally adequate explanations for the fire. A robot that issues this sort of explanation will quickly lose its owner's trust.

NECESSARY CAUSES, SUFFICIENT CAUSES, AND CLIMATE CHANGE

In August 2003, the most intense heat wave in five centuries struck western Europe, concentrating its most severe effects on France. The French government blamed the heat wave for nearly 15,000 deaths, many of them among elderly people who lived by themselves and did not have air-conditioning. Were they victims of global warming or of bad luck—of living in the wrong place at the wrong time?

Before 2003, climate scientists had avoided speculating on such questions. The conventional wisdom was something like this: "Although this is the kind of phenomenon that global warming might make more frequent, it is impossible to attribute this particular event to past emissions of greenhouse gases."

Myles Allen, a physicist at the University of Oxford and author of the above quote, suggested a way to do better: use a metric called fraction of attributable risk (FAR) to quantify the effect of climate change. The FAR requires us to know two numbers: p_0, the probability of a heat wave like the 2003 heat wave before climate change (e.g., before 1800), and p_1, the probability after climate change. For example, if the probability doubles, then we can say that half of the risk is due to climate change. If it triples, then two-thirds of the risk is due to climate change.

Because the FAR is defined purely from data, it does not necessarily have any causal meaning. It turns out, however, that under two mild causal assumptions, it is identical to the probability of necessity. First, we need to assume that the treatment (greenhouse gases) and outcome (heat waves) are not confounded: there is no common cause of each. This is very reasonable, because as far as we know, the only cause of the increase in greenhouse gases is ourselves. Second, we need to assume monotonicity. We discussed this assumption briefly in the last chapter; in this context, it means that the treatment never has the opposite effect from what we expect: that is, greenhouse gases can never protect us from a heat wave.

Provided the assumptions of no confounding and no protection hold, the rung-one metric of FAR is promoted to rung three, where it becomes *PN*. But Allen did not know the causal interpretation of the FAR—it is probably not common knowledge among meteorologists—and this forced him to present his results using somewhat tortuous language.

But what data can we use to estimate the FAR (or *PN*)? We have observed only one such heat wave. We can't do a controlled experiment, because that would require us to control the level of carbon dioxide as if we were flicking a switch. Fortunately, climate scientists have a secret weapon: they can conduct an *in silico* experiment—a computer simulation.

Allen and Peter Stott of the Met Office (the British weather service) took up the challenge, and in 2004 they became the first climate scientists to commit themselves to a causal statement about an individual weather event. Or did they? Judge for yourself. This is what they wrote: "It is very likely that over half the risk of European summer temperature anomalies exceeding a threshold of 1.6° C. is attributable to human influence."

Although I commend Allen and Stott's bravery, it is a pity that their important finding was buried in such a thicket of impenetrable language. Let me unpack this statement and then try to explain why they had to express it in such a convoluted way. First, "temperature anomaly exceeding a threshold of 1.6° C." was their way of defining the outcome. They chose this threshold because the average temperature in Europe that summer was more than 1.6° C above normal, which had never previously happened in recorded history. Their choice balanced the competing objectives of picking an outcome that is sufficiently extreme to capture the effect of global warming but not too closely tailored to the specifics of the 2003 event. Instead of using, for example, the average temperature in France during August, they chose the broader criterion of the average temperature in Europe over the entire summer.

Next, what did they mean by "very likely" and "half the risk"? In mathematical terms, Allen and Stott meant that there was a 90 percent chance that the FAR was over 50 percent. Or, equivalently, there is a 90 percent chance that summers like 2003 are more than twice as likely with current levels of carbon dioxide as they would be with preindustrial levels. Notice that there are two layers of probability here: we are talking about a probability of a probability! No wonder our mind boggles and our eyes swim when we read such a statement. The reason for the double whammy is that the heat wave is subject to two kinds of uncertainty. First, there is uncertainty over the amount of long-term climate change. This is the uncertainty that goes into the first 90 percent figure. Even if we know the amount of long-term climate change exactly, there is uncertainty about the weather in any given year. That is the kind of variability that is built into the 50 percent fraction of attributable risk.

So we have to grant that Allen and Stott were trying to communicate a complicated idea. Nevertheless, one thing is missing from their conclusion: causality. Their statement does not contain even a hint of causation—or maybe just a hint, in the ambiguous and inscrutable phrase "attributable to human influence."

Now compare this with a causal version of the same conclusion: "CO_2 emissions are very likely to have been a necessary cause of the 2003 heat wave." Which sentence, theirs or ours, will you still remember tomorrow? Which one could you explain to your next-door neighbor?

I am not personally an expert on climate change, so I got this example from one of my collaborators, Alexis Hannart of the Franco-Argentine Institute on the Study of Climate and Its Impacts in Buenos Aires, who has been a big proponent of causal analysis in climate science. Hannart draws the causal graph in Figure 8.4. Because Greenhouse Gases is a top-level node in the climate model, with no arrows going into it, he argues that there is no confounding between it and Climate Response. Likewise, he vouches for the no-protection assumption (i.e., greenhouse gases cannot protect us from heat waves).

Hannart goes beyond Allen and Stott and uses our formulas to compute the probability of sufficiency (PS) and of necessity (PN). In the case of the 2003 European heat wave, he finds that PS was extremely low, about 0.0072, meaning that there was no way to predict that this event would happen in this particular year. On the other hand, the probability of necessity PN was 0.9, in agreement with Allen and Stott's results. This means that it is highly likely that, without greenhouse gases, the heat wave would not have happened.

The apparently low value of PS has to be put into a larger context. We don't just want to know the probability of a heat wave this year; we would like to know the probability of a recurrence of such a severe heat wave over a longer time frame—say in the next ten or fifty years. As the time frame lengthens, PN decreases because other possible mechanisms for a heat wave might come into play. On the other hand, PS increases because we are in effect giving the dice more chances to come up snake eyes. So, for

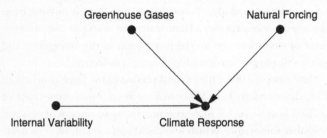

FIGURE 8.4. Causal diagram for the climate change example.

example, Hannart computes that there is an 80 percent probability that climate change will be a sufficient cause of another European heat wave like the 2003 one (or worse) in a two-hundred-year period. That might not sound too terrifying, but that's assuming the greenhouse gas levels of today. In reality, CO_2 levels are certain to continue rising, which can only increase PS and shorten the window of time until the next heat wave.

Can ordinary people learn to understand the difference between necessary and sufficient causes? This is a nontrivial question. Even scientists sometimes struggle. In fact, two conflicting studies came out that analyzed the 2010 heat wave in Russia, when Russia had its hottest summer ever and peat fires darkened the skies of Moscow. One group concluded that natural variability caused the heat wave; another concluded that climate change caused it. In all likelihood, the disagreement occurred because the two groups defined their outcome differently. One group apparently based its argument on PN and got a high likelihood that climate change was the cause, while the other used PS and got a low likelihood. The second group attributed the heat wave to a persistent high-pressure or "blocking pattern" over Russia—which sounds to me like a sufficient cause—and found that greenhouse gases had little to do with this phenomenon. But any study that uses PS as a metric, over a short period, is setting a high bar for proving causation.

Before leaving this example, I would like to comment again on the computer models. Most other scientists have to work very hard to get counterfactual information, for example by painfully combining data from observational and experimental studies. Climate scientists can get counterfactuals very easily from their computer models: just enter in a new number for the carbon dioxide concentration and let the program run. "Easily" is, of course, relative. Behind the simple causal diagram of Figure 8.4 lies a fabulously complex response function, given by the millions of lines of computer code that go into a climate simulation.

This brings up a natural question: How much can we trust the computer simulations? The question has political ramifications, especially here in the United States. However, I will try to give an apolitical answer. I would consider the response function in this example much more credible than the linear models that one sees so often in natural and social sciences. Linear models are often chosen for no good reason other than convenience. By comparison, the climate models reflect more than a century of study by physicists, meteorologists, and climate scientists. They represent the best efforts of a community of scientists to understand the processes that govern our weather and climate. By any normal scientific standards, the climate models are strong and compelling evidence, but with one caveat. Though they are excellent at forecasting the weather a few days ahead, they have never been verified in a prospective trial over century-long timescales, so they could still contain systematic errors that we don't know about.

A WORLD OF COUNTERFACTUALS

I hope that by now it is obvious that counterfactuals are an essential part of how humans learn about the world and how our actions affect it. While we can never walk down both the paths that diverge in a wood, in a great many cases we can know, with some degree of confidence, what lies down each.

Beyond doubt, the variety and richness of causal queries that we can pose to our "inference engine" are greatly enhanced when we can include counterfactuals in the mix. Another very popular kind of query, which I have not discussed here, called the effect of treatment on the treated (ETT), is used to evaluate whether people who gain access to a treatment are those who would benefit most from it. This measure is in many cases superior to the conventional measure of a treatment's effectiveness, the average causal effect (ACE). The ACE, which you can get from a randomized controlled trial, averages treatment efficacy over the entire population. But what if, in actual implementation, those recruited for a treatment program are the ones least likely to benefit from it? To assess the overall effectiveness of the program, ETT measures how adversely treated patients would be affected had they not been treated—a counterfactual measure of critical significance in practical decision making. My former student Ilya Shpitser (now at Johns Hopkins) has now done for ETT what the *do*-calculus did for ACE—provided a complete understanding of when it is estimable from data, given a causal diagram.

Undoubtedly the most popular application of counterfactuals in science today is called mediation analysis. For that reason, I devote a separate chapter to it (Chapter 9). Oddly, many people, especially if using classical mediation analysis techniques, may not realize that they are talking about a counterfactual effect.

In a scientific context, a mediator, or mediating variable, is one that transmits the effect of the treatment to the outcome. We have seen many mediation examples in this book, such as Smoking → Tar → Cancer (where Tar is the mediator). The main question of interest in such cases is whether the mediating variable accounts for the entire effect of the treatment variable or some part of the effect does not require a mediator. We would represent such an effect by a separate arrow leading directly from the treatment to the outcome, such as Smoking → Cancer.

Mediation analysis aims to disentangle the direct effect (which does not pass through the mediator) from the indirect effect (the part that passes through the mediator). The importance is easy to see. If smoking causes lung cancer only through the formation of tar deposits, then we could eliminate the excess cancer risk by giving smokers tar-free cigarettes, such as e-cigarettes. On the other hand, if smoking causes cancer directly or through a different mediator, then e-cigarettes might not solve the problem. At present this medical question is unresolved.

At this point it is probably not obvious to you that direct and indirect effects involve counterfactual statements. It was definitely not obvious to me! In fact, it was one of the biggest surprises of my career. The next chapter tells this story and gives many real-life applications of mediation analysis.

In 1912, a cairn of snow and a cross of skis mark the final resting place of Captain Robert Falcon Scott (right) and the last two men from his ill-fated expedition to the South Pole. Among numerous hardships, Scott's men suffered from scurvy. This part of the tragedy could have been averted if scientists had understood the mechanism by which citrus fruits prevent the disease. (*Source:* left, photograph by Tryggve Gran (presumed); right, photograph by Herbert Ponting. Courtesy of Canterbury Museum, New Zealand.)

9

MEDIATION: THE SEARCH
FOR A MECHANISM

For want of a nail the shoe was lost.
For want of a shoe the horse was lost. . . .
For want of a battle the kingdom was lost.
And all for the want of a nail.

—Anonymous

IN ordinary language, the question "Why?" has at least two versions. The first is straightforward: you see an effect, and you want to know the cause. Your grandfather is lying in the hospital, and you ask, "Why? How could he have had a heart attack when he seemed so healthy?"

But there is a second version of the "Why?" question, which we ask when we want to better understand the connection between a known cause and a known effect. For instance, we observe that Drug *B* prevents heart attacks. Or, like James Lind, we observe that citrus fruits prevent scurvy. The human mind is restless and always wants to know more. Before long we start asking the second version of the question: "Why? What is the mechanism by which citrus fruits prevent scurvy?" This chapter focuses on this second version of "why."

The search for mechanisms is critical to science, as well as to everyday life, because different mechanisms call for different actions when circumstances change. Suppose we run out of oranges. Knowing the mechanism by which oranges work, we can still prevent scurvy. We simply need another source of vitamin C. If we didn't know the mechanism, we might be tempted to try bananas.

The word that scientists use for the second type of "Why?" question is "mediation." You might read in a journal a statement like this: "The effect of Drug B on heart attacks is mediated by its effect on blood pressure." This statement encodes a simple causal model: Drug $A \to$ Blood Pressure $\to$ Heart Attack. In this case, the drug reduces high blood pressure, which in turn reduces the risk of heart attack. (Biologists typically use a different symbol, $A \dashv B$, when cause A inhibits effect B, but in the causality literature it is customary to use $A \to B$ both for positive and negative causes.) Likewise, we can summarize the effect of citrus fruits on scurvy by the causal model Citrus Fruits $\to$ Vitamin C $\to$ Scurvy.

We want to ask certain typical questions about a mediator: Does it account for the entire effect? Does Drug B work exclusively through blood pressure or perhaps through other mechanisms as well? The placebo effect is a common type of mediator in medicine: if a drug acts only through the patient's belief in its benefit, most doctors will consider it ineffective. Mediation is also an important concept in the law. If we ask whether a company discriminated against women when it paid them lower salaries, we are asking a mediation question. The answer depends on whether the observed salary disparity is produced directly in response to the applicant's sex or indirectly, through a mediator such as qualification, over which the employer has no control.

All the above questions require a sensitive ability to tease apart *total effects*, *direct effects* (which do not pass through a mediator), and *indirect effects* (which do). Even defining these terms has been a major challenge for scientists over the past century. Inhibited by the taboos against uttering the word "causation," some tried to define mediation using a causality-free vocabulary. Others dismissed mediation analysis altogether and declared the concepts of direct and indirect effects as "more deceptive than helpful to clear statistical thinking."

For me, too, mediation was a struggle—ultimately one of the most rewarding of my career, because I was wrong at first, and as I was learning from my mistake, I came up with an unexpected solution. For a while, I was of the opinion that indirect effects have no operational implications because, unlike direct effects, they cannot be defined in the language of interventions. It was a personal breakthrough when I realized that they can be defined in terms of counterfactuals and that they can also have important policy implications. They can be quantified only after we have reached the third rung of the Ladder of Causation, and that is why I have placed them at the end of this book. Mediation has flourished in its new habitat and enabled us to quantify, often from the bare data, the portion of the effect mediated by any desired path.

Understandably, due to their counterfactual dressing, indirect effects remain somewhat enigmatic even among champions of the Causal Revolution. I believe that their overwhelming usefulness, however, will eventually overcome any lingering doubts over the metaphysics of counterfactuals. Perhaps they could be compared to irrational and imaginary numbers: they made people uncomfortable at first (hence the name "irrational"), but eventually their usefulness transformed discomfort into delight.

To illustrate this point, I will give several examples of how researchers in various disciplines have gleaned useful insights from mediation analysis. One researcher studied an education reform called "Algebra for All," which at first seemed a failure but later turned into a success. A study of tourniquet use in the Iraq and Afghanistan wars failed to show that it had any benefit; careful mediation analysis explains why the benefit may have been masked in the study.

In summary, over the last fifteen years, the Causal Revolution has uncovered clear and simple rules for quantifying how much of a given effect is direct and how much is indirect. It has transformed mediation from a poorly understood concept with doubtful legitimacy into a popular and widely applicable tool for scientific analysis.

SCURVY: THE WRONG MEDIATOR

I would like to begin with a truly appalling historical example that highlights the importance of understanding the mediator.

One of the earliest examples of a controlled experiment was sea captain James Lind's study of scurvy, published in 1747. In Lind's time scurvy was a terrifying disease, estimated to have killed 2 million sailors between 1500 and 1800. Lind established, as conclusively as anybody could at that time, that a diet of citrus fruit prevented sailors from developing this dread disease. By the early 1800s, scurvy had become a problem of the past for the British navy, as all its ships took to the seas with an adequate supply of citrus fruit. This is usually the point at which history books end the story, celebrating a great triumph of the scientific method.

It seems very surprising, then, that this completely preventable disease made an unexpected comeback a century later, when British expeditions started to explore the polar regions. The British Arctic Expedition of 1875, the Jackson-Harmsworth Expedition to the Arctic in 1894, and most notably the two expeditions of Robert Falcon Scott to Antarctica in 1903 and 1911 all suffered greatly from scurvy.

How could this have happened? In two words: ignorance and arrogance—always a potent combination. By 1900 the leading physicians in Britain had forgotten the lessons of a century before. Scott's physician on the 1903 expedition, Dr. Reginald Koettlitz, attributed scurvy to tainted meat. Further, he added, "the benefit of the so-called 'antiscorbutics' [i.e., scurvy preventatives, such as lime juice] is a delusion." In his 1911 expedition, Scott stocked dried meat that had been scrupulously inspected for signs of decay but no citrus fruits or juices (see Figure 9.1). The trust he placed in the doctor's opinion may have contributed to the tragedy that followed. All of the five men who made it to the South Pole died, two of an unspecified illness that was most likely scurvy. One team member turned back before the pole and made it back alive, but with a severe case of scurvy.

With hindsight, Koettlitz's advice borders on criminal malpractice. How could the lesson of James Lind have been so thoroughly

FIGURE 9.1. Daily rations for the men on Scott's trek to the pole: chocolate, pemmican (a preserved meat dish), sugar, biscuits, butter, tea. Conspicuously absent: any fruit containing vitamin C. (*Source:* Photograph by Herbert Ponting, courtesy of Canterbury Museum, New Zealand.)

forgotten—or worse, dismissed—a century later? The explanation, in part, is that doctors did not really understand how citrus fruits worked against scurvy. In other words, they did not know the mediator.

From Lind's day onward, it had always been believed (but never proved) that citrus fruits prevented scurvy as a result of their acidity. In other words, doctors understood the process to be governed by the following causal diagram:

Citrus Fruits → Acidity → Scurvy

From this point of view, any acid would do. Even Coca-Cola would work (although it had not yet been invented). At first sailors used Spanish lemons; then, for economic reasons, they substituted West Indian limes, which were as acidic as the Spanish lemons but contained only a quarter of the vitamin C. To make things worse, they started "purifying" the lime juice by cooking it, which may have broken down whatever vitamin C it still contained. In other words, they were disabling the mediator.

When the sailors on the 1875 Arctic expedition fell ill with scurvy despite taking lime juice, the medical community was thrown into utter confusion. Those sailors who had eaten freshly killed meat did not get scurvy, while those who had eaten tinned meat did. Koettlitz and others blamed improperly preserved meat as the culprit. Sir Almroth Wright concocted a theory that bacteria in the (supposedly) tainted meat caused "ptomaine poisoning," which then led to scurvy. Meanwhile the theory that citrus fruits could prevent scurvy was consigned to the dustbin.

The situation was not straightened out until the true mediator was discovered. In 1912, a Polish biochemist named Casimir Funk proposed the existence of micronutrients that he called "vitamines" (the *e* was intentional). By 1930 Albert Szent-Gyorgyi had isolated the particular nutrient that prevented scurvy. It was not any old acid but one acid in particular, now known as vitamin C or ascorbic acid (a nod to its "antiscorbutic" past). Szent-Gyorgyi received the Nobel Prize for his discovery in 1937. Thanks to Szent-Gyorgyi, we now know the actual causal path: Citrus Fruits → Vitamin C → Scurvy.

I think that it is fair to predict that scientists will never "forget" this causal path again. And I think the reader will agree that mediation analysis is more than an abstract mathematical exercise.

NATURE VERSUS NURTURE: THE TRAGEDY OF BARBARA BURKS

To the best of my knowledge, the first person to explicitly represent a mediator with a diagram was a Stanford graduate student named Barbara Burks, in 1926. This very little-known pioneer in women's science is one of the true heroes of this book. There is reason to believe that she actually invented path diagrams independently of Sewall Wright. And in regard to mediation, she was ahead of Wright and decades ahead of her time.

Burks's main research interest, throughout her unfortunately brief career, was the role of nature versus nurture in determining human intelligence. Her advisor at Stanford was Lewis Terman, a psychologist famous for developing the Stanford-Binet IQ test and a firm believer that intelligence was inherited, not acquired. Bear in mind that this was the heyday of the eugenics movement, now discredited but at that time legitimized by the active research of people like Francis Galton, Karl Pearson, and Terman.

The nature-versus-nurture debate is, of course, a very old one that continued long after Burks. Her unique contribution was to boil it down to a causal diagram (see Figure 9.2), which she used to ask (and answer) the query "How much of the causal effect is due to the direct path Parental Intelligence → Child's Intelligence (nature), and how much is due to the indirect path Parental Intelligence → Social Status → Child's Intelligence (nurture)?"

In this diagram, Burks has used some double-headed arrows, either to represent mutual causation or simply out of uncertainty about the direction of causation. For simplicity we are going to assume that the main effect of both arrows goes from left to right, which makes Social Status a mediator, so that the parents' intelligence elevates their social standing, and this in turn gives the child

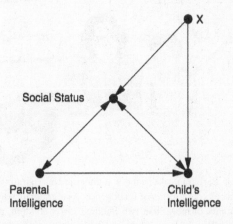

FIGURE 9.2. The nature-versus-nurture debate, as framed by Barbara Burks.

a better opportunity to develop his or her intelligence. The variable X represents "other unmeasured remote causes."

In her dissertation Burks collected data from extensive home visits to 204 families with foster children, who would presumably get only the benefits of nurture and none of the benefits of nature from their foster parents (see Figure 9.3). She gave IQ tests to all of them and to a control group of 105 families without foster children. In addition, she gave them questionnaires that she used to grade various aspects of the child's social environment. Using her data and path analysis, she computed the direct effect of parental IQ on children's IQ and found that only 35 percent, or about one-third, of IQ variation is inherited. In other words, parents with an IQ fifteen points above average would typically have children five points above average.

FIGURE 9.3. Barbara Burks (right) was interested in separating the "nature" and "nurture" components of intelligence. As a graduate student, she visited the homes of more than two hundred foster children, gave them IQ tests, and collected data on their social environment. She was the first researcher other than Sewall Wright to use path diagrams, and in some ways she anticipated Wright. (*Source:* Drawing by Dakota Harr.)

As a disciple of Terman, Burks must have been disappointed to see such a small effect. (In fact, her estimates have held up quite well over time.) So she questioned the then accepted method of analysis, which was to control for Social Status. "The true measure of contribution of a cause to an effect is mutilated," she wrote, "if we have rendered constant variables which may *in part or in whole be caused by either of the two factors whose true relationship is to be measured, or by still other unmeasured remote causes which also affect either of the two isolated factors*" (emphasis in the original). In other words, if you are interested in the total effect of Parental Intelligence on Child's Intelligence, you should not adjust for (render constant) any variable on the pathway between them.

But Burks didn't stop there. Her italicized criterion, translated into modern language, reads that a bias will be introduced if we condition on variables that are (a) effects of either Parental Intelligence or Child's Intelligence, or (b) effects of unmeasured causes of either Parental Intelligence or Child's Intelligence (such as X in Figure 9.2).

These criteria were far ahead of their time and unlike anything that Sewall Wright had written. In fact, criterion (b) is one of the earliest examples ever of collider bias. If we look at Figure 9.2, we see that Social Status is a collider (Parental Intelligence → Social Status ← X). Therefore, controlling for Social Status opens the back-door path Parental Intelligence → Social Status ← X → Child's Intelligence. Any resulting estimate of the indirect and direct effects will be biased. Because statisticians before (and after) Burks did not think in terms of arrows and diagrams, they were totally immersed in the myth that, while simple correlation has no causal implications, controlled correlation (or partial regression coefficients, see p. 222) is a step in the direction of causal explanation.

Burks was not the first person to discover the collider effect, but one can argue that she was the first to characterize it generally in graphical terms. Her criterion (b) applies perfectly to the examples of M-bias in Chapter 4. Hers is the first warning ever against conditioning on a pretreatment factor, a habit deemed safe by all twentieth-century statisticians and oddly still considered safe by some.

Now put yourself in Barbara Burks's shoes. You've just discovered that all your colleagues have been controlling for the wrong variables. You have two strikes against you: you're only a student, and you're a woman. What do you do? Do you put your head down, pretend to accept the conventional wisdom, and communicate with your colleagues in their inadequate vocabulary?

Not Barbara Burks! She titled her first published paper "On the Inadequacy of the Partial and Multiple Correlation Technique" and started it out by saying, "Logical considerations lead to the conclusion that the techniques of partial and multiple correlation are fraught with dangers that seriously restrict their applicability." Fighting words from someone who doesn't have a PhD yet! As Terman wrote, "Her ability was somewhat tempered by her tendency to rub people the wrong way. I think the trouble lay partly in the fact that she was more aggressive in standing up for her own ideas than many teachers and male graduate students liked." Evidently Burks was ahead of her time in more ways than one.

Burks may actually have invented path diagrams independently of Sewall Wright, who preceded her by only six years. We can say for sure that she didn't learn them in any class. Figure 9.2 is the first appearance of a path diagram outside Sewall Wright's work and the first ever in the social or behavioral sciences. True, she credits Wright at the very end of her 1926 paper, but she does so in a manner that looks like a last-minute addition. I have a hunch that she found out about Wright's diagrams only after she had drawn her own, possibly after being tipped off by Terman or an astute reviewer.

It is fascinating to wonder what Burks might have become, had she not been a victim of her times. After obtaining her doctorate she never managed to get a job as a professor at a university, for which she was certainly qualified. She had to make do with less secure research positions, for example at the Carnegie Institution. In 1942 she got engaged, which one might have expected to mark an upturn in her fortunes; instead, she went into a deep depression. "I am convinced that, whether right or not, she was sure some sinister change was going on in her brain, from which she could never recover," her mother, Frances Burks, wrote to Terman. "So in tenderest love to us all she chose to spare us the grief of sharing with her the spectacle of such a tragic decline." On May 25, 1943, at age forty, she jumped to her death from the George Washington Bridge in New York.

But ideas have a way of surviving tragedies. When sociologists Hubert Blalock and Otis Duncan resuscitated path analysis in the 1960s, Burks's paper served as the source of their inspiration. Duncan explained that one of his mentors, William Fielding Ogburn, had briefly mentioned path coefficients in his 1946 lecture on partial correlations. "Ogburn had a report of a brief paper by Wright, the one that dealt with Burks' material, and I acquired this reprint," Duncan said.

So there we have it! Burks's 1926 paper got Wright interested in the inappropriate use of partial correlations. Wright's response found its way into Ogburn's lecture twenty years later and implanted itself into Duncan's mind. Twenty years after that, when Duncan read Blalock's work on path diagrams, it called back this half-forgotten memory from his student years. It's truly amazing to see how this fragile butterfly of an idea fluttered almost unnoticed through two generations before reemerging triumphantly into the light.

IN SEARCH OF A LANGUAGE (THE BERKELEY ADMISSIONS PARADOX)

Despite Burks's early work, half a century later statisticians were struggling even to express the idea of, let alone estimate, direct and indirect effects. A case in point is a well-known paradox, related to Simpson's paradox but with a twist.

In 1973 Eugene Hammel, an associate dean at the University of California, noticed a worrisome trend in the university's admission rates for men and women. His data showed that 44 percent of the men who applied to graduate school at Berkeley had been accepted, compared to only 35 percent of the women. Gender discrimination was coming to wide public attention, and Hammel didn't want to wait for someone else to start asking questions. He decided to investigate the reasons for the disparity.

Graduate admissions decisions, at Berkeley as at other universities, are made by individual departments rather than by the university as a whole. So it made sense to look at the admissions data
department by department to isolate the culprit. But when he did
so, Hammel discovered an amazing fact. Department after department, the admissions decisions were consistently more favorable to
women than to men. How could this be?

At this point Hammel did something smart: he called a statistician. Peter Bickel, when asked to look at the data, immediately
recognized a form of Simpson's paradox. As we saw in Chapter 6,
Simpson's paradox refers to a trend that seems to go one direction
in each layer of a population (women are accepted at a higher rate
in each department) but in the opposite direction for the whole
population (men are accepted at a higher rate in the university as
a whole). We also saw in Chapter 6 that the correct resolution of
the paradox depends very much on the question you want to answer. In this case the question is clear: Is the university (or someone
within the university) discriminating against women?

When I first told my wife about this example, her reaction was,
"It's impossible. If each department discriminates one way, the
school cannot discriminate the other way." And she's right! The
paradox offends our understanding of discrimination, which is a
causal concept, involving preferential response to an applicant's
reported sex. If all actors prefer one sex over the other, the group
as a whole must show that same preference. If the data seem to
say otherwise, it must mean that we are not processing the data
properly, in accordance with the logic of causation. Only with such
logic, and with a clear causal story, can we determine the university's innocence or guilt.

In fact, Bickel and Hammel found a causal story that completely
satisfied them. They wrote an article, published in *Science* magazine in 1975, proposing a simple explanation: women were rejected
in greater numbers because they applied to harder departments to
get into.

To be specific, a higher proportion of females than males applied to departments in the humanities and social sciences. There
they faced a double whammy: the number of students applying to

get in was greater, and the number of places for those students was smaller. On the other hand, females did not apply as often to departments like mechanical engineering, which were easier to get into. These departments had more money and more spaces for graduate students—in short, a higher acceptance rate.

Why did women apply to departments that are harder to get into? Perhaps they were discouraged from applying to technical fields because they had more math requirements or were perceived as more "masculine." Perhaps they had been discriminated against at earlier stages of their education: society tended to push women away from technical fields, as Barbara Burks's story shows far too clearly. But these circumstances were not under Berkeley's control and hence would not constitute discrimination by the university. Bickel and Hammel concluded, "The campus as a whole did not engage in discrimination against women applicants."

At least in passing, I would like to take note of the precision of Bickel's language in this paper. He carefully distinguishes between two terms that, in common English, are often taken as synonyms: "bias" and "discrimination." He defines bias as "a pattern of association between a particular decision and a particular sex of applicant." Note the words "pattern" and "association." They tell us that bias is a phenomenon on rung one of the Ladder of Causation. On the other hand, he defines discrimination as "the exercise of decision influenced by the sex of the applicant when that is immaterial to the qualifications for entry." Words like "exercise of decision," "influence," and "immaterial" are redolent of causation, even if Bickel could not bring himself to utter that word in 1975. Discrimination, unlike bias, belongs on rung two or three of the Ladder of Causation.

In his analysis, Bickel felt that the data should be stratified by department because the departments were the decision-making units. Was this the right call? To answer that question, we start by drawing a causal diagram (Figure 9.4). It is also very illuminating to look at the definition of discrimination in US case law. It uses counterfactual terminology, a clear signal that we have climbed to level three of the Ladder of Causation. In *Carson v. Bethlehem Steel Corp.* (1996), the Seventh Circuit Court wrote, "The central

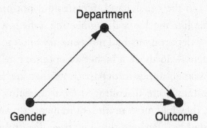

FIGURE 9.4. Causal diagram for Berkeley admission paradox—
simple version.

question in any employment-discrimination case is whether the employer would have taken the same action had the employee been of
a different race (age, sex, religion, national origin, etc.) and everything else had been the same." This definition clearly expresses the
idea that we should disable or "freeze" all causal pathways that
lead from gender to admission through any other variable (e.g.,
qualification, choice of department, etc.). In other words, discrimination equals the direct effect of gender on the admission outcome.

We have seen before that conditioning on a mediator is incorrect if we want to estimate the total effect of one variable on another. But in a case of discrimination, according to the court, it is
not the total effect but the direct effect that matters. Thus Bickel
and Hammel are vindicated: under the assumptions shown in Figure 9.4, they were right to partition the data by departments, and
their result provides a valid estimate of the direct effect of Gender
on Outcome. They succeeded even though the language of direct
and indirect effects was not available to Bickel in 1973.

However, the most interesting part of this story is not the original paper that Bickel and Hammel wrote but the discussion that
followed it. After their paper was published, William Kruskal of
the University of Chicago wrote a letter to Bickel arguing that their
explanation did not really exonerate Berkeley. In fact, Kruskal

queried whether any purely observational study (as opposed to a randomized experiment—say, using fake application forms) could ever do so.

To me their exchange of letters is fascinating. It is not very often that we can witness two great minds struggling with a concept (causation) for which they lacked an adequate vocabulary. Bickel would later go on to earn a MacArthur Foundation "genius" grant in 1984. But in 1975 he was at the beginning of his career, and it must have been both an honor and a challenge for him to match wits with Kruskal, a giant of the American statistics community.

In his letter to Bickel, Kruskal pointed out that the relation between Department and Outcome could have an unmeasured confounder, such as State of Residence. He worked out a numerical example for a hypothetical university with two sex-discriminating departments that produce exactly the same data as in Bickel's example. He did this by assuming that both departments accept all in-state males and out-of-state females and reject all out-of-state males and in-state females and that this is their only decision criterion. Clearly this admissions policy is a blatant, textbook example of discrimination. But because the total numbers of applicants of each gender accepted and rejected were exactly the same as in Bickel's example, Bickel would have to conclude that there was no discrimination. According to Kruskal, the departments appear innocent because Bickel has controlled for only one variable instead of two.

Kruskal put his finger exactly on the weak spot in Bickel's paper: the lack of a clearly justified criterion for determining which variables to control for. Kruskal did not offer a solution, and in fact his letter despairs of ever finding one.

Unlike Kruskal, we can draw a diagram and see exactly what the problem is. Figure 9.5 shows the causal diagram representing Kruskal's counterexample. Does it look slightly familiar? It should! It is exactly the same diagram that Barbara Burks drew in 1926, but with different variables. One is tempted to say, "Great minds think alike," but perhaps it would be more appropriate to say that great problems attract great minds.

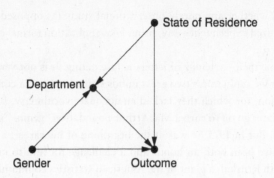

FIGURE 9.5. Causal diagram for Berkeley admissions paradox—
Kruskal's version.

Kruskal argued that the analysis in this situation should control for both the department and the state of residence, and a look
at Figure 9.5 explains why this is so. To disable all but the direct
path, we need to stratify by department. This closes the indirect
path Gender → Department → Outcome. But in so doing, we open
the spurious path Gender → Department ← State of Residence →
Outcome, because of the collider at Department. If we control for
State of Residence as well, we close this path, and therefore any
correlation remaining must be due to the (discriminatory) direct
path Gender → Outcome. Lacking diagrams, Kruskal had to convince Bickel with numbers, and in fact his numbers showed the
same thing. If we do not adjust for any variables, then females
have a lower admission rate. If we adjust for Department, then females appear to have a higher admission rate. If we adjust for both
Department and State of Residence, then once again the numbers
show a lower admission rate for females.

From arguments like this, you can see why the concept of mediation aroused (and still arouses) such suspicions. It seems un-
stable and hard to pin down. First the admission rates are biased
against women, then against men, then against women. In his re-
ply to Kruskal, Bickel continued to maintain that conditioning on
a decision-making unit (Department) is somehow different from

conditioning on a criterion for a decision (State of Residence). But he did not sound at all confident about it. He asks plaintively, "I see a nonstatistical question here: What do we mean by bias?" Why does the bias sign change depending on the way we measure it? In fact he had the right idea when he distinguished between bias and discrimination. Bias is a slippery statistical notion, which may disappear if you slice the data a different way. Discrimination, as a causal concept, reflects reality and must remain stable.

The missing phrase in both their vocabularies was "hold constant." To disable the indirect path from Gender to Outcome, we must hold constant the variable Department and then tweak the variable Gender. When we hold the department constant, we prevent (figuratively speaking) the applicants from choosing which department to apply to. Because statisticians do not have a word for this concept, they do something superficially similar: they condition on Department. That was exactly what Bickel had done: he looked at the data department-by-department and concluded that there was no evidence of discrimination against women. That procedure is valid when Department and Outcome are unconfounded; in that case, seeing is the same as doing. But Kruskal correctly asked, "What if there is a confounder, State of Residence?" He probably didn't realize that he was following in the footsteps of Burks, who had drawn essentially the same diagram.

I cannot stress enough how often this blunder has been repeated over the years—conditioning on the mediator instead of holding the mediator constant. For that reason I call it the Mediation Fallacy. Admittedly, the blunder is harmless if there is no confounding of the mediator and the outcome. However, if there is confounding, it can completely reverse the analysis, as Kruskal's numerical example showed. It can lead the investigator to conclude there is no discrimination when in fact there is.

Burks and Kruskal were unusual in recognizing the Mediation Fallacy as a blunder, although they didn't exactly offer a solution. R. A. Fisher fell victim to the same blunder in 1936, and eighty years later statisticians are still struggling with the problem. Fortunately there has been huge progress since the time of Fisher. Epidemiologists, for example, know now that one has to watch out for confounders between mediator and outcome. Yet those who eschew the language of diagrams (some economists still do) complain and confess that it is a torture to explain what this warning means.

Thankfully, the problem that Kruskal once called "perhaps insoluble" was solved two decades ago. I have this strange feeling that Kruskal would have enjoyed the solution, and in my fantasy I imagine showing him the power of the *do*-calculus and the algorithmization of counterfactuals. Unfortunately, he retired in 1990, just when the rules of *do*-calculus were being shaped, and he died in 2005.

I'm sure that some readers are wondering: What finally happened in the Berkeley case? The answer is, nothing. Hammel and Bickel were convinced that Berkeley had nothing to worry about, and indeed no lawsuits or federal investigations ever materialized. The data hinted at reverse discrimination against males, and in fact there was explicit evidence of this: "In most of the cases involving favored status for women it appears that the admissions committees were seeking to overcome long-established shortages of women in their fields," Bickel wrote. Just three years later, a lawsuit over affirmative action on another campus of the University of California went all the way to the Supreme Court. Had the Supreme Court struck down affirmative action, such "favored status for women" might have become illegal. However, the Supreme Court upheld affirmative action, and the Berkeley case became a historical footnote.

A wise man leaves the final word not with the Supreme Court but with his wife. Why did mine have such a strong intuitive conviction that it is utterly impossible for a school to discriminate while each of its departments acts fairly? It is a theorem of causal calculus similar to the sure-thing principle. The sure-thing principle, as Leonard Savage originally stated it, pertains to total effects, while this theorem holds for direct effects. The very definition of a direct effect on a global level relies on aggregating direct effects in the subpopulations.

To put it succinctly, local fairness everywhere implies global fairness. My wife was right.

DAISY, THE KITTENS AND INDIRECT EFFECTS

So far we have discussed the concepts of direct and indirect effects in a vague and intuitive way, but I have not given them a precise scientific meaning. It is long past time for us to rectify this omission.

Let's start with the direct effect, because it is undoubtedly easier, and we can define a version of it using the *do*-calculus (i.e., at rung two of the Ladder of Causation). We'll consider first the simplest case, which includes three variables: a treatment X, an outcome Y, and a mediator M. We get the direct effect of X on Y when we "wiggle" X without allowing M to change. In the context of the Berkeley admissions paradox example, we force everybody to apply to the history department—that is, we $do(M = 0)$. We randomly assign some people to report their sex (on the application) as male ($do(X = 1)$) and some to report it as female ($do(X = 0)$), regardless of their actual genders. Then we observe the difference in admission rates between the two reporting groups. The result is called the controlled direct effect, or CDE(0). In symbols,

$$CDE(0) =$$
$$P(Y = 1 \mid do(X = 1), do(M = 0)) - P(Y = 1 \mid do(X = 0), do(M = 0)) \quad (9.1)$$

The "0" in CDE(0) indicates that we forced the mediator to take on the value zero. We could also do the same experiment, forcing everybody to apply to engineering: $do(M = 1)$. We would denote the resulting controlled direct effect as CDE(1).

Already we see one difference between direct effects and total effects: we have two different versions of the controlled direct effect, CDE(0) and CDE(1). Which one is right? One option is simply to report both versions. Indeed, it is not unthinkable that one department will discriminate against females and the other against males, and it would be interesting to find out who does what. That was, after all, Hammel's original intention.

However, I would not recommend running this experiment, and here is why. Imagine an applicant named Joe whose lifetime dream is to study engineering and who happened to be (randomly) assigned to apply to the history department. Having sat on a few

admissions committees, I can categorically vow that Joe's application would look awfully strange to the committee. His A+ in elec-
tromagnetic waves and B− in European nationalism would totally
distort the committee's decision, regardless of whether he marked
"male" or "female" on his application. The proportion of males
and females admitted under these distortions would hardly reflect
the admissions policy compared to applicants who normally apply
to the history department.

Luckily, an alternative avoids the pitfalls of this overcontrolled
experiment. We instruct the applicants to report a randomized gender but to apply to the department they would have preferred. We
call this the natural direct effect (NDE), because every applicant
ends up in a department of his or her choice. The "would have"
phrasing is a clue that NDE's formal definition requires counterfactuals. For readers who enjoy mathematics, here is the definition
expressed as a formula:

$$\text{NDE} = P(Y_{M = M_0} = 1 \mid do(X = 1)) - P(Y_{M = M_0} = 1 \mid do(X = 0))$$

$$(9.2)$$

The interesting term is the first, which stands for the probability
that a female student selecting a department of her choice ($M = M_0$)
would be admitted if she faked her sex to read "male" ($do(X = 1)$).
Here the choice of department is governed by the actual sex while
admission is decided by the reported (fake) sex. Since the former
cannot be mandated, we cannot translate this term to one involving do-operators; we need to invoke the counterfactual subscript.

Now you know how we define the controlled direct effect and
the natural direct effect, but how do we compute them? The task is
simple for the controlled direct effect; because it can be expressed
as a do-expression, we need only use the laws of do-calculus to reduce the do-expressions to see-expressions (i.e., conditional probabilities, which can be estimated from observational data).

The natural direct effect poses a greater challenge, though, because it cannot be defined in a do-expression. It requires the language of counterfactuals, and hence it cannot be estimated using
the do-calculus. It was one of the greatest thrills in my life when

I managed to strip the formula for the NDE from all of its counterfactual subscripts. The result, called the Mediation Formula, makes the NDE a truly practical tool because we can estimate it from observational data.

Indirect effects, unlike direct effects, have no "controlled" version because there is no way to disable the direct path by holding some variable constant. But they do have a "natural" version, the natural indirect effect (NIE), which is defined (like NDE) using counterfactuals. To motivate the definition, I will consider a somewhat playful example that my coauthor suggested.

My coauthor and his wife adopted a dog named Daisy, a rambunctious poodle-and-Chihuahua mix with a mind of her own. Daisy was not as easy to house-train as their previous dog, and after several weeks she was still having occasional "accidents" inside the house. But then something very odd happened. Dana and his wife brought home three foster kittens from the animal shelter, and the "accidents" stopped. The foster kittens remained with them for three weeks, and Daisy did not break her training a single time during that period.

Was it just coincidence, or had the kittens somehow inspired Daisy to civilized behavior? Dana's wife suggested that the kittens might have given Daisy a sense of belonging to a "pack," and she would not want to mess up the area where the pack lived. This theory was reinforced when, a few days after the kittens went back to the shelter, Daisy started urinating in the house again, as if she had never heard of good manners.

But then it occurred to Dana that something else had changed when the kittens arrived and departed. While the kittens had been there, Daisy had to be either separated from them or carefully supervised. So she spent long periods in her crate or being closely watched by a human, even leashed to a human. Both interventions, crating and leashing, also happen to be recognized methods for housebreaking.

When the kittens left, the Mackenzies stopped the intensive supervision, and the uncouth behavior returned. Dana hypothesized that the effect of the kittens was not direct (as in the pack theory) but indirect, mediated by crating and supervision. Figure 9.6

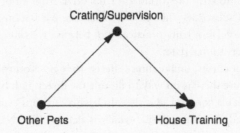

FIGURE 9.6. Causal diagram for Daisy's house training.

shows a causal graph. At this point, Dana and his wife tried an experiment. They treated Daisy as they would have with kittens around, keeping her in a crate and supervising her carefully outside the crate. If the accidents stopped, they could reasonably conclude that the mediator was responsible. If they didn't stop, then the direct effect (the pack psychology) would become more plausible.

In the hierarchy of scientific evidence, their experiment would be considered very shaky—certainly not one that could ever be published in a scientific journal. A real experiment would have to be carried out on more than just one dog and in both the presence and absence of the kittens. Nevertheless, it is the causal logic behind the experiment that concerns us here. We are intending to recreate what would have happened had the kittens not been present and had the mediator been set to the value it would take with the kittens present. In other words, we remove the kittens (intervention number one) and supervise the dog as we would if the kittens were present (intervention number two).

When you look carefully at the above paragraph, you might notice two "would haves," which are counterfactual conditions. The kittens were present when the dog changed her behavior—but we ask what would have happened if they had not been present. Likewise, if the kittens had not been present, Dana would not have supervised Daisy—but we ask what would have happened if he had.

You can see why statisticians struggled for so long to define indirect effects. If even a single counterfactual was outlandish, then double-nested counterfactuals were completely beyond the pale. Nevertheless, this definition conforms closely with our natural intuition about causation. Our intuition is so compelling that Dana's wife, with no special training, readily understood the logic of the proposed experiment.

For readers who are comfortable with formulas, here is how to define the NIE that we have just described in words:

$$\text{NIE} = P(Y_{M = M_1} = 1 \mid do(X = 0)) - P(Y_{M = M_0} = 1 \mid do(X = 0))$$

$$(9.3)$$

The first P term is the outcome of the Daisy experiment: the probability of successful house training ($Y = 1$), given that we do not introduce other pets ($X = 0$) but set the mediator to the value it would have if we had introduced them ($M = M_1$). We contrast this with the probability of successful house training under "normal" conditions, with no other pets. Note that the counterfactual, M_1, has to be computed for each animal on a case-by-case basis: different dogs might have different needs for Crating/Supervision. This puts the indirect effect out of reach of the do-calculus. It may also render the experiment unfeasible, because the experimenter may not know $M_1(u)$ for a particular dog u. Nevertheless, assuming there is no confounding between M and Y, the natural indirect effect can still be computed. It is possible to remove all the counterfactuals from the NIE and arrive at a Mediation Formula for it, like the one for the NDE. This quantity, which requires information from the third rung of the Ladder of Causation, can nevertheless be reduced to an expression that can be computed with rung-one data. Such a reduction is only possible because we have made an assumption of no confounding, which, owing to the deterministic nature of the equations in a structural causal model, is on rung three.

To finish Daisy's story, the experiment was inconclusive. It's questionable whether Dana and his wife monitored Daisy as carefully as they would have if they had been keeping her away from kittens. (So it's not clear that M was truly set to M_1.) With patience and time—it took several months—Daisy eventually learned to "do her business" outside. Even so, Daisy's story holds some useful lessons. Simply by being attuned to the possibility of a mediator, Dana was able to conjecture another causal mechanism. That mechanism had an important practical consequence: he and his wife did not have to keep the house filled with a foster kitten "pack" for the rest of Daisy's life.

MEDIATION IN LINEAR WONDERLAND

When you first hear about counterfactuals, you might wonder if such an elaborate machinery is really needed to express an indirect effect. Surely, you might argue, an indirect effect is simply what is left over after you take away the direct effect. Alternatively, we could write,

$$\text{Total Effect} = \text{Direct Effect} + \text{Indirect Effect} \qquad (9.4)$$

The short answer is that this does not work in models that involve interactions (sometimes called moderation). For example, imagine a drug that causes the body to secrete an enzyme that acts as a catalyst: it combines with the drug to cure a disease. The total effect of the drug is, of course, positive. But the direct effect is zero, because if we disable the mediator (for example, by preventing the body from stimulating the enzyme), the drug will not work. The indirect effect is also zero, because if we don't receive the drug and do artificially get the enzyme, then the disease will not be cured. The enzyme itself has no curing power. Thus Equation 9.4 does not hold: the total effect is positive but the direct and indirect effects are zero.

However, Equation 9.4 does hold automatically in one situation, with no apparent need to invoke counterfactuals. That is the case of a linear causal model, of the sort that we saw in Chapter 8. As discussed there, linear models do not allow interactions, which can be both a virtue and a drawback. It is a virtue in the sense that it makes mediation analysis much easier, but it is a drawback if we want to describe a real-world causal process that does involve interactions.

Because mediation analysis is so much easier for linear models, let's see how it is done and what the pitfalls are. Suppose we have a causal diagram that looks like Figure 9.7. Because we are working with a linear model, we can represent the strength of each effect with a single number. The labels (path coefficients) indicate that increasing the Treatment variable by one unit will increase the Mediator variable by two units. Similarly, a one-unit increase in Mediator will increase Outcome by three units, and a one-unit increase in Treatment will increase Outcome by seven units. These are all direct effects. Here we come to the first reason why linear models are so simple: direct effects do not depend on the level of the mediator. That is, the controlled direct effect CDE(m) is the same for all values m, and we can simply speak of "the" direct effect.

What would be the total effect of an intervention that causes Treatment to increase by one unit? First, this intervention directly causes Outcome to increase by seven units (if we hold Mediator constant). It also causes Mediator to increase by two units. Finally, because each one-unit increase in Mediator directly causes a three-unit increase in Outcome, a two-unit increase in Mediator will lead to an additional six-unit increase in Outcome. So the net increase in Outcome, from both causal pathways, will be thirteen units. The

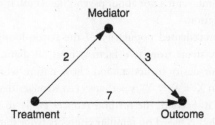

FIGURE 9.7. Example of a linear model (path diagram) with mediation.

first seven units correspond to the direct effect, and the remaining six units correspond to the indirect effect. Easy as pie!

In general, if there is more than one indirect pathway from X to Y, we evaluate the indirect effect along each pathway by taking the product of all the path coefficients along that pathway. Then we get the total indirect effect by adding up all the indirect causal pathways. Finally, the total effect of X on Y is the sum of the direct and indirect effects. This "sum of products" rule has been used since Sewall Wright invented path analysis, and, formally speaking, it indeed follows from the *do*-operator definition of total effect.

In 1986, Reuben Baron and David Kenny articulated a set of principles for detecting and evaluating mediation in a system of equations. The essential principles are, first, that the variables are all related by linear equations, which are estimated by fitting them to the data. Second, direct and indirect effects are computed by fitting two equations to the data: one with the mediator included and one with the mediator excluded. Significant change in the coefficients when the mediator is introduced is taken as evidence of mediation.

The simplicity and plausibility of the Baron-Kenny method took the social sciences by storm. As of 2014, their article ranks thirty-third on the list of most frequently cited scientific papers of all time. As of 2017, Google Scholar reports that 73,000 scholarly articles have cited Baron and Kenny. Just think about that! They've been cited more times than Albert Einstein, more than Sigmund Freud, more than almost any other famous scientist you can think of. Their article ranks second among all papers in psychology and psychiatry, and yet it's not about psychology at all. It's about non-causal mediation.

The unprecedented popularity of the Baron-Kenny approach undoubtedly stems from two factors. First, mediation is in high demand. Our desire to understand "how nature works" (i.e., to find the M in $X \rightarrow M \rightarrow Y$) is perhaps even stronger than our desire to quantify it. Second, the method reduces easily to a cookbook procedure that is based on familiar concepts from statistics, a discipline that has long claimed to have exclusive ownership of objectivity and empirical validity. So hardly anyone noticed the grand leap forward involved, the fact that a causal quantity (mediation) was defined and assessed by purely statistical means.

However, cracks in this regression-based edifice began to appear in the early 2000s, when practitioners tried to generalize the sum-of-products rule to nonlinear systems. That rule involves two assumptions—effects along distinct paths are additive, and path coefficients along one path multiply—and both of them lead to wrong answers in nonlinear models, as we will see below.

It has taken a long time, but the practitioners of mediation analysis have finally woken up. In 2001, my late friend and colleague Rod McDonald wrote, "I think the best way to discuss the question of detecting or showing moderation or mediation in a regression is to set aside the entire literature on these topics and start from scratch." The latest literature on mediation seems to heed McDonald's advice; counterfactual and graphical methods are pursued much more actively than the regression approach. And in 2014, the father of the Baron-Kenny approach, David Kenny, posted a new section on his website called "causal mediation analysis." Though I would not call him a convert yet, Kenny clearly recognizes that times are changing and that mediation analysis is entering a new era.

For now, let's look at one very simple example of how our expectations go wrong when we leave Linear Wonderland. Consider Figure 9.8, a slight modification of Figure 9.7, where a job applicant will decide to take a job if and only if the salary offered exceeds a certain threshold value, in our case ten. The salary offer is determined, as shown in the diagram, by 7 × Education + 3 × Skill. Note that the functions determining Skill and Salary are still assumed to be linear, but the relationship of Salary to Outcome is nonlinear, because it has a threshold effect.

Let us compute, for this model, the total, direct, and indirect effects associated with increasing Education by one unit. The total effect is clearly equal to one, because as Education shifts from zero to one, Salary goes from zero to $(7 \times 1) + (3 \times 2) = 13$, which is above the threshold of ten, making Outcome switch from zero to one.

Remember that the natural indirect effect is the expected change in the outcome, given that we make no change to Education but

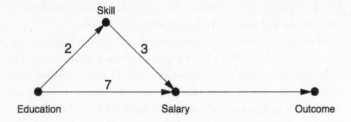

FIGURE 9.8. Mediation combined with a threshold effect.

set Skill at the level it would take if we had increased Education by one. It's easy to see that in this case, Salary goes from zero to 2×3 = 6. This is below the threshold of ten, so the applicant will turn the offer down. Thus NIE = 0.

Now what about the direct effect? As mentioned before, we have the problem of figuring out what value to hold the mediator at. If we hold Skill at the level it had before we changed Education, then Salary will increase from zero to seven, making Outcome = 0. Thus, CDE(0) = 0. On the other hand, if we hold Skill at the level it attains after the change in Education (namely two), Salary will increase from six to thirteen. This changes the Outcome from zero to one, because thirteen is above the applicant's threshold for accepting the job offer. So CDE(2) = 1.

Thus, the direct effect is either zero or one depending on the constant value we choose for the mediator. Unlike in Linear Wonderland, the choice of a value for the mediator makes a difference, and we have a dilemma. If we want to preserve the additive principle, Total Effect = Direct Effect + Indirect Effect, we need to use CDE(2) as our definition of the causal effect. But this seems arbitrary and even somewhat unnatural. If we are contemplating a change in Education and we want to know its direct effect, we would most likely want to keep Skill at the level it already has. In other words, it makes more intuitive sense to use CDE(0) as our direct effect. Not only that, this agrees with the natural direct effect in this example. But then we lose additivity: Total Effect ≠ Direct Effect + Indirect Effect.

However—quite surprisingly—a somewhat modified version of additivity does hold true, not only in this example but in general. Readers who don't mind doing a little computation might be interested in computing the NIE of going back from $X = 1$ to $X = 0$. In this case the salary offer drops from thirteen to seven, and the Outcome drops from one to zero (i.e., the applicant does not accept the offer). So computed in the reverse direction, NIE $= -1$. The cool and amazing fact is that

$$\text{Total Effect } (X = 0 \rightarrow X = 1) =$$
$$\text{NDE } (X = 0 \rightarrow X = 1) - \text{NIE } (X = 1 \rightarrow X = 0)$$

or in this case, $1 = 0 - (-1)$. This is the "natural effects" version of the additivity principle, only it is a subtractivity principle! I was extremely happy to see this version of additivity emerging from the analysis, despite the nonlinearity of the equations.

A staggering amount of ink has been spilled on the "right" way to generalize direct and indirect effects from linear to nonlinear models. Unfortunately, most of the articles go at the problem backward. Instead of rethinking from scratch what we mean by direct and indirect effects, they start from the supposition that we only have to tweak the linear definitions a little bit. For example, in Linear Wonderland we saw that the indirect effect is given by a product of two path coefficients. So some researchers tried to define the indirect effect in the form of a product of two quantities, one measuring the effect of X on M, the other the effect of M on Y. This came to be known as the "product of coefficients" method. But we also saw that in Linear Wonderland the indirect effect is given by the difference between the total effect and the direct effect. So another, equally dedicated group of researchers defined the indirect effect as a difference of two quantities, one measuring the total effect, the other the direct effect. This came to be known as the "difference in coefficients" method.

Which of these is right? Neither! Both groups of researchers confused the procedure with the meaning. The procedure is mathematical; the meaning is causal. In fact, the problem goes even deeper: the indirect effect never had a meaning for regression analysts outside the bubble of linear models. The indirect effect's only meaning was as the outcome of an algebraic procedure ("multiply the path coefficients"). Once that procedure was taken away from them, they were cast adrift, like a boat without an anchor.

One reader of my book *Causality* described this lost feeling beautifully in a letter to me. Melanie Wall, now at Columbia University, used to teach a modeling course to biostatistics and public health students. One time, she explained to her students as usual how to compute the indirect effect by taking the product of direct path coefficients. A student asked her what the indirect effect meant. "I gave the answer that I always give, that the indirect effect is the effect that a change in X has on Y through its relationship with the mediator, Z," Wall told me.

But the student was persistent. He remembered how the teacher had explained the direct effect as the effect remaining after holding the mediator fixed, and he asked, "Then what is being held constant when we interpret an indirect effect?"

Wall didn't know what to say. "I'm not sure I have a good answer for you," she said. "How about I get back to you?"

This was in October 2001, just four months after I had presented a paper on causal mediation at the Uncertainty in Artificial Intelligence conference in Seattle. Needless to say, I was eager to impress Melanie with my newly acquired solution to her puzzle, and I wrote to her the same answer I have given you here: "The indirect effect of X on Y is the increase we would see in Y while holding X constant and increasing M to whatever value M would attain under a unit increase in X."

I am not sure if Melanie was impressed with my answer, but her inquisitive student got me thinking, quite seriously, about how science progresses in our times. Here we are, I thought, forty years after Blalock and Duncan introduced path analysis to social science. Dozens of textbooks and hundreds of research papers are published every year on direct and indirect effects, some with oxymoronic titles like "Regression-Based Approach to Mediation." Each generation passes along to the next the received wisdom that the indirect effect is just the product of two other effects, or the difference between the total and direct effects. Nobody dares to ask the simple question "But what does the indirect effect mean in the first place?" Just like the boy in Hans Christian Andersen's fable "The Emperor's New Clothes," we needed an innocent student with unabashed chutzpah to shatter our faith in the oracular role of scientific consensus.

EMBRACE THE "WOULD-HAVES"

At this point I should tell my own conversion story, because for quite a while I was stymied by the same question that puzzled Melanie Wall's student.

I wrote in Chapter 4 about Jamie Robins (Figure 9.9), a pioneering statistician and epidemiologist at Harvard University who, together with Sander Greenland at the University of California, Los Angeles, is largely responsible for the widespread adoption of graphical models in epidemiology today. We collaborated for a couple of years, from 1993 to 1995, and he got me thinking about the problem of sequential intervention plans, which was one of his principal research interests.

FIGURE 9.9. Jamie Robins, a pioneer of causal inference in epidemiology. (*Source:* Photograph by Kris Snibbe, courtesy of Harvard University Photo Services.)

Years earlier, as an expert in occupational health and safety, Robins had been asked to testify in court about the likelihood that chemical exposure in the workplace had caused a worker's death. He was dismayed to discover that statisticians and epidemiologists had no tools to answer such questions. This was still the era when causal language was taboo in statistics. It was only allowed in the case of a randomized controlled trial, and for ethical reasons one could never conduct such a trial on the effects of exposure to formaldehyde.

Usually a factory worker is exposed to a harmful chemical not just once but over a long period. For that reason, Robins became keenly interested in exposures or treatments that vary over time. Such exposures can also be beneficial: for example, AIDS treatment is given over the course of many years, with different plans of action depending on how a patient's CD4 count responds. How can you sort out the causal effect of treatment when it may occur in many stages and the intermediate variables (which you might want to use as controls) depend on earlier stages of treatment? This has been one of the defining questions of Robins's career.

After Jamie flew out to California to meet me on hearing about the "napkin problem" (Chapter 7), he was keenly interested in applying graphical methods to the sequential treatment plans that were his métier. Together we came up with a sequential back-door criterion for estimating the causal effect of such a treatment stream. I learned some important lessons from this collaboration. In particular, he showed me that two actions are sometimes easier to analyze than one because an action corresponds to erasing arrows on a graph, which makes it sparser.

Our back-door criterion dealt with a long-term treatment consisting of some arbitrarily large number of do-operations. But even two operations will produce some interesting mathematics—including the controlled direct effect, which consists of one action that "wiggles" the value of the treatment, while another action fixes the value of the mediator. More importantly, the idea of defining direct effects in terms of do-operations liberated them from the confines of linear models and grounded them in causal calculus.

But I didn't really get interested in mediation until later, when I saw that people were still making elementary mistakes, such as the Mediation Fallacy mentioned earlier. I was also frustrated that the action-based definition of the direct effect did not extend to the indirect effect. As Melanie Wall's student said, we have no variable or set of variables to intervene on to disable the direct path and let the indirect path stay active. For this reason the indirect effect seemed to me like a figment of the imagination, devoid of independent meaning except to remind us that the total effect may differ from the direct effect. I even said so in the first edition (2000) of my book *Causality*. This was one of the three greatest blunders of my career.

In retrospect, I was blinded by the success of the *do*-calculus, which had led me to believe that the only way to disable a causal path was to take a variable and set it to one particular value. This is not so; if I have a causal model, I can manipulate it in many creative ways, by dictating who listens to whom, when, and how. In particular, I can fix the primary variable for the purpose of suppressing its direct effect and, hypothetically yet simultaneously, energize the primary variable for the purpose of transmitting its effect through the mediator. That allows me to set the treatment variable (e.g., kittens) at zero and to set the mediator at the value it would have had if I had set kittens to one. My model of the data-generating process then tells me how to compute the effect of the split intervention.

I am indebted to one reader of the first edition, Jacques Hagenaars (author of *Categorical Longitudinal Data*), for urging me not to give up on the indirect effect. "Many experts in social science agree on the input and output, but differ exactly with respect to the mechanism," he wrote to me. But I was stuck for almost two years on the dilemma I wrote about in the last section: How can I disable the direct effect?

All these struggles came to sudden resolution, almost like a divine revelation, when I read the legal definition of discrimination that I quoted earlier in this chapter: "had the employee been of a different race . . . and everything else had been the same." Here we have it—the crux of the issue! It's a make-believe game. We deal with each individual on his or her own merits, and we keep all characteristics of the individual constant at whatever level they had prior to the change in the treatment variable.

How does this solve our dilemma? It means, first of all, that we have to redefine both the direct effect and the indirect effect. For the direct effect, we let the mediator choose the value it would have—for each individual—in the absence of treatment, and we fix it there. Now we wiggle the treatment and register the difference. This is different from the controlled direct effect I discussed earlier, where the mediator is fixed at one value for everyone. Because we let the mediator choose its "natural" value, I called it the natural direct effect. Similarly, for the natural indirect effect I first deny treatment to everyone, and then I let the mediator choose the value it would have, for each individual, in the presence of treatment. Finally I record the difference.

I don't know if the legal words in the definition of discrimination would have moved you, or anyone else, in the same way. But by 2000 I could already speak counterfactuals like a native. Having learned how to read them in causal models, I realized that they were nothing but quantities computed by innocent operations on equations or diagrams. As such, they stood ready to be encapsulated in a mathematical formula. All I had to do was embrace the "would-haves."

In a second, I realized that every direct and indirect effect could be translated into a counterfactual expression. Once I saw how to do that, it was a snap to derive a formula that tells you how to estimate the natural direct and indirect effects from data and when it is permissible. Importantly, the formula makes no assumptions about the specific functional form of the relationship between X, M, and Y. We have escaped from Linear Wonderland.

I called the new rule the Mediation Formula, though there are actually two formulas, one for the natural direct effect and one for the natural indirect effect. Subject to transparent assumptions, explicitly displayed in the graph, it tells you how they can be estimated from data. For example, in a situation like Figure 9.4, where there is no confounding between any of the variables, and M is the mediator between treatment X and outcome Y:

$$\text{NIE} = \Sigma_m \, [P(M = m \mid X = 1) - P(M = m \mid X = 0)] \times$$
$$\times \, P(Y = 1 \mid X = 0, M = m) \qquad (9.5)$$

The interpretation of this formula is illuminating. The expression in brackets stands for the effect of X on M, and the following expression stands for the effect of M on Y (when $X = 0$). So it reveals the origin of the product-of-coefficients idea, cast as a product of two nonlinear effects. Note also that unlike Equation 9.3, Equation 9.5 has no subscripts and no *do*-operators, so it can be estimated from rung-one data.

Whether you are a scientist in a laboratory or a child riding a bicycle, it is always a thrill to find you can do something today that you could not do yesterday. And that is how I felt when the Mediation Formula first appeared on paper. I could see at a glance everything about direct and indirect effects: what is needed to make them large or small, when we can estimate them from observational or interventional data, and when we can deem a mediator "responsible" for transmitting observed changes to the outcome variable. The relationship between cause and effect can be linear or nonlinear, numerical or logical. Previously each of these cases had to be handled in a different way, if they were discussed at all. Now a single formula would apply to all of them. Given the right data and the right model, we could determine if an employer was guilty of discrimination or what kinds of confounders would prevent us from making that determination. From Barbara Burks's data, we could estimate how much of the child's IQ comes from nature and how much from nurture. We could even calculate the percentage of the total effect *explained* by mediation and the percentage *owed* to mediation—two complementary concepts that collapse to one in linear models.

After I wrote down the counterfactual definition of the direct and indirect effects, I learned that I was not the first to hit on the idea. Robins and Greenland got there before me, all the way back in 1992. But their paper describes the concept of the natural effect in words, without committing it to a mathematical formula.

More seriously, they took a pessimistic view of the whole idea of natural effects and stated that such effects cannot be estimated from experimental studies and certainly not from observational studies. This statement prevented other researchers from seeing the potential of natural effects. It is hard to tell if Robins and Greenland would have switched to a more optimistic view had they taken the extra step of expressing the natural effect as a formula in counterfactual language. For me, this extra step was crucial.

There is possibly another reason for their pessimistic view, which I do not agree with but will try to explain. They examined the counterfactual definition of the natural effect and saw that it combines information from two different worlds, one in which you hold the treatment constant at zero and another in which you change the mediator to what it would have been if you had set the treatment to one. Because you cannot replicate this "cross-worlds" condition in any experiment, they believed it was out of bounds.

This is a philosophical difference between their school and mine. They believe that the legitimacy of causal inference lies in replicating a randomized experiment as closely as possible, on the assumption that this is the only route to the scientific truth. I believe that there may be other routes, which derive their legitimacy from a combination of data and established (or assumed) scientific knowledge. To that end, there may be methods more powerful than a randomized experiment, based on rung-three assumptions, and I do not hesitate to use them. Where they gave researchers a red light, I gave them a green light, which was the Mediation Formula: if you feel comfortable with these assumptions, here is what you can do! Unfortunately, Robins and Greenland's red light kept the field of mediation at a standstill for nine full years.

Many people find formulas daunting, seeing them as a way of concealing rather than revealing information. But to a mathematician, or to a person who is adequately trained in the mathematical way of thinking, exactly the reverse is true. A formula reveals everything: it leaves nothing to doubt or ambiguity. When reading a scientific article, I often catch myself jumping from formula to formula, skipping the words altogether. To me, a formula is a baked idea. Words are ideas in the oven.

A formula serves two purposes, one practical and one social. From the practical point of view, students or colleagues can read it as they would a recipe. The recipe may be simple or complex, but at the end of the day it promises that if you follow the steps, you will know the natural direct and indirect effects—provided, of course, your causal model accurately reflects the real world.

The second purpose is subtler. I had a friend in Israel who was a famous artist. I visited his studio to acquire one of his paintings, and his canvases were all over the place—a hundred under the bed, dozens in the kitchen. They were priced at between $300 and $500 each, and I had a hard time deciding. Finally, I pointed to one on the wall and said, "I like this one." "This one is $5,000," he said. "How come?" I asked, partly surprised and partly protesting. He answered, "This one is framed." It took me a few minutes to figure out, but then I understood what he meant. It wasn't valuable because it was framed; it was framed because it was valuable. Out of all the hundreds of paintings in his apartment, that one was his personal choice. It best expressed what he had labored to express in the others, and it was thus anointed with a seal of completeness—a frame.

That is the second purpose of a formula. It is a social contract. It puts a frame around an idea and says, "This is something I believe is important. This is something that deserves sharing."

That is why I have chosen to put a frame around the Mediation Formula. It deserves sharing because, to me and many like me, it represents the end to an age-old dilemma. And it is important, because it offers a practical tool for identifying mechanisms and assessing their importance. This is the social promise that the Mediation Formula expresses.

Since then, once the realization took hold that nonlinear mediation analysis is possible, research in the field has taken off. If you go to a database of academic articles and search for titles with the words "mediation analysis," you will find almost nothing before 2004. Then there were seven papers a year, then ten, then twenty; now there are more than a hundred papers a year. I'd like to end this chapter with three examples, which I hope will illustrate the variety of possibilities of mediation analysis.

CASE STUDIES OF MEDIATION

"Algebra for All": A Program and Its Side Effects

Like many big-city public school systems, the Chicago Public Schools face problems that sometimes seem intractable: high poverty rates, low budgets, and big achievement gaps between black, Latino, white, and Asian students. In 1988, then US secretary of education William Bennett called Chicago's public schools the worst in the nation.

But in the 1990s, under new leadership, the Chicago Public Schools undertook a number of reforms and moved from "worst in the nation" to "innovator for the nation." Some of the superintendents responsible for these changes gained nationwide prominence, such as Arne Duncan, who became secretary of education under President Barack Obama.

One innovation that actually predated Duncan was a policy, adopted in 1997, eliminating remedial courses in high school and requiring all ninth graders to take college-prep courses like English I and Algebra I. The math part of this policy was called "Algebra for All."

Was "Algebra for All" a success? That question, it turned out, was surprisingly difficult to answer. There was both good news and bad news. The good news was that test scores did improve. Math scores rose by 7.8 points over three years, a statistically significant change that is equivalent to about 75 percent of students scoring above the mean that existed before the policy change.

But before we can talk about causality, we have to rule out confounders, and in this case there is an important one. By 1997, the qualifications of incoming ninth-grade students were already improving thanks to earlier changes in the K–8 curriculum. So we are not comparing apples to apples. Because these children began ninth grade with better math skills than students had in 1994, the higher scores could be due to the already instituted K–8 changes, not to "Algebra for All."

Guanglei Hong, a professor of human development at the University of Chicago, studied the data and found no significant improvement in test scores once this confounder was taken into account. At this point it would have been easy for Hong to jump to the conclusion that "Algebra for All" was not a success. But she didn't, because there was another factor—this time a mediator, not a confounder—to take into account.

As any good teacher knows, students' success depends not only on what you teach them but on how you teach them. When the "Algebra for All" policy was introduced, more than the curriculum changed. The lower-achieving students found themselves in classrooms with higher-achieving students and could not keep up. This led to all sorts of negative consequences: discouragement, class cutting, and, of course, lower test scores. Also, in a mixed-ability classroom, the low-achieving students may have received less attention from their teachers than they would have in a remedial class. Finally, the teachers themselves may have struggled with the new demands placed on them. The teachers experienced in teaching Algebra I probably were not experienced in teaching low-ability students, and the teachers experienced with low-ability students may not have been as qualified to teach algebra. All of these were unanticipated side effects of "Algebra for All." Mediation analysis is ideally suited for evaluating the influence of side effects.

Hong hypothesized, therefore, that classroom environment had changed and had strongly affected the outcome of the intervention. In other words, she postulated the causal diagram shown in Figure 9.10. Environment (which Hong measured by the median skill level of all the students in the classroom) functions as a mediator between the "Algebra for All" intervention and the students' learning outcomes. The question, as usual in mediation analysis, is how much of the effect of the policy was direct and how much was indirect. Interestingly, the two effects worked in opposing directions. Hong found that the direct effect was positive: the new policy directly led to a roughly 2.7-point increase in test scores. This was

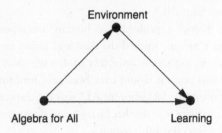

FIGURE 9.10. Causal diagram for "Algebra for All" experiment.

at least a change in the right direction, and it was statistically significant (meaning that such an improvement would be unlikely to
happen by chance). However, because of the changes in classroom
environment, the indirect effect had almost completely cancelled
out this improvement, reducing test scores by 2.3 points.

Hong concluded that the implementation of "Algebra for All"
had seriously undermined the policy. Maintaining the curricu-
lar change but returning to the prepolicy classroom environment
should result in a modest increase in student test scores (and hopefully, student learning).

Serendipitously, that is exactly what happened. In 2003 the Chicago Public Schools (now led by Duncan) instituted a new reform
called "Double-Dose Algebra." This reform would still require all
students to take algebra, but students who scored below the national median in eighth grade would take two classes of algebra a
day instead of one. This repaired the adverse side effect of the previous reform. Now, at least once a day, lower-achieving students
got a classroom environment closer to the one they enjoyed before
the "Algebra for All" reform. The "Double-Dose Algebra" reform
was generally deemed a success and continues to this day.

I consider the story of "Algebra for All" a success for mediation
analysis as well, because the analysis explains both the unimpres-
sive results of the original policy and the improved results under

the modified policy. Even though causal inference came along too late to affect the policy in real time, it does answer our "Why?" questions after the fact: Why did the original reform have little effect? Why did the second reform work better? In this way it can guide policy for the future.

I want to point out one other interesting thing about Hong's work. She was well aware of the Baron-Kenny approach to direct and indirect effects that I have called the Linear Wonderland. In her paper she actually performed the same analysis twice: once using a variation of the Mediation Formula, the other using the "conventional procedures" (her term) of Baron and Kenny. The Baron-Kenny method failed to detect the indirect effect. The reason is most likely just what I discussed before: linear methods cannot spot interactions between the treatment and the mediator. Perhaps the combination of more difficult material and a less supportive classroom environment caused the low-achieving students to become discouraged. Is this plausible? I think so. Algebra is a hard subject. Perhaps its difficulty made the extra attention from the teachers under the double-dose policy that much more valuable.

The Smoking Gene: Mediation and Interaction

In Chapter 5 I wrote about the scientific and political war over smoking in the 1950s and 1960s. The skeptics of that era, who included R. A. Fisher and Jacob Yerushalmy, argued that the apparent link between smoking and cancer might be a statistical artifact due to a confounding variable. Yerushalmy thought in terms of a smoking personality type, while Fisher suggested the possibility of a gene that would predispose people both toward smoking and toward developing lung cancer.

Ironically, genomics researchers discovered in 2008 that Fisher was right: there is a "smoking gene" that operates exactly in the way he suggested. This discovery came about through a new genomic analysis technique called a genome-wide association study (GWAS for short, pronounced "gee-wahss.") This is a prototypical "big-data" method that allows researchers to comb through the whole genome statistically, looking for genes that happen to show up more often in people with a certain disease, such as diabetes or schizophrenia or lung cancer.

It is important to notice the word "association" in the term GWAS. This method does not prove causality; it only identifies genes associated with a certain disease in the given sample. It is a data-driven rather than hypothesis-driven method, and this presents problems for causal inference.

Although previous hypothesis-driven gene studies had failed to find any clear evidence of genes related to smoking or to lung cancer, things changed overnight in 2008. In that year researchers identified a gene located in a region of the fifteenth chromosome that codes for nicotine receptors in lung cells. It has an official name, rs16969968, but that is a mouthful even for genomics experts. So they started calling it "the Big One" or "Mr. Big" because of its extremely strong association with lung cancer. "In the smoking field, if you say Mr. Big, people will know what you are talking about," says Laura Bierut, a smoking expert at Washington University in St. Louis. I'll just call it the smoking gene.

At this point I think I hear the cantankerous ghost of R. A. Fisher rattling his chains in the basement and demanding a retraction of all the things I wrote in Chapter 5. Yes, the smoking gene is associated with lung cancer. It has two variants, one common and one less common. People who inherit two copies of the less common variant (about one-ninth of the population) have about a 77 percent greater risk of getting lung cancer. The smoking gene also seems related to smoking behavior. People who have the risky variant seem to need more nicotine to feel satisfied and have more difficulty stopping. However, there is also some good news: these people respond better to nicotine replacement therapy than people without the smoking gene.

The discovery of the smoking gene should not change anybody's mind about the overwhelmingly more important causal factor in lung cancer, which is smoking. We know that smoking is associated with more than a tenfold increase in the risk of contracting lung cancer. By comparison, even a double dose of the smoking gene less than doubles your risk. This is serious business, no doubt,

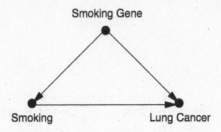

FIGURE 9.11. Causal diagram for the smoking gene example.

but it does not compare to the danger you face (for no good reason) if you are a regular smoker.

As always, it helps to visualize the discussion with a causal diagram. Fisher thought of the (at that time purely hypothetical) smoking gene as a confounder of smoking and cancer (Figure 9.11). But as a confounder, it is not nearly strong enough to account for the overwhelmingly strong effect of smoking on the risk of lung cancer. This is, in essence, the argument that Jerome Cornfield made in his 1959 paper that settled the argument about the genetic hypothesis.

We can easily rewrite the same causal diagram as shown in Figure 9.12. When we look at the diagram this way, we see that smoking behavior is a mediator between the smoking gene and lung cancer. This tiny change in perspective completely revamps the scientific debate. Instead of asking whether smoking causes cancer (a question we know the answer to), we ask instead how the gene works. Does it make people smoke more and inhale harder? Or does it somehow make lung cells more vulnerable to cancer? Which is stronger, the indirect effect or the direct effect?

The answer makes a difference for treatment. If the effect is direct, then people who have the high-risk gene should perhaps receive extra screening for lung cancer. On the other hand, if the effect is indirect, smoking behavior becomes crucial. We should counsel such patients about their increased risk and the importance

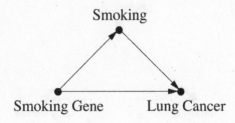

FIGURE 9.12. Figure 9.11, slightly rearranged.

of not smoking in the first place. If they already smoke, we may need to intervene more aggressively, perhaps with nicotine replacement therapy.

Tyler VanderWeele, an epidemiologist at Harvard University, read the first report about the smoking gene in *Nature*, and he contacted a research group at Harvard led by David Christiani. Since 1992, Christiani had asked his lung cancer patients, as well as their friends and family, to fill out questionnaires and provide DNA samples to help the research effort. By the mid-2000s he had collected data on 1,800 patients with cancer as well as 1,400 people without lung cancer, who served as controls. The DNA samples were still chilling in a freezer when VanderWeele called.

The results of VanderWeele's analysis were surprising at first. He found that the increased risk of lung cancer due to the indirect effect was only 1 to 3 percent. The people with the high-risk variant of the gene smoked only one additional cigarette per day on average, which was not enough to be clinically relevant. However, their bodies responded differently to smoking. The effect of the smoking gene on lung cancer was large and significant, but only for those people who smoked.

This poses an interesting conundrum in reporting the results. In this case, CDE(0) would be essentially zero: if you don't smoke, the gene won't hurt you. On the other hand, if we set the mediator to one pack a day or two packs a day, which I would denote CDE(1) or CDE(2), then the effect of the gene is strong. The natural direct effect averages these controlled effects. So the natural direct effect, NDE, is positive, and that is how VanderWeele reported it.

This example is a textbook case of interaction. In the end, VanderWeele's analysis proves three important things about the smoking gene. First, it does not significantly increase cigarette consumption. Second, it does not cause lung cancer through a smoking-independent path. Third, for those people who do smoke, it significantly increases the risk of lung cancer. The interaction between the gene and the subject's behavior is everything.

As is the case with any new result, of course more research is needed. Bierut points out one problem with VanderWeele and Christiani's analysis: they had only one measure of smoking behavior— the number of cigarettes per day. The gene could possibly cause people to inhale more deeply to get a larger dose of nicotine per puff. The Harvard study simply didn't have data to test this theory.

Even if some uncertainty remains, the research on the smoking gene provides a glimpse into the future of personalized medicine. It seems quite clear that in this case the important thing is how the gene and behavior interact. We still don't know for sure whether the gene changes behavior (as Bierut suggests) or merely interacts with behavior that would have happened anyway (as Vander-Weele's analysis suggests). Nevertheless, we may be able to use genetic status to give people better information about the risks they face. In the future, causal models capable of detecting interactions between genes and behavior or genes and environment are sure to be an important tool in the epidemiologist's kit.

Tourniquets: A Hidden Fallacy

When John Kragh, an army surgeon, arrived for his first day of duty in a Baghdad hospital in 2006, he received an immediate awakening to the new realities of wartime medicine. Seeing a clipboard with the current day's cases, he remarked to the nurse on duty, "Hey, that's interesting—you've had an emergency tourniquet used during your shift."

The nurse replied, "That's not interesting. We have one every shift."

In his first five minutes on the job, Kragh had stumbled upon a sea change in trauma care that took place during the Iraq and Afghanistan wars. Though used for centuries, both on the battlefield and in the operating room, tourniquets have always been somewhat controversial. A tourniquet left on too long will lead to loss of a limb. Also, tourniquets have often been improvised under duress, from straps or other handy materials, so their effectiveness is unsurprisingly a hit-or-miss affair. After World War II they were considered a treatment of last resort, and their use was officially discouraged.

The Iraq and Afghanistan wars radically changed that policy. Two things happened: more of the severe injuries required tourniquet use and better tourniquet designs became available. In 2005, the surgeon general of the US Army recommended that premanufactured tourniquets be provided to all soldiers. By 2006, as Kragh noted, the arrival of injured soldiers at the hospital with a tourniquet around an arm or leg was an everyday occurrence—a situation unprecedented in medical history.

From 2002 to 2012, Kragh estimates, tourniquets saved 2,000 military lives. Soldiers on the front lines noticed. According to US Army surgeon David Welling, "Combat troops are reportedly going out on dangerous patrol missions with tourniquets already in place on extremities, as they wish to be fully ready to respond to extremity bleeding, if and when the mine or the improvised explosive devices (IED) should go off."

Judging from the anecdotal evidence and the popularity of tourniquets with frontline soldiers, their value should be beyond question by now. However, few, if any, large-scale studies of tourniquet use had ever been performed. In civilian life, the kinds of injuries that necessitate them are too rare, and in military life the chaos of war makes it difficult to conduct a proper scientific study. But Kragh saw the opportunity to document the effects of their use. He and the nurses collected data on every case that came through the hospital doors, and the former tourniquet newbie became known as "the tourniquet guy."

The study results, published in 2015, were not what Kragh ex‑
pected. According to the data, the patients who had tourniquets ap‑
plied before arriving at the hospital did not survive at a higher rate
than those with similar injuries who had not received tourniquets.
Of course, Kragh reasoned, the ones with tourniquets possibly had
more severe injuries to begin with. But even when he controlled for
this factor by comparing cases of equal severity, the tourniquets
did not appear to improve survival rates (see Table 9.1).

TABLE 9.1. Data on survival with and without tourniquets.

Injury Severity	Survived/Total (No Tourniquet)	Survival Rate (No Tourniquet)	Survived/Total (with Tourniquet)	Survival Rate (with Tourniquet)
3 (Serious)	502/555	90%	416/465	89%
4 (Severe)	96/111	86%	212/248	85%
5 (Critical)	16/27	59%	4/7	57%
Total	614/693	89%	632/720	88%

This is not a Simpson's paradox situation. It doesn't matter
whether we aggregate the data or stratify it; in every severity cat‑
egory, as well as in the aggregate, survival was slightly greater for
soldiers who did not get tourniquets. (The difference in survival
rates was, however, too small to be statistically significant.)

What went wrong? One possibility, of course, is that tourni‑
quets aren't better. Our belief in them could be a case of confir‑
mation bias. When a soldier gets a tourniquet and survives, his
doctors and his buddies will say, "That tourniquet saved his life."
But if the soldier doesn't get a tourniquet and survives, nobody will
say, "Not putting on a tourniquet saved his life." So tourniquets
might get more credit than their due, and nonintervention doesn't
get any credit.

But there was another possible bias in this study, which Kragh
himself pointed out: the doctors only collected data on those
soldiers who survived long enough to get to the hospital in the
first place. To see why this matters, let's draw a causal diagram
(Figure 9.13).

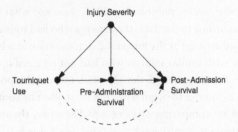

FIGURE 9.13. Causal diagram for tourniquet example. The dashed line is a hypothetical causal effect (not supported by the data).

In this figure, we can see that Injury Severity is a confounder of all three variables, the treatment (Tourniquet Use), the mediator (Pre-Administration Survival), and the outcome (Post-Admission Survival). It is therefore appropriate and necessary to condition on Injury Severity, as Kragh did in his paper.

However, because Kragh studied only the patients who actually survived long enough to get to the hospital, he was also conditioning on the mediator, Pre-Administration Survival. In effect, he was blocking the indirect path from tourniquet use to post-admission survival, and therefore he was computing the direct effect, indicated by the dashed arrow in Figure 9.13. That effect was essentially zero. Nevertheless, there still could be an indirect effect. If tourniquets enabled more soldiers to survive until they got to the hospital, then the tourniquet would be a very favorable intervention. This would mean that the job of a tourniquet is to get the patient to the hospital alive; once it has done that, it has no further value. Unfortunately, nothing in the data (Table 9.1) can either confirm or refute this hypothesis.

William Kruskal once lamented that there is no Homer to sing the praise of statisticians. I would like to sing the praise of Kragh, who under the most adverse conditions imaginable had the presence of mind to collect data and subject the standard treatment to a scientific test. His example is a shining light for anyone who wants to practice evidence-based medicine. It's a particularly bitter irony that his study could not succeed because he had no way to collect data on soldiers who didn't survive to the hospital. We may wish that he could have proved once and for all that tourniquets save lives. Kragh himself wrote in an email, "I have no doubt that tourniquets are a desirable intervention." But in the end he had to report a "null result," the kind that doesn't make headlines. Even so, he deserves credit for sound scientific instincts.

10

BIG DATA, ARTIFICIAL INTELLIGENCE, AND THE BIG QUESTIONS

All is pre-determined, yet permission is always granted.
—MAIMONIDES (MOSHE BEN MAIMON) (1138–1204)

WHEN I began my journey into causation, I was following the tracks of an anomaly. With Bayesian networks, we had taught machines to think in shades of gray, and this was an important step toward humanlike thinking. But we still couldn't teach machines to understand causes and effects. We couldn't explain to a computer why turning the dial of a barometer won't cause rain. Nor could we teach it what to expect when one of the riflemen on a firing squad changes his mind and decides not to shoot. Without the ability to envision alternate realities and contrast them with the currently existing reality, a machine cannot pass the mini-Turing test; it cannot answer the most basic question that makes us human: "Why?" I took this as an anomaly because I did not anticipate such natural and intuitive questions to reside beyond the reach of the most advanced reasoning systems of the time.

Only later did I realize that the same anomaly was afflicting more than just the field of artificial intelligence (AI). The very people who should care the most about "Why?" questions—namely, scientists—were laboring under a statistical culture that denied them the right to ask those questions. Of course they asked them anyway, informally, but they had to cast them as associational questions whenever they wanted to subject them to mathematical analysis.

The pursuit of this anomaly brought me into contact with people in a variety of fields, like Clark Glymour and his team (Richard Scheines and Peter Spirtes) from philosophy, Joseph Halpern from computer science, Jamie Robins and Sander Greenland from epidemiology, Chris Winship from sociology, and Don Rubin and Philip Dawid from statistics, who were thinking about the same problem. Together we lit the spark of a Causal Revolution, which has spread like a chain of firecrackers from one discipline to the next: epidemiology, psychology, genetics, ecology, geology, climate science, and so on. With every passing year I see a greater and greater willingness among scientists to speak and write about causes and effects, not with apologies and downcast eyes but with confidence and assertiveness. A new paradigm has evolved according to which it is okay to base your claims on assumptions as long as you make your assumptions transparent so that you and others can judge how plausible they are and how sensitive your claims are to their violation. The Causal Revolution has perhaps not led to any particular gadget that has changed our lives, but it has led to a transformation in attitudes that will inevitably lead to healthier science.

I often think of this transformation as "the second gift of AI to humanity," and it has been our main focus in this book. But as we bring the story to a conclusion, it is time for us to go back and inquire about the first gift, which has taken an unexpectedly long time to materialize. Are we in fact getting any closer to the day when computers or robots can understand causal conversations? Can we make artificial intelligences with as much imagination as a three-year-old human? I share some thoughts, but give no definitive conclusions, in this final chapter.

CAUSAL MODELS AND "BIG DATA"

Throughout science, business, government, and even sports, the amount of raw data we have about our world has grown at a staggering rate in recent years. The change is perhaps most visible to those of us who use the Internet and social media. In 2014, the last year for which I've seen data, Facebook reportedly was warehousing 300 petabytes of data about its 2 billion active users, or 150 megabytes of data per user. The games people play, the products they like to buy, the names of all their Facebook friends, and of course all their cat videos—all of them are out there in a glorious ocean of ones and zeros.

Less obvious to the public, but just as important, is the rise of huge databases in science. For example, the 1000 Genomes Project collected two hundred terabytes of information in what it calls "the largest public catalogue of human variation and genotype data." NASA's Mikulski Archive for Space Telescopes has collected 2.5 petabytes of data from several deep-space surveys. But Big Data hasn't only affected high-profile sciences; it's made inroads into every science. A generation ago, a marine biologist might have spent months doing a census of his or her favorite species. Now the same biologist has immediate access online to millions of data points on fish, eggs, stomach contents, or anything else he or she wants. Instead of just doing a census, the biologist can tell a story.

Most relevant for us is the question of what comes next. How do we extract meaning from all these numbers, bits, and pixels? The data may be immense, but the questions we ask are simple. Is there a gene that causes lung cancer? What kinds of solar systems are likely to harbor Earth-like planets? What factors are causing the population of our favorite fish to decrease, and what can we do about it?

In certain circles there is an almost religious faith that we can find the answers to these questions in the data itself, if only we are sufficiently clever at data mining. However, readers of this book will know that this hype is likely to be misguided. The questions I have just asked are all causal, and causal questions can never be answered from data alone. They require us to formulate a model of the process that generates the data, or at least some aspects of that process. Anytime you see a paper or a study that analyzes the data in a model-free way, you can be certain that the output of the study will merely summarize, and perhaps transform, but not interpret the data.

This is not to say that data mining is useless. It may be an essential first step to search for interesting patterns of association and pose more precise interpretive questions. Instead of asking, "Are there any lung-cancer-causing genes?" we can now start scanning the genome for genes with a high correlation with lung cancer (such as the "Mr. Big" gene mentioned in Chapter 9). Then we can ask, "Does this gene cause lung cancer? (And how?)" We never could have asked about the "Mr. Big" gene if we did not have data mining. To get any farther, though, we need to develop a causal model specifying (for example) what variables we think the gene affects, what confounders might exist, and what other causal pathways might bring about the result. Data interpretation means hypothesizing on how things operate in the real world.

Another role of Big Data in causal inference problems lies in the last stage of the inference engine described in the Introduction (step 8), which takes us from the estimand to the estimate. This step of statistical estimation is not trivial when the number of variables is large, and only big-data and modern machine-learning techniques can help us to overcome the curse of dimensionality. Likewise, Big Data and causal inference together play a crucial role in the emerging area of personalized medicine. Here, we seek to make inferences from the past behavior of a set of individuals who are similar in as many characteristics as possible to the individual in question. Causal inference permits us to screen off the irrelevant characteristics and to recruit these individuals from diverse studies, while Big Data allows us to gather enough information about them.

It's easy to understand why some people would see data mining as the finish rather than the first step. It promises a solution using available technology. It saves us, as well as future machines, the work of having to consider and articulate substantive assumptions about how the world operates. In some fields our knowledge may be in such an embryonic state that we have no clue how to begin drawing a model of the world. But Big Data will not solve this problem. The most important part of the answer must come from such a model, whether sketched by us or hypothesized and fine-tuned by machines.

Lest I seem too critical of the big-data enterprise, I would like to mention one new opportunity for symbiosis between Big Data and causal inference. This is called transportability.

Thanks to Big Data, not only can we access an enormous number of individuals in any given study, but we can also access an enormous number of studies, conducted in different locations and under different conditions. Often we want to combine the results of these studies and translate them to new populations that may be different even in ways we have not anticipated.

The process of translating the results of a study from one setting to another is fundamental to science. In fact, scientific progress would grind to a halt were it not for the ability to generalize results from laboratory experiments to the real world—for example, from test tubes to animals to humans. But until recently each science had to develop its own criteria for sorting out valid from invalid generalizations, and there have been no systematic methods for addressing "transportability" in general.

Within the last five years, my former student (now colleague) Elias Bareinboim and I have succeeded in giving a complete criterion for deciding when results are transportable and when they are not. As usual, the proviso for using this criterion is that you represent the salient features of the data-generating process with a causal diagram, marked with locations of potential disparities. "Transporting" a result does not necessarily mean taking it at face value and applying it to the new environment. The researcher may have to recalibrate it to allow for disparities between the two environments.

Suppose we want to know the effect of an online advertisement (X) on the likelihood that a consumer will purchase the product (Y)—say, a surfboard. We have data from studies in five different places: Los Angeles, Boston, San Francisco, Toronto, and Honolulu. Now we want to estimate how effective the advertisement will be in Arkansas. Unfortunately, each population and each study differs slightly. For example, the Los Angeles population is younger than our target population, and the San Francisco population differs in click-through rate. Figure 10.1 shows the unique characteristics of each population and each study. Can we combine the data from these remote and disparate studies to estimate the

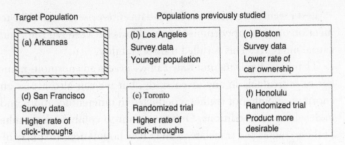

FIGURE 10.1. The transportability problem.

ad's effectiveness in Arkansas? Can we do it without taking any data in Arkansas? Or perhaps by measuring merely a small set of variables or conducting a pilot observational study?

Figure 10.2 translates these differences into graphical form. The variable Z represents age, which is a confounder; young people may be more likely to see the ad and more likely to buy the product even

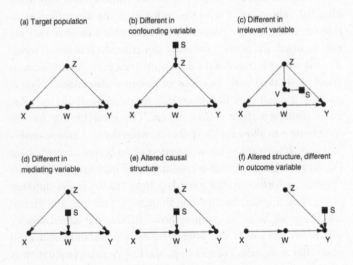

X = Advertisement, Y = Purchase Decision, Z = Age, W = Click-through Rate, V = Car Ownership, S = Indicator Variable

FIGURE 10.2. Differences between the studied populations, expressed in graphical form.

if they don't see the ad. The variable W represents clicking on a link to get more information. This is a mediator, a step that must take place in order to convert "seeing the advertisement" into "buying the product." The letter S, in each case, stands for a "difference-producing" variable, a hypothetical variable that points to the characteristic by which the two populations differ. For example, in Los Angeles (b), the indicator S points to Z, age. In each of the other cities the indicator points to the distinguishing feature of the population mentioned in Figure 10.1.

For the advertising agency, the good news is that a computer can now manage this complicated "data fusion" problem and, guided by the *do*-calculus, tell us which studies we can use to answer our query and by what means, as well as what information we need to collect in Arkansas to support the conclusion. In some cases the effect may transport directly, with no further work and without our even setting foot in Arkansas. For example, the effect of the ad in Arkansas should be the same as in Boston, because according to the diagram, Boston (c) differs from Arkansas only in the variable V, which does not affect either treatment X or outcome Y.

We need to reweight the data in some of the other studies—for instance, to account for the different age structure of the population in the Los Angeles study (b). Interestingly, the experimental study in Toronto (e) is sufficient for estimating our query in Arkansas despite the disparity at W, if we can only measure X, W, and Y in Arkansas.

Remarkably, we have found examples in which no transport is feasible from any one of the available studies; yet the target quantity is nevertheless estimable from their combination. Also, even studies that are not transportable are not entirely useless. Take, for example, the Honolulu (f) study in Figure 10.2, which is not transportable due to the arrow $S \rightarrow Y$. The arrow $X \rightarrow W$, on the other hand, is not contaminated by S, and so the data available from Honolulu can be used to estimate $P(W \mid X)$. By combining this with estimates of $P(W \mid X)$ from other studies, we can increase the precision of this subexpression. By carefully combining such subexpressions, we may be able to synthesize an accurate overall estimate of the target quantity.

Although in simple cases these results are intuitively reasonable, when the diagrams get more complicated, we need the help of a formal method. The *do*-calculus provides a general criterion for determining transportability in such cases. The rule is quite simple: if you can perform a valid sequence of *do*-operations (using the rules from Chapter 7) that transforms the target quantity into another expression in which any factor involving *S* is free of *do*-operators, then the estimate is transportable. The logic is simple; any such factor can be estimated from the available data, uncontaminated by the disparity factor *S*.

Elias Bareinboim has managed to do the same thing for the problem of transportability that Ilya Shpitser did for the problem of interventions. He has developed an algorithm that can automatically determine for you whether the effect you are seeking is transportable, using graphical criteria alone. In other words, it can tell you whether the required separation of *S* from the *do*-operators can be accomplished or not.

Bareinboim's results are exciting because they change what was formerly seen as a threat to validity into an opportunity to leverage the many studies in which participation cannot be mandated and where we therefore cannot guarantee that the study population would be the same as the population of interest. Instead of seeing the difference between populations as a threat to the "external validity" of a study, we now have a methodology for establishing validity in situations that would have appeared hopeless before. It is precisely because we live in the era of Big Data that we have access to information on many studies and on many of the auxiliary variables (like *Z* and *W*) that will allow us to transport results from one population to another.

I will mention in passing that Bareinboim has also proved analogous results for another problem that has long bedeviled statisticians: selection bias. This kind of bias occurs when the sample group being studied differs from the target population in some relevant way. This sounds a lot like the transportability problem—and it is, except for one very important modification: instead of drawing an arrow from the indicator variable *S* to the affected variable, we draw the arrow toward S. We can think of S as standing for "selection" (into the study). For example, if our study observes only hospitalized patients, as in the Berkson bias example, we would draw an arrow from Hospitalization to S, indicating that hospitalization is a cause of selection for our study. In Chapter 6 we saw this situation only as a threat to the validity of our study. But now, we can look at it as an opportunity. If we understand the mechanism by which we recruit subjects for the study, we can recover from bias by collecting data on the right set of deconfounders and using an appropriate reweighting or adjustment formula. Bareinboim's work allows us to exploit causal logic and Big Data to perform miracles that were previously inconceivable.

Words like "miracles" and "inconceivable" are rare in scientific discourse, and the reader may wonder if I am being a little too enthusiastic. But I use them for a good reason. The concept of external validity as a threat to experimental science has been around for at least half a century, ever since Donald Campbell and Julian Stanley recognized and defined the term in 1963. I have talked to dozens of experts and prominent authors who have written about this topic. To my amazement, not one of them was able to tackle any of the toy problems presented in Figure 10.2. I call them "toy problems" because they are easy to describe, easy to solve, and easy to verify if a given solution is correct.

At present, the culture of "external validity" is totally preoccupied with listing and categorizing the threats to validity rather than fighting them. It is in fact so paralyzed by threats that it looks with suspicion and disbelief on the very idea that threats can be disarmed. The experts, who are novices to graphical models, find it easier to configure additional threats than to attempt to remedy any one of them. Language like "miracles," so I hope, should jolt my colleagues into looking at such problems as intellectual challenges rather than reasons for despair.

I wish that I could present the reader with successful case studies of a complex transportability task and recovery from selection bias, but the techniques are still too new to have penetrated into general usage. I am very confident, though, that researchers will discover the power of Bareinboim's algorithms before long, and then external validity, like confounding before it, will cease to have its mystical and terrifying power.

STRONG AI AND FREE WILL

The ink was scarcely dry on Alan Turing's great paper, "Computing Machinery and Intelligence," when science fiction writers and futurologists began toying with the prospect of machines that think. Sometimes they envisioned these machines as benign or even noble figures, like the whirry, chirpy R2D2 and the oddly British android C3PO from *Star Wars*. Other times the machines are much more sinister, plotting the destruction of the human species, as in the *Terminator* movies, or enslaving humans in a virtual reality, as in *The Matrix*.

In all these cases, the AIs say more about the anxieties of the writers or the capabilities of the movie's special effects department than they do about actual artificial intelligence research. Artificial intelligence has turned out to be a more elusive goal than Turing ever suspected, even though the sheer computational power of our computers has no doubt exceeded his expectations.

In Chapter 3 I wrote about some of the reasons for this slow progress. In the 1970s and early 1980s, artificial intelligence research was hampered by its focus on rule-based systems. But rule-based systems proved to be on the wrong track. They were very brittle. Any slight change to their working assumptions required that they be rewritten. They could not cope well with uncertainty or with contradictory data. Finally, they were not scientifically transparent; you could not prove mathematically that they would behave in a certain way, and you could not pinpoint exactly what needed repair when they didn't. Not all AI researchers objected to the lack of transparency. The field at the time was divided into "neats" (who wanted transparent systems with guarantees of behavior) and "scruffies" (who just wanted something that worked). I was always a "neat."

I was lucky to come along at a time when the field was ready for a new approach. Bayesian networks were probabilistic; they could cope with a world full of conflicting and uncertain data. Unlike the rule-based systems, they were modular and easily implemented on a distributed computing platform, which made them fast. Finally, as was important to me (and other "neats"), Bayesian networks dealt with probabilities in a mathematically sound way. This guaranteed that if anything went wrong, the bug was in the program, not in our thinking.

Even with all these advantages, Bayesian networks still could not understand causes and effects. By design, in a Bayesian network, information flows in both directions, causal and diagnostic: smoke increases the likelihood of fire, and fire increases the likelihood of smoke. In fact, a Bayesian network can't even tell what the "causal direction" is. The pursuit of this anomaly—this wonderful anomaly, as it turned out—drew me away from the field of machine learning and toward the study of causation. I could not reconcile myself to the idea that future robots would not be able to communicate with us in our native language of cause and effect. Once in causality land, I was naturally drawn toward the vast spectrum of other sciences where causal asymmetry is of the utmost importance.

So, for the past twenty-five years, I have been somewhat of an expatriate from the land of automated reasoning and machine learning. Nevertheless, from my distant vantage point I can still see the current trends and fashions.

In recent years, the most remarkable progress in AI has taken place in an area called "deep learning," which uses methods like convolutional neural networks. These networks do not follow the rules of probability; they do not deal with uncertainty in a rigorous or transparent way. Still less do they incorporate any explicit representation of the environment in which they operate. Instead, the architecture of the network is left free to evolve on its own. When finished training a new network, the programmer has no idea what computations it is performing or why they work. If the network fails, she has no idea how to fix it.

Perhaps the prototypical example is AlphaGo, a convolutional neural-network-based program that plays the ancient Asian game of Go, developed by DeepMind, a subsidiary of Google. Among human games of perfect information, Go had always been considered the toughest nut for AI. Though computers conquered humans in chess in 1997, they were not considered a match even for the lowest-level professional Go players as recently as 2015. The Go community thought that computers were still a decade or more away from giving humans a real battle.

That changed almost overnight with the advent of AlphaGo. Most Go players first heard about the program in late 2015, when it trounced a human professional 5–0. In March 2016, AlphaGo defeated Lee Sedol, for years considered the strongest human player, 4–1. A few months later it played sixty online games against top human players without losing a single one, and in 2017 it was officially retired after beating the current world champion, Ke Jie. The one game it lost to Sedol is the only one it will ever lose to a human.

All of this is exciting, and the results leave no doubt: deep learning works for certain tasks. But it is the antithesis of transparency. Even AlphaGo's programmers cannot tell you why the program plays so well. They knew from experience that deep networks have been successful at tasks in computer vision and speech recognition. Nevertheless, our understanding of deep learning is completely empirical and comes with no guarantees. The AlphaGo team could not have predicted at the outset that the program would beat the best human in a year, or two, or five. They simply experimented, and it did.

Some people will argue that transparency is not really needed. We do not understand in detail how the human brain works, and yet it runs well, and we forgive our meager understanding. So, they argue, why not unleash deep-learning systems and create a new kind of intelligence without understanding how it works? I cannot say they are wrong. The "scruffies," at this moment in time, have taken the lead. Nevertheless, I can say that I personally don't like opaque systems, and that is why I do not choose to do research on them.

My personal taste aside, there is another factor to add to this analogy with the human brain. Yes, we forgive our meager understanding of how human brains work, but we can still communicate with other humans, learn from them, instruct them, and motivate them in our own native language of cause and effect. We can do that because our brains work the same way. If our robots will all be as opaque as AlphaGo, we will not be able to hold a meaningful conversation with them, and that would be quite unfortunate.

When my house robot turns on the vacuum cleaner while I am still asleep (Figure 10.3) and I tell it, "You shouldn't have woken me up," I want it to understand that the vacuuming was at fault, but I don't want it to interpret the complaint as an instruction never to vacuum the upstairs again. It should understand what you and I perfectly understand: vacuum cleaners make noise, noise wakes people up, and that makes some people unhappy. In other words, our robot will have to understand cause-and-effect relations—in fact, counterfactual relations, such as those encoded in the phrase "You shouldn't have."

Indeed, observe the rich content of this short sentence of instructions. We should not need to tell the robot that the same applies

FIGURE 10.3. A smart robot contemplating the causal ramifications of his/her actions. (*Source:* Drawing by Maayan Harel.)

to vacuum cleaning downstairs or anywhere else in the house, but not when I am awake or not at home, when the vacuum cleaner is equipped with a silencer, and so forth. Can a deep-learning program understand the richness of this instruction? That is why I am not satisfied with the apparently superb performance of opaque systems. Transparency enables effective communication.

One aspect of deep learning does interest me: the theoretical limitations of these systems, primarily limitations that stem from their inability to go beyond rung one of the Ladder of Causation. This limitation does not hinder the performance of AlphaGo in the narrow world of go games, since the board description together with the rules of the game constitutes an adequate causal model of the go-world. Yet it hinders learning systems that operate in environments governed by rich webs of causal forces, while having access merely to surface manifestations of those forces. Medicine, economics, education, climatology, and social affairs are typical examples of such environments. Like the prisoners in Plato's famous cave, deep-learning systems explore the shadows on the cave wall and learn to accurately predict their movements. They lack the understanding that the observed shadows are mere projections of three-dimensional objects moving in a three-dimensional space. Strong AI requires this understanding.

Deep-learning researchers are not unaware of these basic limitations. For example, economists using machine learning have noted that their methods do not answer key questions of interest, such as estimating the impact of untried policies and actions. Typical examples are introducing new price structures or subsidies or changing the minimum wage. In technical terms, machine-learning methods today provide us with an efficient way of going from finite sample estimates to probability distributions, and we still need to get from distributions to cause-effect relations.

When we start talking about strong AI, causal models move from a luxury to a necessity. To me, a strong AI should be a machine that can reflect on its actions and learn from past mistakes. It should be able to understand the statement "I should have acted differently," whether it is told as much by a human or arrives at that conclusion itself. The counterfactual interpretation of this statement reads, "I have done $X = x$, and the outcome was $Y = y$. But if I had acted differently, say $X = x'$, then the outcome would have been better, perhaps $Y = y'$." As we have seen, the estimation of such probabilities has been completely automated, given enough data and an adequately specified causal model.

In fact, I think that a very important target for machine learning is the simpler probability $P(Y_{X = x'} = y' \mid X = x)$, where the machine observes an event $X = x$ but not the outcome Y, and then asks for the outcome under an alternative event $X = x'$. If it can compute this quantity, the machine can treat its intended action as an observed event $(X = x)$ and ask, "What if I change my mind and do $X = x'$ instead?" This expression is mathematically the same as the effect of treatment on the treated (mentioned in Chapter 8), and we have lots of results indicating how to estimate it.

Intent is a very important part of personal decision making. If a former smoker feels himself tempted to light up a cigarette, he should think very hard about the reasons behind that intention and ask whether a contrary action might in fact lead to a better outcome. The ability to conceive of one's own intent and then use it as a piece of evidence in causal reasoning is a level of self-awareness (if not consciousness) that no machine I know of has achieved. I would like to be able to lead a machine into temptation and have it say, "No."

Any discussion of intent leads to another major issue for strong AI: free will. If we are asking a machine to have the intent to do $X = x$, become aware of it, and choose to do $X = x'$ instead, we seem to be asking it to have free will. But how can a robot have free will if it just follows instructions stored in its program?

Berkeley philosopher John Searle has labeled the free will problem "a scandal in philosophy," partly due to the zero progress made on the problem since antiquity and partly because we cannot brush it off as an optical illusion. Our entire conception of "self" presupposes that we have such a thing as choices. For example, there seems to be no way to reconcile my vivid, unmistakable sensation of having an option (say, to touch or not to touch my nose) with my understanding of reality that presupposes causal determinism: all our actions are triggered by electrical neural signals emanating from the brain.

While many philosophical problems have disappeared over time in the light of scientific progress, free will remains stubbornly enigmatic, as fresh as it appeared to Aristotle and Maimonides. Moreover, while human free will has sometimes been justified on spiritual or theological grounds, these explanations would not apply to a programmed machine. So any appearance of robotic free will must be a gimmick—at least this is the conventional dogma.

Not all philosophers are convinced that there really is a clash between free will and determinism. A group called "compatibilists," among whom I count myself, consider it only an apparent clash between two levels of description: the neural level at which processes appear deterministic (barring quantum indeterminism) and the cognitive level at which we have a vivid sensation of options. Such apparent clashes are not infrequent in science. For example, the equations of physics are time reversible on a microscopic level, yet appear irreversible on the macroscopic level of description; the smoke never flows back into the chimney. But that opens up new questions: Granted that free will is (or may be) an illusion, why is it so important to us as humans to have this illusion? Why did evolution labor to endow us with this conception? Gimmick or no gimmick, should we program the next generation of computers to have this illusion? What for? What computational benefits does it entail?

I think that understanding the benefits of the illusion of free will is the key to the stubbornly enigmatic problem of reconciling it with determinism. The problem will dissolve before our eyes once we endow a deterministic machine with the same benefits.

Together with this functional issue, we must also cope with questions of simulation. If neural signals from the brain trigger all our actions, then our brains must be fairly busy decorating some actions with the title "willed" or "intentional" and others with "unintentional." What precisely is this labeling process? What neural path would earn a given signal the label "willed"?

In many cases, voluntary actions are recognized by a trace they leave in short-term memory, with the trace reflecting a purpose or motivation. For example, "Why did you do it?" "Because I wanted to impress you." Or, as Eve innocently answered, "The serpent deceived me, and I ate." But in many other cases an intentional action is taken, and yet no reason or motives come to mind. Rationalization of actions may be a reconstructive, post-action process. For example, a soccer player may explain why he decided to pass the ball to Joe instead of Charlie, but it is rarely the case that those reasons consciously triggered the action. In the heat of the game, thousands of input signals compete for the player's attention. The crucial decision is which signals to prioritize, and the reasons can hardly be recalled and articulated.

AI researchers are therefore trying to answer two questions— about function and simulation—with the first driving the second. Once we understand what computational function free will serves in our lives, then we can attend to equipping machines with such functions. It becomes an engineering problem, albeit a hard one.

To me, certain aspects of the functional question stand out clearly. The illusion of free will gives us the ability to speak about our intents and to subject them to rational thinking, possibly using counterfactual logic. When the coach pulls us out of a soccer game and says, "You should have passed the ball to Charlie," consider all the complex meanings embedded in these eight words.

First, the purpose of such a "should have" instruction is to swiftly transmit valuable information from the coach to the player: in the future, when faced with a similar situation, choose action B rather than action A. But the "similar situations" are far too numerous to list and are hardly known even to the coach himself. Instead of listing the features of these "similar situations," the coach points to the player's action, which is representative of his intent at decision time. By proclaiming the action inadequate, the coach is asking the player to identify the software packages that led to his decision and then reset priorities among those packages so that "pass to Charlie" becomes the preferred action. There is profound wisdom in this instruction because who, if not the player himself, would know the identities of those packages? They are nameless neural paths that cannot be referenced by the coach or any external observer. Asking the player to take an action different from the one taken amounts to encouraging an intent-specific analysis, like the one we mentioned above. Thinking in terms of intents, therefore, offers us a shorthand to convert complicated causal instructions into simple ones.

I would conjecture, then, that a team of robots would play better soccer if they were programmed to communicate as if they had free will. No matter how technically proficient the individual robots are at soccer, their team's performance will improve when they can speak to each other as if they are not preprogrammed robots but autonomous agents believing they have options.

Although it remains to be seen whether the illusion of free will enhances robot-to-robot communication, there is much less uncertainty about robot-to-human communication. In order to communicate naturally with humans, strong AIs will certainly need to understand the vocabulary of options and intents, and thus they will need to emulate the illusion of free will. As I explained above, they may also find it advantageous to "believe" in their own free will themselves, to the extent of being able to observe their intent and act differently.

The ability to reason about one's own beliefs, intents, and desires has been a major challenge to AI researchers and defines the notion of "agency." Philosophers, on the other hand, have studied these abilities as part of the classical question of consciousness. Questions such as "Can machines have consciousness?" or "What makes a software agent different from an ordinary program?" have engaged the best minds of many generations, and I would not pretend to answer them in full. I believe, nevertheless, that the algorithmization of counterfactuals is a major step toward understanding these questions and making consciousness and agency a computational reality. The methods described for equipping a machine with a symbolic representation of its environment and the capacity to imagine a hypothetical perturbation of that environment can be extended to include the machine itself as part of the environment. No machine can process a complete copy of its own software, but it can have a blueprint summary of its major software components. Other components can then reason about that blueprint and mimic a state of self-awareness.

To create the perception of agency, we must also equip this software package with a memory to record past activations, to which it can refer when asked, "Why did you do that?" Actions that pass certain patterns of path activation will receive reasoned explanations, such as "Because the alternative proved less attractive." Others will end up with evasive and useless answers, such as "I wish I knew why" or "Because that's the way you programmed me."

In summary, I believe that the software package that can give a thinking machine the benefits of agency would consist of at least three parts: a causal model of the world; a causal model of its own software, however superficial; and a memory that records how intents in its mind correspond to events in the outside world.

This may even be how our own causal education as infants begins. We may have something like an "intention generator" in our minds, which tells us that we are supposed to take action $X = x$. But children love to experiment—to defy their parents', their teachers', even their own initial intentions—and to do something different, just for fun. Fully aware that we are supposed to do $X = x$, we playfully do $X = x'$ instead. We watch what happens, repeat the process, and keep a record of how good our intention generator is. Finally, when we start to adjust our own software, that is when we begin to take moral responsibility for our actions. This responsibility may be an illusion at the level of neural activation but not at the level of self-awareness software.

Encouraged by these possibilities, I believe that strong AI with causal understanding and agency capabilities is a realizable promise, and this raises the question that science fiction writers have been asking since the 1950s: Should we be worried? Is strong AI a Pandora's box that we should not open?

Recently public figures like Elon Musk and Stephen Hawking have gone on record saying that we should be worried. On Twitter, Musk said that AIs were "potentially more dangerous than nukes." In 2015, John Brockman's website Edge.org posed as its annual question, that year asking, "What do you think about machines that think?" It drew 186 thoughtful and provocative answers (since collected into a book titled *What to Think About Machines That Think*).

Brockman's intentionally vague question can be subdivided into at least five related ones:

1. Have we already made machines that think?
2. Can we make machines that think?
3. Will we make machines that think?
4. Should we make machines that think?

And finally, the unstated question that lies at the heart of our anxieties:

5. Can we make machines that are capable of distinguishing good from evil?

The answer to the first question is no, but I believe that the answer to all of the others is yes. We certainly have not yet made machines that think in any humanlike interpretation of the word. So far we can only simulate human thinking in narrowly defined domains that have only the most primitive causal structures. There we can actually make machines that outperform humans, but this should be no surprise because these domains reward the one thing that computers do well: compute.

The answer to the second question is almost certainly yes, if we define thinking as being able to pass the Turing test. I say that on the basis of what we have learned from the mini-Turing test. The ability to answer queries at all three levels of the Ladder of Causation provides the seeds of "agency" software so that the machine can think about its own intentions and reflect on its own mistakes. The algorithms for answering causal and counterfactual queries already exist (thanks in large part to my students), and they are only waiting for industrious AI researchers to implement them.

The third question depends, of course, on human events that are difficult to predict. But historically, humans have seldom refrained from making or doing things that they are technologically capable of. Partly this is because we do not know we are technologically capable of something until we actually do it, whether it's cloning animals or sending astronauts to the moon. The detonation of the atomic bomb, however, was a turning point: many people think this technology should not have been developed.

Since World War II, a good example of scientists pulling back from the feasible was the 1975 Asilomar conference on DNA recombination, a new technology seen by the media in somewhat apocalyptic terms. The scientists working in the field managed to come to a consensus on good-sense safety practices, and the agreement they reached then has held up well over the ensuing four decades. Recombinant DNA is now a common, mature technology.

In 2017, the Future of Life Institute convened a similar Asilomar conference on artificial intelligence and agreed on a set of twenty-three principles for future research in "beneficial AI." While most of the guidelines are not relevant to the topics discussed in this book, the recommendations on ethics and values are definitely worthy of attention. For example, recommendations 6, "AI systems should be safe and secure throughout their operational lifetime, and verifiably so," and 7, "If an AI system causes harm, it should be possible to ascertain why," clearly speak to the importance of transparency. Recommendation 10, "Highly autonomous AI systems should be designed so that their goals and behaviors can be assured to align with human values throughout their operation," is rather vague as stated but could be given operational meaning if these systems were required to be able to declare their own intents and communicate with humans about causes and effects.

My answer to the fourth question is also yes, based on the answer to the fifth. I believe that we will be able to make machines that can distinguish good from evil, at least as reliably as humans and hopefully more so. The first requirement of a moral machine is the ability to reflect on its own actions, which falls under counterfactual analysis. Once we program self-awareness, however limited, empathy and fairness follow, for it is based on the same computational principles, with another agent added to the equation.

There is a big difference in spirit between the causal approach to building the moral robot and an approach that has been studied and rehashed over and over in science fiction since the 1950s: Asimov's laws of robotics. Isaac Asimov proposed three absolute laws, starting with "A robot may not injure a human being or, through inaction, allow a human being to come to harm." But as science fiction has shown over and over again, Asimov's laws always lead to contradictions. To AI scientists, this comes as no surprise: rule-based systems never turn out well. But it does not follow that building a moral robot is impossible. It means that the approach cannot be prescriptive and rule based. It means that we should equip thinking machines with the same cognitive abilities that we have, which include empathy, long-term prediction, and self-restraint, and then allow them to make their own decisions.

Once we have built a moral robot, many apocalyptic visions start to recede into irrelevance. There is no reason to refrain from building machines that are better able to distinguish good from evil than we are, better able to resist temptation, better able to assign guilt and credit. At this point, like chess and Go players, we may even start to learn from our own creation. We will be able to depend on our machines for a clear-eyed and causally sound sense of justice. We will be able to learn how our own free will software works and how it manages to hide its secrets from us. Such a thinking machine would be a wonderful companion for our species and would truly qualify as AI's first and best gift to humanity.

ACKNOWLEDGMENTS

To enumerate the entire cast of students, friends, colleagues, and teachers who have contributed ideas to this book would amount to writing another book. Still, a few players deserve special mention, from my personal persective. I would like to thank Phil Dawid, for giving me my first audition on the pages of *Biometrika*; Jamie Robins and Sander Greenland, for turning epidemiology into a graph-speaking community; the late Dennis Lindley, for giving me the assurance that even seasoned statisticians can recognize flaws in their field and rally for its reform; Chris Winship, Steven Morgan, and Felix Elwert for ushering social science into the age of causation; and, finally, Peter Spirtes, Clark Glymour, and Richard Scheines, for their help in pushing me over the cliff of probabilities into the stormy waters of causation.

Digging deeper into my ancient history, I must thank Joseph Hermony, Dr. Shimshon Lange, Professor Franz Ollendorff, and other dedicated science teachers who inspired me from grade school to college. They instilled in many of us first-generation Israelis a sense of mission and historical responsibility to pursue scientific explorations as mankind's most noble and fun challenge.

This book would have remained a relic of wishful thinking if it were not for my co-author, Dana Mackenzie, who took my wishful thinking seriously and made it a reality. He not only corrected my foreign accent but also took me to distant lands, from the Navy ships of Captain James Lind to the Antarctic expedition of Captain Robert Scott, adding knowledge, stories, structure, and clarity to a mess of mathematical equations that were awaiting an organizing narrative.

I owe a great debt to members of the Cognitive Systems Laboratory at UCLA whose work and ideas over the past three and a half decades formed the scientific basis of this book: Alex Balke, Elias Bareinboim,

Blai Bonet, Carlo Brito, Avin Chen, Bryant Chen, David Chickering, Adnan Darwiche, Rina Dechter, Andrew Forney, David Galles, Hector Geffner, Dan Geiger, Moises Goldszmidt, David Heckerman, Mark Hopkins, Jin Kim, Manabu Kuroki, Trent Kyono, Karthika Mohan, Azaria Paz, George Rebane, Ilya Shpitser, Jin Tian, Thomas Verma, and Ingrid Zukerman.

Funding agencies receive ritualized thanks in scholarly publications but far too little real credit, considering their crucial role in recognizing seeds of ideas before they become fashionable. I must acknowledge the steady and unfailing support of the National Science Foundation and the Office of Naval Research, through the Machine Learning and Intelligence program headed by Behzad Kamgar-Parsi.

Dana and I would like to thank our agent, John Brockman, who gave us timely encouragement and the benefit of his professional expertise. Our editor at Basic Books, TJ Kelleher, asked us just the right questions and persuaded Basic Books that a story this ambitious could not be told in 200 pages. Our illustrators, Maayan Harel and Dakota Harr, managed to cope with our sometimes conflicting instructions and brought abstract subjects to life with humor and beauty. Kaoru Mulvihill at UCLA deserves much credit for proofing several versions of the manuscript and illustrating the hordes of graphs and diagrams.

Dana will be forever grateful to John Wilkes, who founded the Science Communication Program at UC Santa Cruz, which is still going strong and is the best possible route into a career as a science writer. Dana would also like to thank his wife, Kay, who encouraged him to pursue his childhood dream of being a writer, even when it meant pulling up stakes, crossing the country, and starting over.

Finally, my deepest debt is owed to my family, for their patience, understanding, and support. Especially to my wife, Ruth, my moral compass, for her endless love and wisdom. To my late son, Danny, for showing me the silent audacity of truth. To my daughters Tamara and Michelle for trusting my perennial promise that the book will eventually be done. And to my grandchildren, Leora, Tori, Adam, Ari and Evan, for giving a purpose to my long journeys and for always dissolving my "why" questions away.

NOTES

NOTES TO INTRODUCTION

5   **Students are never allowed:** With possibly one exception: if we have performed a randomized controlled trial, as discussed in Chapter 4.

NOTES TO CHAPTER ONE

42   **then the opposite is true:** In other words, when evaluating an intervention in a causal model, we make the minimum changes possible to enforce its immediate effect. So we "break" the model where it comes to A but not B.

45   **We should thank the language:** I should also mention here that counterfactuals allow us to talk about causality in individual cases: What would have happened to Mr. Smith, who was not vaccinated and died of smallpox, if he had been vaccinated? Such questions, the backbone of personalized medicine, cannot be answered from rung-two information.

48   **Yet we can answer:** To be more precise, in geometry, undefined terms like "point" and "line" are primitives. The primitive in causal inference is the relation of "listening to," indicated by an arrow.

NOTES TO CHAPTER TWO

83   **And now the algebraic magic:** For anyone who takes the trouble to read Wright's paper, let me warn you that he does not

compute his path coefficients in grams per day. He computes them in "standard units" and then converts to grams per day at the end.

NOTES TO CHAPTER FIVE

169 **"Cigarette smoking is causally related":** The evidence for women was less clear at that time, primarily because women had smoked much less than men in the early decades of the century.

NOTES TO CHAPTER EIGHT

263 **And Abraham drew near:** As before, I have used the King James translation but made small changes to align it more closely with the Hebrew.

275 **The ease and familiarity of such:** The 2013 Joint Statistical Meetings dedicated a whole session to the topic "Causal Inference as a Missing Data Problem"—Rubin's traditional mantra. One provocative paper at that session was titled "What Is Not a Missing Data Problem?" This title sums up my thoughts precisely.

277 **This difference in commitment:** Readers who are seeing this distinction for the first time should not feel alone; there are well over 100,000 regression analysts in the United States who are confused by this very issue, together with most authors of statistical textbooks. Things will only change when readers of this book take those authors to task.

283 **Unfortunately, Rubin does not consider:** "Pearl's work is clearly interesting, and many researchers find his arguments that path diagrams are a natural and convenient way to express assumptions about causal structures appealing. In our own work, perhaps influenced by the type of examples arising in social and medical sciences, we have not found this approach to aid the drawing of causal inferences" (Imbens and Rubin 2013, p. 25).

284 **One obstacle I faced was cyclic models:** These are models with arrows that form a loop. I have avoided discussing them in this book, but such models are quite important in economics, for example.

285   **Even today modern-day economists:** Between 1995 and 1998, I
      presented the following toy puzzle to hundreds of econometrics
      students and faculty across the United States:

Consider the classical supply-and-demand equations that every
economics student solves in Economics 101.

1. What is the expected value of the demand Q if the price is *reported* to be $P = p_0$?
2. What is the expected value of the demand Q if the price is *set*
   to $P = p_0$?
3. Given that the current price is $P = p_0$, what would the expected
   value of the demand Q be if we were to set the price at $P = p_1$?

The reader should recognize these queries as coming from
the three levels of the Ladder of Causation: predictions, actions,
and counterfactuals. As I expected, respondents had no trouble
answering question 1, one person (a distinguished professor)
was able to solve question 2, and nobody managed to answer
question 3.

286   **The Model Penal Code expresses:** This is a set of standard legal
      principles proposed by the American Law Institute in 1962 to
      bring uniformity to the various state legal codes. It does not
      have full legal force in any state, but according to Wikipedia, as
      of 2016, more than two-thirds of the states have enacted parts
      of the Model Penal Code.

NOTES TO CHAPTER NINE

304   **Those sailors who had eaten:** The reason is that polar bear liv-
      ers do contain vitamin C.
308   **"On the Inadequacy of the Partial":** The title refers to partial
      correlation, a standard method of controlling for a confounder
      that we discussed in Chapter 7.
321   **here is how to define the NIE:** In the original delivery room,
      NIE was expressed using nested subscripts, as in $Y_{(0,M_1)}$. I hope
      the reader will find the mixture of counterfactual subscripts and
      do-operators above more transparent.

340   **In that year researchers identified:** To be technically correct it
      should be called a "single nucleotide polymorphism," or SNP.
      It is a single letter in the genetic code, while a gene is more like a
      word or a sentence. However, in order not to burden the reader
      with unfamiliar terminology, I will simply refer to it as a gene.

BIBLIOGRAPHY

INTRODUCTION: MIND OVER DATA

Annotated Bibliography

The history of probability and statistics from antiquity to modern days is covered in depth by Hacking (1990); Stigler (1986, 1999, 2016). A less technical account is given in Salsburg (2002). Comprehensive accounts of the history of causal thought are unfortunately lacking, though interesting material can be found in Hoover (2008); Kleinberg (2015); Losee (2012); Mumford and Anjum (2014). The prohibition on causal talk can be seen in almost every standard statistical text, for example, Freedman, Pisani, and Purves (2007) or Efron and Hastie (2016). For an analysis of this prohibition as a linguistic impediment, see Pearl (2009, Chapters 5 and 11), and as a cultural barrier, see Pearl (2000b).

Recent accounts of the achievements and limitations of Big Data and machine learning are Darwiche (2017); Pearl (2017); Mayer-Schönberger and Cukier (2013); Domingos (2015); Marcus (July 30, 2017). Toulmin (1961) provides historical context to this debate.

Readers interested in "model discovery" and more technical treatments of the *do*-operator can consult Pearl (1994, 2000a, Chapters 2–3); Spirtes, Glymour, and Scheines (2000). For a gentler introduction, see Pearl, Glymour, and Jewell (2016). This last source is recommended for readers with college-level mathematical skills but no background in statistics or computer science. It also provides basic introduction to conditional probabilities, Bayes's rule, regression, and graphs.

Earlier versions of the inference engine shown in Figure 1.1 can be found in Pearl (2012); Pearl and Bareinboim (2014).

References

Darwiche, A. (2017). Human-level intelligence or animal-like abilities? Tech. rep., Department of Computer Science, University of California, Los Angeles, CA. Submitted to Communications of the ACM. Accessed online at https://arXiv:1707.04327.

Domingos, P. (2015). *The Master Algorithm: How the Quest for the Ultimate Learning Machine Will Remake Our World.* Basic Books, New York, NY.

Efron, B., and Hastie, T. (2016). *Computer Age Statistical Inference.* Cambridge University Press, New York, NY.

Freedman, D., Pisani, R., and Purves, R. (2007). *Statistics.* 4th ed. W. W. Norton & Company, New York, NY.

Hacking, I. (1990). *The Taming of Chance (Ideas in Context).* Cambridge University Press, Cambridge, UK.

Hoover, K. (2008). Causality in economics and econometrics. In *The New Palgrave Dictionary of Economics* (S. Durlauf and L. Blume, eds.), 2nd ed. Palgrave Macmillan, New York, NY.

Kleinberg, S. (2015). *Why: A Guide to Finding and Using Causes.* O'Reilly Media, Sebastopol, CA.

Losee, J. (2012). *Theories of Causality: From Antiquity to the Present.* Routledge, New York, NY.

Marcus, G. (July 30, 2017). Artificial intelligence is stuck. Here's how to move it forward. *New York Times*, SR6.

Mayer-Schönberger, V., and Cukier, K. (2013). *Big Data: A Revolution That Will Transform How We Live, Work, and Think.* Houghton Mifflin Harcourt Publishing, New York, NY.

Morgan, S., and Winship, C. (2015). *Counterfactuals and Causal Inference: Methods and Principles for Social Research (Analytical Methods for Social Research).* 2nd ed. Cambridge University Press, New York, NY.

Mumford, S., and Anjum, R. L. (2014). *Causation: A Very Short Introduction (Very Short Introductions).* Oxford University Press, New York, NY.

Pearl, J. (1988). *Probabilistic Reasoning in Intelligent Systems.* Morgan Kaufmann, San Mateo, CA.

Pearl, J. (1994). A probabilistic calculus of actions. In *Uncertainty in Artificial Intelligence 10* (R. L. de Mantaras and D. Poole, eds.). Morgan Kaufmann, San Mateo, CA, 454–462.

Pearl, J. (1995). Causal diagrams for empirical research. *Biometrika* 82: 669–710.

Pearl, J. (2000a). *Causality: Models, Reasoning, and Inference.* Cambridge University Press, New York, NY.

Pearl, J. (2000b). Comment on A. P. Dawid's Causal inference without counterfactuals. *Journal of the American Statistical Association* 95: 428–431.

Pearl, J. (2009). *Causality: Models, Reasoning, and Inference.* 2nd ed. Cambridge University Press, New York, NY.

Pearl, J. (2012). The causal foundations of structural equation modeling. In *Handbook of Structural Equation Modeling* (R. Hoyle, ed.). Guilford Press, New York, NY, 68–91.

Pearl, J. (2017). Advances in deep neural networks, at ACM Turing 50 Celebration. Available at: https://www.youtube.com/watch?v =mFYM9j8bGtg (June 23, 2017).

Pearl, J., and Bareinboim, E. (2014). External validity: From *do*-calculus to transportability across populations. *Statistical Science* 29: 579–595.

Pearl, J., Glymour, M., and Jewell, N. (2016). *Causal Inference in Statistics: A Primer.* Wiley, New York, NY.

Provine, W. B. (1986). *Sewall Wright and Evolutionary Biology.* University of Chicago Press, Chicago, IL.

Salsburg, D. (2002). *The Lady Tasting Tea: How Statistics Revolutionized Science in the Twentieth Century.* Henry Holt and Company, LLC, New York, NY.

Spirtes, P., Glymour, C., and Scheines, R. (2000). *Causation, Prediction, and Search.* 2nd ed. MIT Press, Cambridge, MA.

Stigler, S. M. (1986). *The History of Statistics: The Measurement of Uncertainty Before 1900.* Belknap Press of Harvard University Press, Cambridge, MA.

Stigler, S. M. (1999). *Statistics on the Table: The History of Statistical Concepts and Methods.* Harvard University Press, Cambridge, MA.

Stigler, S. M. (2016). *The Seven Pillars of Statistical Wisdom.* Harvard University Press, Cambridge, MA.

Toulmin, S. (1961). *Foresight and Understanding: An Enquiry into the Aims of Science.* University of Indiana Press, Bloomington, IN.

Virgil. (29 BC). Georgics. Verse 490, Book 2.

CHAPTER 1. THE LADDER OF CAUSATION

Annotated Bibliography

A technical account of the distinctions between the three levels of the Ladder of Causation can be found in Chapter 1 of Pearl (2000).

Our comparisons between the Ladder of Causation and human cognitive development were inspired by Harari (2015) and by the recent findings by Kind et al. (2014). Kind's article contains details about the Lion Man and the site where it was found. Related research on the development of causal understanding in babies can be found in Weisberg and Gopnik (2013).

The Turing test was first proposed as an imitation game in 1950 (Turing, 1950). Searle's "Chinese Room" argument appeared in Searle (1980) and has been widely discussed in the years since. See Russell and Norvig (2003); Preston and Bishop (2002); Pinker (1997).

The use of model modification to represent intervention has its conceptual roots with the economist Trygve Haavelmo (1943); see Pearl (2015) for a detailed account. Spirtes, Glymour, and Scheines (1993) gave it a graphical representation in terms of arrow deletion. Balke and Pearl (1994a, 1994b) extended it to simulate counterfactual reasoning, as demonstrated in the firing squad example.

A comprehensive summary of probabilistic causality is given in Hitchcock (2016). Key ideas can be found in Reichenbach (1956); Suppes (1970); Cartwright (1983); Spohn (2012). My analyses of probabilistic causality and probability raising are presented in Pearl (2000; 2009, Section 7.5; 2011).

References

Balke, A., and Pearl, J. (1994a). Counterfactual probabilities: Computational methods, bounds, and applications. In *Uncertainty in Artificial Intelligence 10* (R. L. de Mantaras and D. Poole, eds.). Morgan Kaufmann, San Mateo, CA, 46–54.

Balke, A., and Pearl, J. (1994b). Probabilistic evaluation of counterfactual queries. In *Proceedings of the Twelfth National Conference on Artificial Intelligence*, vol. 1. MIT Press, Menlo Park, CA, 230–237.

Cartwright, N. (1983). *How the Laws of Physics Lie*. Clarendon Press, Oxford, UK.

Haavelmo, T. (1943). The statistical implications of a system of simultaneous equations. *Econometrica* 11: 1–12. Reprinted in D. F. Hendry and M. S. Morgan (Eds.), *The Foundations of Econometric Analysis*, Cambridge University Press, Cambridge, UK, 477–490, 1995.

Harari, Y. N. (2015). *Sapiens: A Brief History of Humankind*. Harper Collins Publishers, New York, NY.

Hitchcock, C. (2016). Probabilistic causation. In *Stanford Encyclopedia of Philosophy (Winter 2016)* (E. N. Zalta, ed.). Metaphysics Research Lab, Stanford, CA. Available at: https://stanford.library .sydney.edu.au/archives/win2016/entries/causation-probabilistic.

Kind, C.-J., Ebinger-Rist, N., Wolf, S., Beutelspacher, T., and Wehrberger, K. (2014). The smile of the Lion Man. Recent excavations in Stadel cave (Baden-Württemberg, south-western Germany) and the restoration of the famous upper palaeolithic figurine. *Quartär* 61: 129–145.

Pearl, J. (2000). *Causality: Models, Reasoning, and Inference*. Cambridge University Press, New York, NY.

Pearl, J. (2009). *Causality: Models, Reasoning, and Inference*. 2nd ed. Cambridge University Press, New York, NY.

Pearl, J. (2011). The structural theory of causation. In *Causality in the Sciences* (P. M. Illari, F. Russo, and J. Williamson, eds.), chap. 33. Clarendon Press, Oxford, UK, 697–727.

Pearl, J. (2015). Trygve Haavelmo and the emergence of causal calculus. *Econometric Theory* 31: 152–179. Special issue on Haavelmo centennial.

Pinker, S. (1997). *How the Mind Works*. W. W. Norton and Company, New York, NY.

Preston, J., and Bishop, M. (2002). *Views into the Chinese Room: New Essays on Searle and Artificial Intelligence*. Oxford University Press, New York, NY.

Reichenbach, H. (1956). *The Direction of Time*. University of California Press, Berkeley, CA.

Russell, S. J., and Norvig, P. (2003). *Artificial Intelligence: A Modern Approach*. 2nd ed. Prentice Hall, Upper Saddle River, NJ.

Searle, J. (1980). Minds, brains, and programs. *Behavioral and Brain Sciences* 3: 417–457.

Spirtes, P., Glymour, C., and Scheines, R. (1993). *Causation, Prediction, and Search*. Springer-Verlag, New York, NY.

Spohn, W. (2012). *The Laws of Belief: Ranking Theory and Its Philosophical Applications*. Oxford University Press, Oxford, UK.

Suppes, P. (1970). *A Probabilistic Theory of Causality*. North-Holland Publishing Co., Amsterdam, Netherlands.

Turing, A. (1950). Computing machinery and intelligence. *Mind 59*: 433–460.

Weisberg, D. S., and Gopnik, A. (2013). Pretense, counterfactuals, and Bayesian causal models: Why what is not real really matters. *Cognitive Science 37*: 1368–1381.

CHAPTER 2. FROM BUCCANEERS TO GUINEA PIGS: THE GENESIS OF CAUSAL INFERENCE

Annotated Bibliography

Galton's explorations of heredity and correlation are described in his books (Galton, 1869, 1883, 1889) and are also documented in Stigler (2012, 2016).

For a basic introduction to the Hardy-Weinberg equilibrium, see Wikipedia (2016a). For the origin of Galileo's quote "E pur si muove," see Wikipedia (2016b). The story of the Paris catacombs and Pearson's shock at correlations induced by "artificial mixtures" can be found in Stigler (2012, p. 9).

Because Wright lived such a long life, he had the rare privilege of seeing a biography (Provine, 1986) come out while he was still alive. Provine's biography is still the best place to learn about Wright's career, and we particularly recommend Chapter 5 on path analysis. Crow's two biographical sketches (Crow, 1982, 1990) also provide a very useful biographical perspective. Wright (1920) is the seminal paper on path diagrams; Wright (1921) is a fuller exposition and the source for the guinea pig birth-weight example. Wright (1983) is Wright's response to Karlin's critique, written when he was over ninety years old.

The fate of path analysis in economics and social science is narrated in Chapter 5 of Pearl (2000) and in Bollen and Pearl (2013). Blalock (1964), Duncan (1966), and Goldberger (1972) introduced Wright's ideas to social science with great enthusiasm, but their theoretical underpinnings were not well articulated. A decade later, when Freedman (1987) challenged path analysts to explain how interventions are modelled, the enthusiasm disappeared, and leading researchers retreated to viewing SEM as an exercise in statistical analysis. This revealing discussion among twelve scholars is documented in the same issue of the *Journal of Educational Statistics* as Freedman's article.

The reluctance of economists to embrace diagrams and structural notation is described in Pearl (2015). The painful consequences for economic education are documented in Chen and Pearl (2013).

A popular exposition of the Bayesian-versus-frequentist debate is given in McGrayne (2011).

More technical discussions can be found in Efron (2013) and Lindley (1987).

References

Blalock, H., Jr. (1964). *Causal Inferences in Nonexperimental Research*. University of North Carolina Press, Chapel Hill, NC.

Bollen, K., and Pearl, J. (2013). Eight myths about causality and structural equation models. In *Handbook of Causal Analysis for Social Research* (S. Morgan, ed.). Springer, Dordrecht, Netherlands, 301–328.

Chen, B., and Pearl, J. (2013). Regression and causation: A critical examination of econometrics textbooks. *Real-World Economics Review* 65: 2–20.

Crow, J. F. (1982). Sewall Wright, the scientist and the man. *Perspectives in Biology and Medicine* 25: 279–294.

Crow, J. F. (1990). Sewall Wright's place in twentieth-century biology. *Journal of the History of Biology* 23: 57–89.

Duncan, O. D. (1966). Path analysis. *American Journal of Sociology* 72: 1–16.

Efron, B. (2013). Bayes' theorem in the 21st century. *Science* 340: 1177–1178.

Freedman, D. (1987). As others see us: A case study in path analysis (with discussion). *Journal of Educational Statistics* 12: 101–223.

Galton, F. (1869). *Hereditary Genius*. Macmillan, London, UK.

Galton, F. (1883). *Inquiries into Human Faculty and Its Development*. Macmillan, London, UK.

Galton, F. (1889). *Natural Inheritance*. Macmillan, London, UK.

Goldberger, A. (1972). Structural equation models in the social sciences. *Econometrica: Journal of the Econometric Society* 40: 979–1001.

Lindley, D. (1987). *Bayesian Statistics: A Review*. CBMS-NSF Regional Conference Series in Applied Mathematics (Book 2). Society for Industrial and Applied Mathematics, Philadelphia, PA.

McGrayne, S. B. (2011). *The Theory That Would Not Die*. Yale University Press, New Haven, CT.

Pearl, J. (2000). *Causality: Models, Reasoning, and Inference*. Cambridge University Press, New York, NY.

Pearl, J. (2015). Trygve Haavelmo and the emergence of causal calculus. *Econometric Theory* 31: 152–179. Special issue on Haavelmo centennial.

Provine, W. B. (1986). *Sewall Wright and Evolutionary Biology*. University of Chicago Press, Chicago, IL.

Stigler, S. M. (2012). Studies in the history of probability and statistics, L: Karl Pearson and the rule of three. *Biometrika* 99: 1–14.

Stigler, S. M. (2016). *The Seven Pillars of Statistical Wisdom*. Harvard University Press, Cambridge, MA.

Wikipedia. (2016a). Hardy-Weinberg principle. Available at: https://en.wikipedia.org/wiki/Hardy-Weinberg-principle (last edited: October 2, 2016).

Wikipedia. (2016b). Galileo Galilei. Available at: https://en.wikipedia.org/wiki/Galileo_Galilei (last edited: October 6, 2017).

Wright, S. (1920). The relative importance of heredity and environment in determining the piebald pattern of guinea-pigs. *Proceedings of the National Academy of Sciences of the United States of America* 6: 320–332.

Wright, S. (1921). Correlation and causation. *Journal of Agricultural Research* 20: 557–585.

Wright, S. (1983). On "Path analysis in genetic epidemiology: A critique." *American Journal of Human Genetics* 35: 757–768.

CHAPTER 3. FROM EVIDENCE TO CAUSES: REVEREND BAYES MEETS MR. HOLMES

Annotated Bibliography

Elementary introductions to Bayes's rule and Bayesian thinking can be found in Lindley (2014) and Pearl, Glymour, and Jewell (2016).

Debates with competing representations of uncertainty are presented in Pearl (1988); see also the extensive list of references given there.

Our mammogram data are based primarily on information from the Breast Cancer Surveillance Consortium (BCSC, 2009) and US Preventive Services Task Force (USPSTF, 2016) and are presented for instructional purposes only.

"Bayesian networks" received their name in 1985 (Pearl, 1985) and were first presented as a model of self-activated memory. Applications to expert systems followed the development of belief updating algorithms for loopy networks (Pearl, 1986; Lauritzen and Spiegelhalter, 1988).

The concept of d-separation, which connects path blocking in a diagram to dependencies in the data, has its roots in the theory of graphoids (Pearl and Paz, 1985). The theory unveils the common properties of graphs (hence the name) and probabilities and explains why these two seemingly alien mathematical objects can support one another in so many ways. See also "Graphoid," Wikipedia.

The amusing example of the bag on the airline flight can be found in Conrady and Jouffe (2015, Chapter 4).

The Malaysia Airlines Flight 17 disaster was well covered in the media; see Clark and Kramer (October 14, 2015) for an update on the investigation a year after the incident. Wiegerinck, Burgers, and Kappen (2013) describes how Bonaparte works. Further details on the identification of Flight 17 victims, including the pedigree shown in Figure 3.7, came from personal correspondence from W. Burgers to D. Mackenzie (August 24, 2016) and from a phone interview with W. Burgers and B. Kappen by D. Mackenzie (August 23, 2016).

The complex and fascinating story of turbo and low-density parity-check codes has not been told in a truly layman-friendly form, but good starting points are Costello and Forney (2007) and Hardesty (2010a, 2010b). The crucial realization that turbo codes work by the belief propagation algorithm stems from McEliece, David, and Cheng (1998).

Efficient codes continue to be a battleground for wireless communications; Carlton (2016) takes a look at the current contenders for "5G" phones (due out in the 2020s).

References

Breast Cancer Surveillance Consortium (BCSC). (2009). Performance measures for 1,838,372 screening mammography examinations from 2004 to 2008 by age. Available at: http://www.bcsc-research .org/statistics/performance/screening/2009/perf_age.html (accessed October 12, 2016).

Carlton, A. (2016). Surprise! Polar codes are coming in from the cold. *Computerworld*. Available at: https://www.computerworld.com /article/3151866/mobile-wireless/surprise-polar-codes-are -coming-in-from-the-cold.html (posted December 22, 2016).

Clark, N., and Kramer, A. (October 14, 2015). Malaysia Airlines Flight 17 most likely hit by Russian-made missile, inquiry says. *New York Times*.

Conrady, S., and Jouffe, L. (2015). *Bayesian Networks and Bayesia Lab: A Practical Introduction for Researchers*. Bayesia USA, Franklin, TN.

Costello, D. J., and Forney, G. D., Jr. (2007). Channel coding: The road to channel capacity. *Proceedings of IEEE 95*: 1150–1177.

Hardesty, L. (2010a). Explained: Gallager codes. *MIT News*. Available at: http://news.mit.edu/2010/gallager-codes-0121 (posted: January 21, 2010).

Hardesty, L. (2010b). Explained: The Shannon limit. *MIT News*. Available at: http://news.mit.edu/2010/explained-shannon-0115 (posted January 19, 2010).

Lauritzen, S., and Spiegelhalter, D. (1988). Local computations with probabilities on graphical structures and their application to expert systems (with discussion). *Journal of the Royal Statistical Society, Series B 50*: 157–224.

Lindley, D. V. (2014). *Understanding Uncertainty*. Rev. ed. John Wiley and Sons, Hoboken, NJ.

McEliece, R. J., David, J. M., and Cheng, J. (1998). Turbo decoding as an instance of Pearl's "belief propagation" algorithm. *IEEE Journal on Selected Areas in Communications 16*: 140–152.

Pearl, J. (1985). Bayesian networks: A model of self-activated memory for evidential reasoning. In *Proceedings, Cognitive Science Society* (CSS-7). UCLA Computer Science Department, Irvine, CA.

Pearl, J. (1986). Fusion, propagation, and structuring in belief networks. *Artificial Intelligence 29*: 241–288.

Pearl, J. (1988). *Probabilistic Reasoning in Intelligent Systems*. Morgan Kaufmann, San Mateo, CA.

Pearl, J., Glymour, M., and Jewell, N. (2016). *Causal Inference in Statistics: A Primer*. Wiley, New York, NY.

Pearl, J., and Paz, A. (1985). GRAPHOIDS: A graph-based logic for reasoning about relevance relations. Tech. Rep. 850038 (R-53-L). Computer Science Department, University of California, Los Angeles. Short version in B. DuBoulay, D. Hogg, and L. Steels (Eds.) *Advances in Artificial Intelligence—II*, Amsterdam, North Holland, 357–363, 1987.

US Preventive Services Task Force (USPSTF) (2016). Final recommendation statement: Breast cancer: Screening. Available at: https://www.uspreventiveservicestaskforce.org/Page/Document /RecommendationStatementFinal/breast-cancer-screening1 (updated: January 2016).

Wikipedia. (2018). Graphoid. Available at: https://en.wikipedia.org /wiki/Graphoid (last edited: January 8, 2018).

Wiegerinck, W., Burgers, W., and Kappen, B. (2013). Bayesian networks, introduction and practical applications. In *Handbook on Neural Information Processing* (M. Bianchini, M. Maggini, and L. C. Jain, eds.). Intelligent Systems Reference Library (Book 49). Springer, Berlin, Germany, 401–431.

CHAPTER 4. CONFOUNDING AND DECONFOUNDING: OR, SLAYING THE LURKING VARIABLE

Annotated Bibliography

The story of Daniel has frequently been cited as the first controlled trial; see, for example, Lilienfeld (1982) or Stigler (2016). The results of the Honolulu walking study were reported in Hakim (1998).

Fisher Box's lengthy quote about "the skillful interrogation of Nature" comes from her excellent biography of her father (Box, 1978, Chapter 6). Fisher, too, wrote about experiments as a dialogue with Nature; see Stigler (2016). Thus I believe we can think of her quote as nearly coming from the patriarch himself, only more beautifully expressed.

It is fascinating to read Weinberg's papers on confounding (Weinberg, 1993; Howards et al., 2012) back-to-back. They are like two snapshots of the history of confounding, one taken just before causal diagrams became widespread and the second taken twenty years later, revisiting the same examples using causal diagrams. Forbes's complicated diagram of the causal network for asthma and smoking can be found in Williamson et al. (2014).

Morabia's "classic epidemiological definition of confounding" can be found in Morabia (2011). The quotes from David Cox come from Cox (1992, pp. 66–67). Other good sources on the history of confounding are Greenland and Robins (2009) and Wikipedia (2016).

The back-door criterion for eliminating confounding bias, together with its adjustment formula, were introduced in Pearl (1993). Its impact on epidemiology can be seen through Greenland, Pearl, and Robins (1999). Extensions to sequential interventions and other nuances are developed in Pearl (2000, 2009) and more gently described in Pearl, Glymour, and Jewell (2016). Software for computing causal effects using *do*-calculus is available in Tikka and Karvanen (2017).

The paper by Greenland and Robins (1986) was revisited by the authors a quarter century later, in light of the extensive developments since that time, including the advent of causal diagrams (Greenland and Robins, 2009).

References

Box, J. F. (1978). *R. A. Fisher: The Life of a Scientist*. John Wiley and Sons, New York, NY.

Cox, D. (1992). *Planning of Experiments*. Wiley-Interscience, New York, NY.

Greenland, S., Pearl, J., and Robins, J. (1999). Causal diagrams for epidemiologic research. *Epidemiology* 10: 37–48.

Greenland, S., and Robins, J. (1986). Identifiability, exchangeability, and epidemiological confounding. *International Journal of Epidemiology* 15: 413–419.

Greenland, S., and Robins, J. (2009). Identifiability, exchangeability, and confounding revisited. *Epidemiologic Perspectives & Innovations* 6. doi:10.1186/1742-5573-6-4.

Hakim, A. (1998). Effects of walking on mortality among nonsmoking retired men. *New England Journal of Medicine* 338: 94–99.

Hernberg, S. (1996). Significance testing of potential confounders and other properties of study groups—Misuse of statistics. *Scandinavian Journal of Work, Environment and Health* 22: 315–316.

Howards, P. P., Schisterman, E. F., Poole, C., Kaufman, J. S., and Weinberg, C. R. (2012). "Toward a clearer definition of confounding" revisited with directed acyclic graphs. *American Journal of Epidemiology* 176: 506–511.

Lilienfeld, A. (1982). Ceteris paribus: The evolution of the clinical trial. *Bulletin of the History of Medicine* 56: 1–18.

Morabia, A. (2011). History of the modern epidemiological concept of confounding. *Journal of Epidemiology and Community Health* 65: 297–300.

Pearl, J. (1993). Comment: Graphical models, causality, and intervention. *Statistical Science* 8: 266–269.

Pearl, J. (2000). *Causality: Models, Reasoning, and Inference.* Cambridge University Press, New York, NY.

Pearl, J. (2009). *Causality: Models, Reasoning, and Inference.* 2nd ed. Cambridge University Press, New York, NY.

Pearl, J., Glymour, M., and Jewell, N. (2016). *Causal Inference in Statistics: A Primer.* Wiley, New York, NY.

Stigler, S. M. (2016). *The Seven Pillars of Statistical Wisdom.* Harvard University Press, Cambridge, MA.

Tikka, J., and Karvanen, J. (2017). Identifying causal effects with the R Package causaleffect. *Journal of Statistical Software* 76, no. 12. doi:10.18637/jss.r076.i12.

Weinberg, C. (1993). Toward a clearer definition of confounding. *American Journal of Epidemiology* 137: 1–8.

Wikipedia. (2016). Confounding. Available at: https://en.wikipedia.org/wiki/Confounding (accessed: September 16, 2016).

Williamson, E., Aitken, Z., Lawrie, J., Dharmage, S., Burgess, H., and Forbes, A. (2014). Introduction to causal diagrams for confounder selection. *Respirology* 19: 303–311.

CHAPTER 5. THE SMOKE-FILLED DEBATE: CLEARING THE AIR

Annotated Bibliography

Two book-length studies, Brandt (2007) and Proctor (2012a), contain all the information any reader could ask for about the smoking–lung cancer debate, short of reading the actual tobacco company documents (which are available online). Shorter surveys of the smoking-cancer debate in the 1950s are Salsburg (2002, Chapter 18), Parascandola (2004), and Proctor (2012b). Stolley (1991) takes a look at the unique role of R. A. Fisher, and Greenhouse (2009) comments on Jerome Cornfield's importance. The shot heard around the world was Doll and Hill (1950), which first implicated smoking in lung cancer; though technical, it is a scientific classic.

For the story of the surgeon general's committee and the emergence of the Hill guidelines for causation, see Blackburn and Labarthe (2012) and Morabia (2013). Hill's own description of his criteria can be found in Hill (1965).

Lilienfeld (2007) is the source of the "Abe and Yak" story with which we began the chapter.

VanderWeele (2014) and Hernández-Díaz, Schisterman, and Hernán (2006) resolve the birth-weight paradox using causal diagrams. An interesting "before-and-after" pair of articles is Wilcox (2001, 2006), written before and after the author learned about causal diagrams; his excitement in the latter article is palpable.

Readers interested in the latest statistics and historical trends in cancer mortality and smoking may consult US Department of Health and Human Services (USDHHS, 2014), American Cancer Society (2017), and Wingo (2003).

References

American Cancer Society. (2017). Cancer facts and figures. Available at: https://www.cancer.org/research/cancer-facts-statistics.html (posted: February 19, 2015).

Blackburn, H., and Labarthe, D. (2012). Stories from the evolution of guidelines for causal inference in epidemiologic associations: 1953–1965. *American Journal of Epidemiology* 176: 1071–1077.

Brandt, A. (2007). *The Cigarette Century*. Basic Books, New York, NY.

Doll, R., and Hill, A. B. (1950). Smoking and carcinoma of the lung. *British Medical Journal* 2: 739–748.

Greenhouse, J. (2009). Commentary: Cornfield, epidemiology, and causality. *International Journal of Epidemiology* 38: 1199–1201.

Hernández-Díaz, S., Schisterman, E., and Hernán, M. (2006). The birth weight "paradox" uncovered? *American Journal of Epidemiology* 164: 1115–1120.

Hill, A. B. (1965). The environment and disease: Association or causation? *Journal of the Royal Society of Medicine* 58: 295–300.

Lilienfeld, D. (2007). Abe and Yak: The interactions of Abraham M. Lilienfeld and Jacob Yerushalmy in the development of modern epidemiology (1945–1973). *Epidemiology* 18: 507–514.

Morabia, A. (2013). Hume, Mill, Hill, and the sui generis epidemiologic approach to causal inference. *American Journal of Epidemiology* 178: 1526–1532.

Parascandola, M. (2004). Two approaches to etiology: The debate over smoking and lung cancer in the 1950s. *Endeavour* 28: 81–86.

Proctor, R. (2012a). *Golden Holocaust: Origins of the Cigarette Catastrophe and the Case for Abolition.* University of California Press, Berkeley, CA.

Proctor, R. (2012b). The history of the discovery of the cigarette–lung cancer link: Evidentiary traditions, corporate denial, and global toll. *Tobacco Control* 21: 87–91.

Salsburg, D. (2002). *The Lady Tasting Tea: How Statistics Revolutionized Science in the Twentieth Century.* Henry Holt and Company, New York, NY.

Stolley, P. (1991). When genius errs: R. A. Fisher and the lung cancer controversy. *American Journal of Epidemiology* 133: 416–425.

US Department of Health and Human Services (USDHHS). (2014). The health consequences of smoking—50 years of progress: A report of the surgeon general. USDHHS and Centers for Disease Control and Prevention, Atlanta, GA.

VanderWeele, T. (2014). Commentary: Resolutions of the birthweight paradox: Competing explanations and analytical insights. *International Journal of Epidemiology* 43: 1368–1373.

Wilcox, A. (2001). On the importance—and the unimportance—of birthweight. *International Journal of Epidemiology* 30: 1233–1241.

Wilcox, A. (2006). The perils of birth weight—A lesson from directed acyclic graphs. *American Journal of Epidemiology* 164: 1121–1123.

Wingo, P. (2003). Long-term trends in cancer mortality in the United States, 1930–1998. *Cancer* 97: 3133–3275.

CHAPTER 6. PARADOXES GALORE!

Annotated Bibliography

The Monty Hall paradox appears in many introductory books on probability theory (e.g., Grinstead and Snell, 1998, p. 136; Lindley,

2014, p. 201). The equivalent "three prisoners dilemma" was used to demonstrate the inadequacy of non-Bayesian approaches in Pearl (1988, pp. 58–62).

Tierney (July 21, 1991) and Crockett (2015) tell the amazing story of vos Savant's column on the Monty Hall paradox; Crockett gives several other entertaining and embarrassing comments that vos Savant received from so-called experts. Tierney's article tells what Monty Hall himself thought of the fuss—an interesting human-interest angle!

An extensive account of the history of Simpson's paradox is given in Pearl (2009, pp. 174–182), including many attempts by statisticians and philosophers to resolve it without invoking causation. A more recent account, geared for educators, is given in Pearl (2014).

Savage (2009), Julious and Mullee (1994), and Appleton, French, and Vanderpump (1996) give the three real-world examples of Simpson's paradox mentioned in the text (relating to baseball, kidney stones, and smoking, respectively).

Savage's sure-thing principle (Savage, 1954) is treated in Pearl (2016b), and its corrected causal version is derived in Pearl (2009, pp. 181–182).

Versions of Lord's paradox (Lord, 1967) are described in Glymour (2006); Hernández-Díaz, Schisterman, and Hernán (2006); Senn (2006); Wainer (1991); Wainer and Brown (2007). A comprehensive analysis can be found in Pearl (2016a).

Paradoxes invoking counterfactuals are not included in this chapter but are no less intriguing. For a sample, see Pearl (2013).

References

Appleton, D., French, J., and Vanderpump, M. (1996). Ignoring a covariate: An example of Simpson's paradox. *American Statistician* 50: 340–341.

Crockett, Z. (2015). The time everyone "corrected" the world's smartest woman. *Priceonomics*. Available at: http://priceonomics.com /the-time-everyone-corrected-the-worlds-smartest (posted: February 19, 2015).

Glymour, M. M. (2006). Using causal diagrams to understand common problems in social epidemiology. In *Methods in Social Epidemiology*. John Wiley and Sons, San Francisco, CA, 393–428.

Grinstead, C. M., and Snell, J. L. (1998). *Introduction to Probability*. 2nd rev. ed. American Mathematical Society, Providence, RI.

Hernández-Díaz, S., Schisterman, E., and Hernán, M. (2006). The birth weight "paradox" uncovered? *American Journal of Epidemiology* 164: 1115–1120.

Julious, S., and Mullee, M. (1994). Confounding and Simpson's paradox. *British Medical Journal* 309: 1480–1481.

Lindley, D. V. (2014). *Understanding Uncertainty*. Rev. ed. John Wiley and Sons, Hoboken, NJ.

Lord, F. M. (1967). A paradox in the interpretation of group comparisons. *Psychological Bulletin* 68: 304–305.

Pearl, J. (1988). *Probabilistic Reasoning in Intelligent Systems*. Morgan Kaufmann, San Mateo, CA.

Pearl, J. (2009). *Causality: Models, Reasoning, and Inference*. 2nd ed. Cambridge University Press, New York, NY.

Pearl, J. (2013). The curse of free-will and paradox of inevitable regret. *Journal of Causal Inference* 1: 255–257.

Pearl, J. (2014). Understanding Simpson's paradox. *American Statistician* 88: 8–13.

Pearl, J. (2016a). Lord's paradox revisited—(Oh Lord! Kumbaya!). *Journal of Causal Inference* 4. doi:10.1515/jci-2016-0021.

Pearl, J. (2016b). The sure-thing principle. *Journal of Causal Inference* 4: 81–86.

Savage, L. (1954). *The Foundations of Statistics*. John Wiley and Sons, New York, NY.

Savage, S. (2009). *The Flaw of Averages: Why We Underestimate Risk in the Face of Uncertainty*. John Wiley and Sons, Hoboken, NJ.

Senn, S. (2006). Change from baseline and analysis of covariance revisited. *Statistics in Medicine* 25: 4334–4344.

Simon, H. (1954). Spurious correlation: A causal interpretation. *Journal of the American Statistical Association* 49: 467–479.

Tierney, J. (July 21, 1991). Behind Monty Hall's doors: Puzzle, debate and answer? *New York Times*.

Wainer, H. (1991). Adjusting for differential base rates: Lord's paradox again. *Psychological Bulletin* 109: 147–151.

Wainer, H., and Brown, L. (2007). Three statistical paradoxes in the interpretation of group differences: Illustrated with medical school admission and licensing data. Rao C, Sinharay S, editors. *Handbook of Statistics 26: Psychometrics* Vol. 26. North Holland: Elsevier B.V., 893–918.

CHAPTER 7. BEYOND ADJUSTMENT:
THE CONQUEST OF MOUNT INTERVENTION

Annotated Bibliography

Extensions of the back-door and front-door adjustments were first reported in Tian and Pearl (2002) based on Tian's *c*-component factorization. These were followed by Shpitser's algorithmization of the *do*-calculus (Shpitser and Pearl, 2006a) and then the completeness results of Shpitser and Pearl (2006b) and Huang and Valtorta (2006).

The economists among our readers should note that the cultural resistance of some economists to graphical tools of analysis (Heckman and Pinto, 2015; Imbens and Rubin, 2015) is not shared by all economists. White and Chalak (2009), for example, have generalized and applied the *do*-calculus to economic systems involving equilibrium and learning. Recent textbooks in the social and behavioral sciences, Morgan and Winship (2007) and Kline (2016), further signal to young researchers that cultural orthodoxy, like the fear of telescopes in the seventeenth century, is not long lasting in the sciences.

John Snow's investigation of cholera was very little appreciated during his lifetime, and his one-paragraph obituary in *Lancet* did not even mention it. Remarkably, the premier British medical journal "corrected" its obituary 155 years later (Hempel, 2013). For more biographical material on Snow, see Hill (1955) and Cameron and Jones (1983). Glynn and Kashin (2018) is one of the first papers to demonstrate empirically that front-door adjustment is superior to back-door adjustment when there are unobserved confounders. Freedman's critique of the smoking–tar–lung cancer example can be found in a chapter of Freedman (2010) titled "On Specifying Graphical Models for Causation."

Introductions to instrumental variables can be found in Greenland (2000) and in many textbooks of econometrics (e.g., Bowden and Turkington, 1984; Wooldridge, 2013).

Generalized instrumental variables, extending the classical definition given in our text, were introduced in Brito and Pearl (2002).

The program DAGitty (available online at http://www.dagitty.net /dags.html) permits users to search the diagram for generalized instrumental variables and reports the resulting estimands (Textor, Hardt, and Knüppel, 2011). Another diagram-based software package for decision making is BayesiaLab (www.bayesia.com).

Bounds on instrumental variable estimates are studied at length in Chapter 8 of Pearl (2009) and are applied to the problem of noncompliance. The LATE approximation is advocated and debated in Imbens (2010).

References

Bareinboim, E., and Pearl, J. (2012). Causal inference by surrogate experiments: z-identifiability. In *Proceedings of the Twenty-Eighth Conference on Uncertainty in Artificial Intelligence* (N. de Freitas and K. Murphy, eds.). AUAI Press, Corvallis, OR.

Bowden, R., and Turkington, D. (1984). *Instrumental Variables*. Cambridge University Press, Cambridge, UK.

Brito, C., and Pearl, J. (2002). Generalized instrumental variables. In *Uncertainty in Artificial Intelligence, Proceedings of the Eighteenth Conference* (A. Darwiche and N. Friedman, eds.). Morgan Kaufmann, San Francisco, CA, 85–93.

Cameron, D., and Jones, I. (1983). John Snow, the Broad Street pump, and modern epidemiology. *International Journal of Epidemiology* 12: 393–396.

Cox, D., and Wermuth, N. (2015). Design and interpretation of studies: Relevant concepts from the past and some extensions. *Observational Studies* 1. Available at: https://arxiv.org/pdf/1505.02452 .pdf.

Freedman, D. (2010). *Statistical Models and Causal Inference: A Dialogue with the Social Sciences*. Cambridge University Press, New York, NY.

Glynn, A., and Kashin, K. (2018). Front-door versus back-door adjustment with unmeasured confounding: Bias formulas for front-door and hybrid adjustments. *Journal of the American Statistical Association*. To appear.

Greenland, S. (2000). An introduction to instrumental variables for epidemiologists. *International Journal of Epidemiology* 29: 722–729.

Heckman, J. J., and Pinto, R. (2015). Causal analysis after Haavelmo. *Econometric Theory* 31: 115–151.

Hempel, S. (2013). Obituary: John Snow. *Lancet* 381: 1269–1270.

Hill, A. B. (1955). Snow—An appreciation. *Journal of Economic Perspectives* 48: 1008–1012.

Huang, Y., and Valtorta, M. (2006). Pearl's calculus of intervention is complete. In *Proceedings of the Twenty-Second Conference on Uncertainty in Artificial Intelligence* (R. Dechter and T. Richardson, eds.). AUAI Press, Corvallis, OR, 217–224.

Imbens, G. W. (2010). Better LATE than nothing: Some comments on Deaton (2009) and Heckman and Urzua (2009). *Journal of Economic Literature* 48: 399–423.

Imbens, G. W., and Rubin, D. B. (2015). *Causal Inference for Statistics, Social, and Biomedical Sciences: An Introduction*. Cambridge University Press, Cambridge, MA.

Kline, R. B. (2016). *Principles and Practice of Structural Equation Modeling*. 3rd ed. Guilford, New York, NY.

Morgan, S., and Winship, C. (2007). *Counterfactuals and Causal Inference: Methods and Principles for Social Research (Analytical Methods for Social Research)*. Cambridge University Press, New York, NY.

Pearl, J. (2009). *Causality: Models, Reasoning, and Inference*. 2nd ed. Cambridge University Press, New York, NY.

Pearl, J. (2013). Reflections on Heckman and Pinto's "Causal analysis after Haavelmo." Tech. Rep. R-420. Department of Computer Science, University of California, Los Angeles, CA. Working paper.

Pearl, J. (2015). Indirect confounding and causal calculus (on three papers by Cox and Wermuth). Tech. Rep. R-457. Department of Computer Science, University of California, Los Angeles, CA.

Shpitser, I., and Pearl, J. (2006a). Identification of conditional interventional distributions. In *Proceedings of the Twenty-Second Conference on Uncertainty in Artificial Intelligence* (R. Dechter and T. Richardson, eds.). AUAI Press, Corvallis, OR, 437–444.

Shpitser, I., and Pearl, J. (2006b). Identification of joint interventional distributions in recursive semi-Markovian causal models. In *Proceedings of the Twenty-First National Conference on Artificial Intelligence*. AAAI Press, Menlo Park, CA, 1219–1226.

Stock, J., and Trebbi, F. (2003). Who invented instrumental variable regression? *Journal of Economic Perspectives* 17: 177–194.

Textor, J., Hardt, J., and Knüppel, S. (2011). DAGitty: A graphical tool for analyzing causal diagrams. *Epidemiology* 22: 745.

Tian, J., and Pearl, J. (2002). A general identification condition for causal effects. In *Proceedings of the Eighteenth National Conference on Artificial Intelligence*. AAAI Press/MIT Press, Menlo Park, CA, 567–573.

Wermuth, N., and Cox, D. (2008). Distortion of effects caused by indirect confounding. *Biometrika* 95: 17–33. (See Pearl [2009, Chapter 4] for a general solution.)

Wermuth, N., and Cox, D. (2014). Graphical Markov models: Overview. ArXiv: 1407.7783.

White, H., and Chalak, K. (2009). Settable systems: An extension of Pearl's causal model with optimization, equilibrium and learning. *Journal of Machine Learning Research* 10: 1759–1799.

Wooldridge, J. (2013). *Introductory Econometrics: A Modern Approach*. 5th ed. South-Western, Mason, OH.

CHAPTER 8. COUNTERFACTUALS: MINING WORLDS THAT COULD HAVE BEEN

Annotated Bibliography

The definition of counterfactuals as derivatives of structural equations was introduced by Balke and Pearl (1994a, 1994b) and was used to estimate probabilities of causation in legal settings. The relationships between this framework and those developed by Rubin and Lewis are discussed at length in Pearl (2000, Chapter 7), where they are shown to be logically equivalent; a problem solved in one framework would yield the same solution in another.

Recent books in social science (e.g., Morgan and Winship, 2015) and in health science (e.g., VanderWeele, 2015) are taking the hybrid, graph-counterfactual approach pursued in our book.

The section on linear counterfactuals is based on Pearl (2009, pp. 389–391), which also provides the solution to the problem posed in note 12. Our discussion of ETT is based on Shpitser and Pearl (2009).

Legal questions of attribution, as well as probabilities of causation, are discussed at length in Greenland (1999), who pioneered the counterfactual approach to such questions. Our treatment of *PN*, *PS*, and *PNS* is based on Tian and Pearl (2000) and Pearl (2009, Chapter 9). A gentle approach to counterfactual attribution, including a tool kit for estimation, is given in Pearl, Glymour, and Jewell (2016). An advanced formal treatment of actual causation can be found in Halpern (2016).

Matching techniques for estimating causal effects are used routinely by potential outcome researchers (Sekhon, 2007), though they usually ignore the pitfalls shown in our education-experience-salary example. My realization that missing-data problems should be viewed in the context of causal modeling was formed through the analysis of Mohan and Pearl (2014).

Cowles (2016) and Reid (1998) tell the story of Neyman's tumultuous years in London, including the anecdote about Fisher and the wooden models. Greiner (2008) is a long and substantive introduction to "but-for" causation in the law. Allen (2003), Stott et al. (2013), Trenberth (2012), and Hannart et al. (2016) address the problem of attribution of weather events to climate change, and Hannart in particular invokes the ideas of necessary and sufficient probability, which bring more clarity to the subject.

References

Allen, M. (2003). Liability for climate change. *Nature* 421: 891–892.

Balke, A., and Pearl, J. (1994a). Counterfactual probabilities: Computational methods, bounds, and applications. In *Uncertainty in Artificial Intelligence 10* (R. L. de Mantaras and D. Poole, eds.). Morgan Kaufmann, San Mateo, CA, 46–54.

Balke, A., and Pearl, J. (1994b). Probabilistic evaluation of counterfactual queries. In *Proceedings of the Twelfth National Conference on Artificial Intelligence*, vol. 1. MIT Press, Menlo Park, CA, 230–237.

Cowles, M. (2016). *Statistics in Psychology: An Historical Perspective*. 2nd ed. Routledge, New York, NY.

Duncan, O. (1975). *Introduction to Structural Equation Models*. Academic Press, New York, NY.

Freedman, D. (1987). As others see us: A case study in path analysis (with discussion). *Journal of Educational Statistics* 12: 101–223.

Greenland, S. (1999). Relation of probability of causation, relative risk, and doubling dose: A methodologic error that has become a social problem. *American Journal of Public Health* 89: 1166–1169.

Greiner, D. J. (2008). Causal inference in civil rights litigation. *Harvard Law Review* 81: 533–598.

Haavelmo, T. (1943). The statistical implications of a system of simultaneous equations. *Econometrica* 11: 1–12. Reprinted in D. F. Hendry and M. S. Morgan (Eds.), *The Foundations of Econometric Analysis*, Cambridge University Press, Cambridge, UK, 477–490, 1995.

Halpern, J. (2016). *Actual Causality*. MIT Press, Cambridge, MA.

Hannart, A., Pearl, J., Otto, F., Naveu, P., and Ghil, M. (2016). Causal counterfactual theory for the attribution of weather and climate-related events. *Bulletin of the American Meteorological Society (BAMS)* 97: 99–110.

Holland, P. (1986). Statistics and causal inference. *Journal of the American Statistical Association* 81: 945–960.

Hume, D. (1739). *A Treatise of Human Nature*. Oxford University Press, Oxford, UK. Reprinted 1888.

Hume, D. (1748). *An Enquiry Concerning Human Understanding*. Reprinted Open Court Press, LaSalle, IL, 1958.

Joffe, M. M., Yang, W. P., and Feldman, H. I. (2010). Selective ignorability assumptions in causal inference. *International Journal of Biostatistics* 6. doi:10.2202/1557-4679.1199.

Lewis, D. (1973a). Causation. *Journal of Philosophy* 70: 556–567. Reprinted with postscript in D. Lewis, *Philosophical Papers*, vol. 2, Oxford University Press, New York, NY, 1986.

Lewis, D. (1973b). *Counterfactuals*. Harvard University Press, Cambridge, MA.

Lewis, M. (2016). *The Undoing Project: A Friendship That Changed Our Minds*. W. W. Norton and Company, New York, NY.

Mohan, K., and Pearl, J. (2014). Graphical models for recovering probabilistic and causal queries from missing data. *Proceedings of Neural Information Processing* 27: 1520–1528.

Morgan, S., and Winship, C. (2015). *Counterfactuals and Causal Inference: Methods and Principles for Social Research (Analytical Methods for Social Research)*. 2nd ed. Cambridge University Press, New York, NY.

Neyman, J. (1923). On the application of probability theory to agricultural experiments. Essay on principles. Section 9. *Statistical Science* 5: 465–480.

Pearl, J. (2000). *Causality: Models, Reasoning, and Inference*. Cambridge University Press, New York, NY.

Pearl, J. (2009). *Causality: Models, Reasoning, and Inference*. 2nd ed. Cambridge University Press, New York, NY.

Pearl, J., Glymour, M., and Jewell, N. (2016). *Causal Inference in Statistics: A Primer*. Wiley, New York, NY.

Reid, C. (1998). *Neyman*. Springer-Verlag, New York, NY.

Rubin, D. (1974). Estimating causal effects of treatments in randomized and nonrandomized studies. *Journal of Educational Psychology* 66: 688–701.

Sekhon, J. (2007). The Neyman-Rubin model of causal inference and estimation via matching methods. In *The Oxford Handbook of Political Methodology* (J. M. Box-Steffensmeier, H. E. Brady, and D. Collier, eds.). Oxford University Press, Oxford, UK.

Shpitser, I., and Pearl, J. (2009). Effects of treatment on the treated: Identification and generalization. In *Proceedings of the Twenty-Fifth Conference on Uncertainty in Artificial Intelligence*. AUAI Press, Montreal, Quebec, 514–521.

Stott, P. A., Allen, M., Christidis, N., Dole, R. M., Hoerling, M., Huntingford, C., Pardeep Pall, J. P., and Stone, D. (2013). Attribution of weather and climate-related events. In *Climate Science for Serving Society: Research, Modeling, and Prediction Priorities* (G. R. Asrar and J. W. Hurrell, eds.). Springer, Dordrecht, Netherlands, 449–484.

Tian, J., and Pearl, J. (2000). Probabilities of causation: Bounds and identification. *Annals of Mathematics and Artificial Intelligence* 28: 287–313.

Trenberth, K. (2012). Framing the way to relate climate extremes to climate change. *Climatic Change* 115: 283–290.

VanderWeele, T. (2015). *Explanation in Causal Inference: Methods for Mediation and Interaction*. Oxford University Press, New York, NY.

CHAPTER 9. MEDIATION: THE SEARCH FOR A MECHANISM

Annotated Bibliography

There are several books dedicated to the topic of mediation. The most up-to-date reference is VanderWeele (2015); MacKinnon (2008) also contains many examples. The dramatic transition from the statistical approach of Baron and Kenny (1986) to the counterfactual-based approach of causal mediation is described in Pearl (2014) and Kline (2015). McDonald's quote (to discuss mediation, "start from scratch") is taken from McDonald (2001).

Natural direct and indirect effects were conceptualized in Robins and Greenland (1992) and deemed problematic. They were later formalized and legitimized in Pearl (2001), leading to the Mediation Formula.

In addition to the comprehensive text of VanderWeele (2015), new results and applications of mediation analysis can be found in De Stavola et al. (2015); Imai, Keele, and Yamamoto (2010); and Muthén and Asparouhov (2015). Shpitser (2013) provides a general criterion for estimating arbitrary path-specific effects in graphs.

The Mediation Fallacy and the fallacy of "conditioning" on a mediator are demonstrated in Pearl (1998) and Cole and Hernán (2002). Fisher's falling for this fallacy is told in Rubin (2005), whereas Rubin's dismissal of mediation analysis as "deceptive" is expressed in Rubin (2004).

The startling story of how the cure for scurvy was "lost" is told in Lewis (1972) and Ceglowski (2010). Barbara Burks's story is told in King, Montañez Ramírez, and Wertheimer (1996); the quotes from Terman and Burks's mother are drawn from the letters (L. Terman to R. Tolman, 1943).

The source paper for the Berkeley admissions paradox is Bickel, Hammel, and O'Connell (1975), and the ensuing correspondence between him and Kruskal is found in Fairley and Mosteller (1977).

VanderWeele (2014) is the source for the "smoking gene" example, and Bierut and Cesarini (2015) tells the story of how the gene was discovered.

The surprising history of tourniquets, before and during the Gulf War, is told in Welling et al. (2012) and Kragh et al. (2013). The latter article is written in a personal and entertaining style that is quite unusual for a scholarly publication. Kragh et al. (2015) describes the research that unfortunately failed to prove that tourniquets improve the chances for survival.

References

Baron, R., and Kenny, D. (1986). The moderator-mediator variable distinction in social psychological research: Conceptual, strategic, and statistical considerations. *Journal of Personality and Social Psychology* 51: 1173–1182.

Bickel, P. J., Hammel, E. A., and O'Connell, J. W. (1975). Sex bias in graduate admissions: Data from Berkeley. *Science* 187: 398–404.

Bierut, L., and Cesarini, D. (2015). How genetic and other biological factors interact with smoking decisions. *Big Data* 3: 198–202.

Burks, B. S. (1926). On the inadequacy of the partial and multiple correlation technique (parts I–II). *Journal of Experimental Psychology* 17: 532–540, 625–630.

Burks, F., to Mrs. Terman. (June 16, 1943). Correspondence. Lewis M. Terman Archives, Stanford University.

Ceglowski, M. (2010). Scott and scurvy. *Idle Words* (blog). Available at: http://www.idlewords.com/2010/03/scott_and_scurvy.htm (posted: March 6, 2010).

Cole, S., and Hernán, M. (2002). Fallibility in estimating direct effects. *International Journal of Epidemiology* 31: 163–165.

De Stavola, B. L., Daniel, R. M., Ploubidis, G. B., and Micali, N. (2015). Mediation analysis with intermediate confounding. *American Journal of Epidemiology* 181: 64–80.

Fairley, W. B., and Mosteller, F. (1977). *Statistics and Public Policy*. Addison-Wesley, Reading, MA.

Imai, K., Keele, L., and Yamamoto, T. (2010). Identification, inference, and sensitivity analysis for causal mediation effects. *Statistical Science* 25: 51–71.

King, D. B., Montañez Ramírez, L., and Wertheimer, M. (1996). Barbara Stoddard Burks: Pioneer behavioral geneticist and humanitarian. In *Portraits of Pioneers in Psychology* (C. W. G. A. Kimble and M. Wertheimer, eds.), vol. 2. Erlbaum Associates, Hillsdale, NJ, 212–225.

Kline, R. B. (2015). The mediation myth. *Chance* 14: 202–213.

Kragh, J. F., Jr., Nam, J. J., Berry, K. A., Mase, V. J., Jr., Aden, J. K., III, Walters, T. J., Dubick, M. A., Baer, D. G., Wade, C. E., and Blackbourne, L. H. (2015). Transfusion for shock in U.S. military war casualties with and without tourniquet use. *Annals of Emergency Medicine* 65: 290–296.

Kragh, J. F., Jr., Walters, T. J., Westmoreland, T., Miller, R. M., Mabry, R. L., Kotwal, R. S., Ritter, B. A., Hodge, D. C., Greydanus, D. J., Cain, J. S., Parsons, D. S., Edgar, E. P., Harcke, T., Baer, D. G., Dubick, M. A., Blackbourne, L. H., Montgomery, H. R., Holcomb, J. B., and Butler, F. K. (2013). Tragedy into drama: An American history of tourniquet use in the current war. *Journal of Special Operations Medicine* 13: 5–25.

Lewis, H. (1972). Medical aspects of polar exploration: Sixtieth anniversary of Scott's last expedition. *Journal of the Royal Society of Medicine* 65: 39–42.

MacKinnon, D. (2008). *Introduction to Statistical Mediation Analysis*. Lawrence Erlbaum Associates, New York, NY.

McDonald, R. (2001). Structural equations modeling. *Journal of Consumer Psychology* 10: 92–93.

Muthén, B., and Asparouhov, T. (2015). Causal effects in mediation modeling. *Structural Equation Modeling* 22: 12–23.

Pearl, J. (1998). Graphs, causality, and structural equation models. *Sociological Methods and Research* 27: 226–284.

Pearl, J. (2001). Direct and indirect effects. In *Proceedings of the Seventeenth Conference on Uncertainty in Artificial Intelligence*. Morgan Kaufmann, San Francisco, CA, 411–420.

Pearl, J. (2014). Interpretation and identification of causal mediation. *Psychological Methods* 19: 459–481.

Robins, J., and Greenland, S. (1992). Identifiability and exchangeability for direct and indirect effects. *Epidemiology* 3: 143–155.

Rubin, D. (2004). Direct and indirect causal effects via potential outcomes. *Scandinavian Journal of Statistics* 31: 161–170.

Rubin, D. (2005). Causal inference using potential outcomes: Design, modeling, decisions. *Journal of the American Statistical Association* 100: 322–331.

Shpitser, I. (2013). Counterfactual graphical models for longitudinal mediation analysis with unobserved confounding. *Cognitive Science* 37: 1011–1035.

Terman, L., to Tolman, R. (August 6, 1943). Correspondence. Lewis M. Terman Archives, Stanford University.

VanderWeele, T. (2014). A unification of mediation and interaction: A four-way decomposition. *Epidemiology* 25: 749–761.

VanderWeele, T. (2015). *Explanation in Causal Inference: Methods for Mediation and Interaction*. Oxford University Press, New York, NY.

Welling, D., MacKay, P., Rasmussen, T., and Rich, N. (2012). A brief history of the tourniquet. *Journal of Vascular Surgery* 55: 286–290.

CHAPTER 10. BIG DATA, ARTIFICIAL INTELLIGENCE, AND THE BIG QUESTIONS

Annotated Bibliography

An accessible source for the perpetual free will debate is Harris (2012). The compatibilist school of philosophers is represented in the writings of Mumford and Anjum (2014) and Dennett (2003).

Artificial intelligence conceptualizations of agency can be found in Russell and Norvig (2003) and Wooldridge (2009). Philosophical views on agency are compiled in Bratman (2007). An intent-based learning system is described in Forney et al. (2017).

The twenty-three principles for "beneficial AI" agreed to at the 2017 Asilomar meeting can be found at Future of Life Institute (2017).

References

Bratman, M. E. (2007). *Structures of Agency: Essays*. Oxford University Press, New York, NY.

Brockman, J. (2015). *What to Think About Machines That Think*. HarperCollins, New York, NY.

Dennett, D. C. (2003). *Freedom Evolves*. Viking Books, New York, NY.

Forney, A., Pearl, J., and Bareinboim, E. (2017). Counterfactual data-fusion for online reinforcement learners. *Proceedings of the 34th International Conference on Machine Learning. Proceedings of Machine Learning Research* 70: 1156–1164.

Future of Life Institute. (2017). Asilomar AI principles. Available at: https://futureoflife.org/ai-principles (accessed December 2, 2017).

Harris, S. (2012). *Free Will*. Free Press, New York, NY.

Mumford, S., and Anjum, R. L. (2014). *Causation: A Very Short Introduction (Very Short Introductions)*. Oxford University Press, New York, NY.

Russell, S. J., and Norvig, P. (2003). *Artificial Intelligence: A Modern Approach*. 2nd ed. Prentice Hall, Upper Saddle River, NJ.

Wooldridge, J. (2009). *Introduction to Multi-agent Systems*. 2nd ed. John Wiley and Sons, New York, NY.

INDEX

ALLEN LANE

an imprint of

PENGUIN BOOKS

Also Published

David Brooks, *The Second Mountain*

Roberto Calasso, *The Unnamable Present*

Lee Smolin, *Einstein's Unfinished Revolution: The Search for What Lies Beyond the Quantum*

Clare Carlisle, *Philosopher of the Heart: The Restless Life of Søren Kierkegaard*

Nicci Gerrard, *What Dementia Teaches Us About Love*

Edward O. Wilson, *Genesis: On the Deep Origin of Societies*

John Barton, *A History of the Bible: The Book and its Faiths*

Carolyn Forché, *What You Have Heard is True: A Memoir of Witness and Resistance*

Elizabeth-Jane Burnett, *The Grassling*

Kate Brown, *Manual for Survival: A Chernobyl Guide to the Future*

Roderick Beaton, *Greece: Biography of a Modern Nation*

Matt Parker, *Humble Pi: A Comedy of Maths Errors*

Ruchir Sharma, *Democracy on the Road*

David Wallace-Wells, *The Uninhabitable Earth: A Story of the Future*

Randolph M. Nesse, *Good Reasons for Bad Feelings: Insights from the Frontier of Evolutionary Psychiatry*

Anand Giridharadas, *Winners Take All: The Elite Charade of Changing the World*

Richard Bassett, *Last Days in Old Europe: Triste '79, Vienna '85, Prague '89*

Paul Davies, *The Demon in the Machine: How Hidden Webs of Information Are Finally Solving the Mystery of Life*

Toby Green, *A Fistful of Shells: West Africa from the Rise of the Slave Trade to the Age of Revolution*

Paul Dolan, *Happy Ever After: Escaping the Myth of The Perfect Life*

Sunil Amrith, *Unruly Waters: How Mountain Rivers and Monsoons Have Shaped South Asia's History*

Christopher Harding, *Japan Story: In Search of a Nation, 1850 to the Present*

Timothy Day, *I Saw Eternity the Other Night: King's College, Cambridge, and an English Singing Style*

Richard Abels, *Aethelred the Unready: The Failed King*

Eric Kaufmann, *Whiteshift: Populism, Immigration and the Future of White Majorities*

Alan Greenspan and Adrian Wooldridge, *Capitalism in America: A History*

Philip Hensher, *The Penguin Book of the Contemporary British Short Story*

Paul Collier, *The Future of Capitalism: Facing the New Anxieties*

Andrew Roberts, *Churchill: Walking With Destiny*